Sabine Kunst   Tanja Kruse   Andrea Burmester  (Eds.)
Sustainable Water and Soil Management

Springer

*Berlin
Heidelberg
New York
Barcelona
Hong Kong
London
Milan
Paris
Tokyo*

Sabine Kunst
Tanja Kruse
Andrea Burmester (Eds.)

# Sustainable Water and Soil Management

With 62 Figures

Springer

Professor Dr. Ing. habil. Dr. phil. Sabine Kunst
Institut für Siedlungswasserwirtschaft
und Abfalltechnik Hannover ISAH
Universität Hannover
30167 Hannover
Germany

Dipl.-Päd. Tanja Kruse
Projektbereich Wasser
Internationale Frauenuniversität
Warmbüchenstrasse 15
30159 Hannover
Germany

Dipl.-Ing. Andrea Burmester
Projektbereich Wasser
Internationale Frauenuniversität
Warmbüchenstrasse 15
30159 Hannover
Germany

ISBN 3-540-42428-8 Springer-Verlag Berlin Heidelberg New York

Library of Congress Cataloging-in-Publication Data
Sustainable water and soil management / Sabine Kunst, Tanja Kruse, Andrea Burmester, eds.
  p.cm. Includes bibliographical references and index.
ISBN 35404244288
  1.Water-supply—Management. 2. Water-supply—Developing countries—Management.
3. Water resources development. 4. Sewage disposal, Rural. 5. Soil management. 6. Women in development. I. Kunst, Sabine. II. Kruse, Tanja, 1971-III. Burmester, Andrea, 1970-

This work is subject to copyright. All rights are reserved, whether the whole or part of the material is concerned, specifically the rights of translation, reprinting, reuse of illustrations, recitation, broadcasting, reproduction on microfilm or in any other way, and storage in data banks. Duplication of this publication or parts thereof is permitted only under the provisions of the German Copyright Law of September 9, 1965, in its current version, and permission for use must always be obtained from Springer-Verlag. Violations are liable for prosecution under the German Copyright Law.

Springer-Verlag is a member of BertelsmannSpringer Science+Business Media GmbH

http://www.springer.de

© Springer-Verlag Berlin Heidelberg 2002
Printed in Germany

The use of general descriptive names, registered names, trademarks, etc. in this publication does not imply, even in the absence of a specific statement, that such names are exempt from the relevant protective laws and regulations and therefore free for general use.

Product liability: The publishers cannot guarantee the accuracy of any information about the application of operative techniques and medications contained in this book. In every individual case the user must check such information by consulting the relevant literature.

Camera ready by authors
Cover design: design & production, Heidelberg
Printed on acid-free paper  SPIN 10833667/3130/as  5 4 3 2 1 0

# Foreword

Aylâ Neusel

The idea of holding an International Women's University *ifu* as part of the EXPO 2000 World Exposition was born in Lower Saxony in the mid-1990s. In 1992, Lower Saxony's then Minister of Science Helga Schuchardt had set up a Women's Research Commission that in 1994 presented its report with the programmatic title "Promoting Women's Interests Means Academic Reform – Women's Research Means a Critique of Science". A spin-off, so to speak, of this commission's was the idea of a women's university as an EXPO project. The 2$^{nd}$ Lower Saxony Women's Research Commission (1995-1997) stated: "From 15 July until 15 October, an International Women's University is to be organised offering an interdisciplinary, international, multimedia, postgraduate study programme".

Initially conceived as a purely research-oriented university, *ifu* evolved into an academic project for women scientists on an international scale. The *ifu* concept was based on the (self-) image of science as an ongoing, evolving, forward-looking research project.

The unique concept of the International Women's University as an academic reform project was founded on three key principles:

1. Problem Orientation of Teaching and Research
   The choice of the globally relevant controversial issues Work – Information – Body – Migration – City – Water and the idea of addressing these issues from the perspective of the natural and engineering sciences, the humanities and social sciences as well as art, consciously focusing on questions of practical relevance, gave rise to a problem-oriented, interdisciplinary approach.
2. Promotion of Women's Interests and Gender Perspective
   As a single-sex academic institution, *ifu* has introduced new effective ways of promoting networking and mentoring among young women scientists on an international scale. Gender perspective is a key element of research. Research topics, theories and methods are subjected to critical scrutiny, fundamental questions are asked about the role of science and academic institutions, and efforts are made to promote innovative approaches in science, academia and practice.
3. Transnationality and Interculturality
   *ifu*'s consistent application of the principle of internationality – both quantitatively in terms of the number of countries represented and qualitatively in terms of the international nature of the student body *and* faculty – is quite unique and has initiated a productive North-South dialogue among the women scientists involved. This comprehensive international discourse, incorporating intercultural forms of teaching and learning and addressing topics of global relevance, has helped participants to broaden their horizons, sharpen their critical faculties and question cultural and scientific certainties.

Between 15 July and 15 October 2000, a total of 747 women junior scientists from 105 different countries studied at the International Women's University. The faculty staff consisted of 313 women lecturers and visiting scholars from 49 countries. In all, then, *ifu* brought together some 1,000 established women scientists from every continent in the world.

The study programme was conducted in English in co-operation with the University of Hannover (WORK, BODY, MIGRATION), the University of Hamburg (INFORMATION), the University of Kassel (CITY), the University of Bremen (subphase of the study programme BODY), at the University of Applied Sciences in Suderburg in co-operation with the University of Hannover (WATER), and with the collaboration of the University of Clausthal (WORK).

More than 60% of the junior scientists came from Africa, Asia, Latin America and Eastern Europe, 20% from Germany and the remainder from the USA, Australia, Canada and West European countries. 79% of junior scientists received a grant. 97% of junior scientists successfully completed their studies and obtained certification to this effect.

A project like the International Women's University is not the brainchild of one individual: it owes its origin and genesis to a group of critical women scientists who succeeded in winning support among policymakers and the public for its implementation. Before reaching maturity, then, the idea passed through many minds and was subject to constant modification.

The present volume is the first in a series of publications presenting the results of *ifu*'s pilot semester to the international scientific community. My special thanks go to Sabine Kunst, Dean of the project area Water, and her scientific collaborators Andrea Burmester and Tanja Kruse for being the first to publish the results of the pilot semester.

# Contents

**List of Contributors** .................................................................................. **XIV**

## *Gendersensitive, Participatory Approach of Water and Soil Management*

**1   The International Women's University – Framework for the Project Area Water** ..............................................................................................1
Andrea Burmester, Tanja Kruse
    1.1    The Future of Higher Education (Aims) ...............................................1
    1.2    The International Women's University and Intercultural Science .......2
    1.3    Junior Women Scientists........................................................................2
        1.3.1    Selection Process in the Project Area Water.............................3
        1.3.2    Profile of the Junior Scientists in the Project Area Water..........4
        1.3.3    Catering to Participants' Needs: Service Centre .......................5
    1.4    *ifu* as a Platform for Global Dialogue ..................................................6

**2   *ifu* – an Intercultural Innovation in Higher Education ?**..............................9
Vathsala Aithal
    2.1    The Intercultural *ifu* ..............................................................................9
        2.1.1    Internationalisation or Parochialisation?...................................10
        2.1.2    The Socio-Political Context ......................................................11
        2.1.3    The Feminist Agenda .................................................................12
        2.1.4    Culture as Social Practice...........................................................14
        2.1.5    The Many Differences................................................................15
        2.1.6    Intercultural Training at *ifu* ......................................................16
        2.1.7    The Interculturality of Knowledge Production .........................17
    2.2    Conclusion ............................................................................................17
    References ......................................................................................................18

**3   The Project Area Water** ..............................................................................21
Sabine Kunst, Andrea Burmester, Tanja Kruse
    3.1    Water is Life – Background Information ............................................21
    3.2    Feminist Perspectives at the Action Level ..........................................23
    3.3    The Concept of the Project Area Water ...............................................24
    3.4    Curriculum of the Project Area Water .................................................26
        3.4.1    Knowledge Transfer...................................................................26
        3.4.2    Practical Projects........................................................................26
    References ......................................................................................................28

## *Aspects of Water and Soil Management*

**4  Rural Development with Special Emphasis on Women, Water and Environment ..................................................................................................29**
Leelamma Devasia
    4.1    An Experiment in the Creation of Knowledge, Skills and Attitude ...29
    4.2    Feminisation of Water Management – an Indian Concept.................31
        4.2.1    India – the Land and People......................................................31
        4.2.2    Rural Women's Participation in Water Management in Maharashtra State....................................................................32
        4.2.3    An Alternative Vision Planned and Directed by Rural Women......................................................................................42
    4.3    An Interdisciplinary and International Approach to Rural Development within *ifu*.........................................................................44
        4.3.1    Women and Rural Development................................................44
        4.3.2    Rural Women - Water and the Environment.............................46
        4.3.3    Skill Development .....................................................................48
        4.3.4    Exposure to Different Realities, Field Trips and Excursions.....51
    4.4    The *ifu* Experiment as a Beginning of a New Endeavour .................52
    References ....................................................................................................55

**5  Water Treatment and Rainwater Harvesting............................................57**
Namrata Pathak
    5.1    Overview.............................................................................................57
    5.2    Water Disinfection Methods ...............................................................60
        5.2.1    Physical and Chemical Methods ...............................................60
        5.2.2    Biological Method .....................................................................62
        5.2.3    Bacterial Contamination ............................................................63
    5.3    Rainwater Harvesting – Two Scenarios..............................................64
    5.4    Description of the Project ...................................................................66
        5.4.1    Presentation of Excursions.........................................................67
        5.4.2    Results.........................................................................................71
    5.5    Rainwater Harvesting Project Plan Developed by the Women Junior Scientists ..................................................................................76
        5.5.1    Rainwater Harvesting for Household Consumption (Philippines by Angelica R. Martinez)......................................76
        5.5.2    Case Study on Rainwater Harvesting (Albania by Gentiana Haxhillazi) ..............................................78
        5.5.3    Rainwater Harvesting - A Proposal for Secondary Schools (Tanzania by Eng. Immaculata Nshange Raphael) ....................80
        5.5.4    Rainwater Harvesting (USA by Margaret Fredricks).................81
        5.5.5    Promotion of Rainwater Harvesting in the Arid Area (Cameroon by Michele Denise Akamba Ava, Maroua Salak) ...83

  5.5.6 Rainwater Harvesting Draft Plan for a Vegetable Garden
     (India by Nandini Sankarampadi, Sanjulata Prasad) .................84
  5.5.7 Rainwater Harvesting Plan for Loyola College
     (Nigeria by Theresa Odejayi, Yetunde Odeyemi,
     Helen Oloyede) ........................................................................86
 5.6 Summary ..............................................................................................88
 References ......................................................................................................89

# 6 Wastewater Treatment ...............................................................................91
Sabine Kunst, Artur Mennerich, Marc Wichern
 6.1 Mechanical Wastewater Treatment ......................................................92
  6.1.1 Overview .................................................................................92
  6.1.2 Rakes and Strainers .................................................................92
  6.1.3 Sand Catchers .........................................................................94
  6.1.4 Preliminary Treatment/Settling Tank ......................................97
 6.2 Biological Wastewater Treatment ......................................................100
  6.2.1 Overview ...............................................................................100
  6.2.2 Legal Requirements for Wastewater Treatment in Europe ......101
 6.3 Models for the Design and Simulation of Wastewater
   Treatment Plants (WWTPs) ................................................................112
  6.3.1 Overview ...............................................................................112
  6.3.2 Dynamic Models ...................................................................118
  6.3.3 Use of Computer Programs ...................................................121
 6.4 Evaluation of Centralised Wastewater Treatment .............................123
  6.4.1 Comparison of Wastewater Treatment Plants .......................127
  6.4.2 Conclusions ..........................................................................131
 References ....................................................................................................135

# 7 Decentralised Wastewater Treatment - Wastewater Treatment in Rural Areas ..................................................................................................137
Katrin Kayser, Sabine Kunst
 7.1 Situation ..............................................................................................137
  7.1.1 Principles and Spheres of Action ..........................................138
  7.1.2 Decentralisation and User Participation ...............................139
 7.2 Nature-Based Wastewater and Sludge Treatment Methods as
   Components of Sustainable Concepts in Rural Regions
   – State of the Art .................................................................................141
  7.2.1 Introduction ...........................................................................141
  7.2.2 Overview - Wastewater Quantities and Wastewater Agents
     in Rural Areas .......................................................................142
  7.2.3 Pre-Treatment ........................................................................144
  7.2.4 Planted Soil Filters ................................................................146
  7.2.5 Wastewater Lagoons .............................................................153
  7.2.6 Sludge Composting in Reed Beds .........................................156
  7.2.7 Practical Examples ................................................................159

7.3 Conclusions for Design Parameters ..................................................164
    7.3.1 Characteristics of Decentralised Wastewater Treatment Systems which are Conducive to Sustainable Development....168
    7.3.2 Impact of Gender Perspectives on Planning Criteria ...............169
7.4 Examples of Planning Ideas...............................................................172
    7.4.1 Wastewater Purification for Remote Villages..........................173
    7.4.2 Planning Ideas for Sensitive Regions in Rural Areas...............176
References ..................................................................................................180

## 8 Alternative Technologies for Sanitation, Recycling and Reuse................183
Sabine Kunst, Namrata Pathak

8.1 Overview............................................................................................183
8.2 The Composting Process....................................................................186
    8.2.1 The Phases ...............................................................................186
    8.2.2 Environmental Factors in Composting.....................................187
    8.2.3 Composting Micro-Organisms.................................................190
    8.2.4 Quality Criteria for Compost as a Product...............................193
8.3 Types of Toilets .................................................................................194
    8.3.1 Water Toilets............................................................................194
    8.3.2 Waterless Toilets......................................................................195
8.4 Composting Toilet Systems ...............................................................196
    8.4.1 Dimensioning Composting Toilets ..........................................196
    8.4.2 Dry Sanitation with Reuse .......................................................198
    8.4.3 Dehydration Toilets .................................................................201
    8.4.4 Decomposition Toilets .............................................................204
    8.4.5 Types of Composting Toilet Systems ......................................207
8.5 SIRDO ................................................................................................211
    8.5.1 Pathogens Elimination .............................................................214
    8.5.2 Social Evaluation .....................................................................216
References ..................................................................................................219

## 9 River Development Planning..................................................................219
Andrea Töppe

9.1 River Protection for the Balance of Nature.......................................219
9.2 Hydraulics and River Protection........................................................222
    9.2.1 Some River Characteristics......................................................222
    9.2.2 River Discharge .......................................................................223
    9.2.3 The River Protection System of Lower Saxony/Germany........229
    9.2.4 Making an Inventory in Situ ....................................................231
9.3 Stahlbach River Development Plan ...................................................236
    9.3.1 The Elbe Catchment Area........................................................237
    9.3.2 The Stahlbach River.................................................................239
    9.3.3 River Stahlbach Development Project (Project Modules) .......239
    9.3.4 Results......................................................................................241
9.4 Summary............................................................................................256
References ..................................................................................................258

**10 Water and Soil Towards Sustainable Land Use** ............................................. 261
Brigitte Urban
    10.1 Overview .................................................................................................. 261
        10.1.1 Soils ............................................................................................. 261
        10.1.2 Soil and Water ............................................................................. 264
        10.1.3 Global Significance ..................................................................... 267
    10.2 Project Water and Soil ............................................................................. 267
        10.2.1 Skills and Aims ........................................................................... 267
        10.2.2 Stahlbach Creek Project .............................................................. 268
        10.2.3 Methodology ............................................................................... 269
    10.3 Results of The Project .............................................................................. 270
        10.3.1 Case Study: River Elbe Ecology Project .................................... 270
        10.3.2 Description and Results of the Three Project Sites .................... 277
    10.4 Summary .................................................................................................. 287
        10.4.1 Intergroup Interferences .............................................................. 287
    References ...................................................................................................... 291

# *Conclusions*

**11 Evaluation "There is No Unanimous Judgement on *ifu*"** ........................ 295
Sigrid Metz-Göckel
    11.1 Evaluation Concept ................................................................................. 295
    11.2 Bridging the Gap Between Mutually Unfamiliar Disciplines and
        Socio-Technical Innovation .................................................................... 297
        11.2.1 Curriculum of the Project Area Water ........................................ 297
        11.2.2 Evaluation of the Curriculum from the Perspective of the
               Junior Scientists ........................................................................... 298
        11.2.3 From the Perspective of the Visiting Scholars: "You Can Feel
               It in the Air" ................................................................................. 303
        11.2.4 Description of the Study Venue – the Environment from the
               Perspective of the Junior scientists .............................................. 306
    11.3 Incongruity of the Perspectives: A Summary ........................................ 307
    References ...................................................................................................... 308

**12 Future Perspectives for Sustainable Water and Soil Management** ......... 309
Sabine Kunst
    12.1 Internationality and Intercultural Work .................................................. 310
    12.2 Interdisciplinary Work and Gender Perspectives ................................... 312
    12.3 Women's International Network for Sustainability: A Post-*ifu*
        Initiative Promoting Equitable and Ecologically Sound
        Alternatives to Mainstream Development ............................................. 315
        Dolly Wittberger
        12.3.1 "Development is Well-Being – Concerning the Individual as
               well as the Community Level – for the Past, Present and
               Future." (Andrea Heckert, U.S./Mexico) ................................... 315

12.3.2 "What Should I Say? Now We Are Developed?"
(Christobel Chakwana, Malawi) ............................................. 316
12.3.3 "...I Would Like to Have a Computer, this Would Empower
Me." (Arig Bakhiet, Sudan) ..................................................... 317
References ................................................................................................ 317

# *Appendices*

**13 Manual for Analysis of Soils and Related Materials ................................. 321**
Brigitte Urban
13.1 Introduction to Soil Exploration and Soil Sampling ..................... 321
13.2 Moisture Content and Dry Weight .................................................. 322
13.3 Determination of Organic Matter .................................................... 324
13.4 Determination of pH ......................................................................... 325
13.5 Salinity of Soils (Electric Conductivity, EC) ................................. 326
13.6. Cress Test (Germinability of Lepidium sativum) .......................... 328
13.7 Determination of Total Amount of Micro-organisms in Solids
(Microbial Number) ......................................................................... 329
13.8 Soil Respiration, Biological Oxygen Demand (BOD) ................. 334
13.9 Respiration Activity of Compost .................................................... 336
13.10 Carbon Content ................................................................................ 338
13.11 Determination of Nitrogen (Kjeldahl Procedure) .......................... 339
13.12 C/N and C/P Ratio ............................................................................ 341
13.13 Determination of Carbonate ............................................................ 342
13.14 Determination of Plant-Available Phosphorus and Potassium ...... 344
13.15 Determination of Plant-Available Potassium and Magnesium
(diluted with Calcium Chloride) ..................................................... 347
13.16 Determination of N – min ($NO_3$ and $NO_2$) ................................... 348
13.17 Determination of N-min ($NH_4$) ..................................................... 351
13.18 Determining Exchangeable Cations at Soil pH ............................ 352
13.19 Nitrohydrochloric Acid Disintegration .......................................... 356
13.20 Sewage Sludge Regulations ............................................................ 357
13.21 Elution with Water ........................................................................... 358
13.22 Soil Moisture Retention Capacity, pF Value ................................ 359
13.23 Soil Texture (Grain Size Distribution) ........................................... 362
13.24 Grain Fractions and Texture Types ................................................ 366
References ................................................................................................ 369

**14 Influencing BOD and N Removal Assessment of Important
Parameters .................................................................................................... 371**
Sabine Kunst
14.1 Batch Tests as a Method for Classifying Nitrification and
Denitrification Activities in Activated Sludge .............................. 371
14.1.1 Batch Tests for Nitrification (Aerobic) ............................ 371
14.1.2 Batch Tests for Denitrification (Anoxic) ......................... 373

    14.2  Respirometry: Determination of the Oxygen Uptake Rate (OUR) ..375
        14.2.1  Determination of the Respiration Rate of Activated Sludge
               by Measuring the $O_2$ Utilisation Rate ......................................376
        14.2.2  Evaluation of the Recorded Data .............................................377
        14.2.3  Dependence of Oxygen Consumption on Toxic or Inhibiting
               Substances in Water ................................................................378
        14.2.4  Further Applications for Oxygen-Consumption
               Measurements ..........................................................................378

**Vitae of Contributors** .......................................................................................381

**Index** ...................................................................................................................383

# List of Contributors

Aithal, Vathsala, M.A.
 Fachbereich Cornelia Goethe Center, Johann Wolfgang Goethe Universität
 Postfach 11 19 32, 60054 Frankfurt am Main, Germany
 Email: aithal@em.uni-frankfurt.de
Burmester, Andrea, Dipl.-Ing.
 Project Area Water, International Womens University *ifu*
 Wesselstraße 24, 30449 Hannover, Germany
 Email: burmester@isah.uni-hannover.de
Devasia, Leelamma, Dr.
 F - 2, Krishna Ganga, Temple Road, Civil Lines
 Nagpur – 440001, India
 Email: dearchu@nagpur.dot.net.in
Kayser, Katrin, Dipl.-Ing.
 Institut für Siedlungswasserwirtschaft und Abfalltechnik Hannover ISAH,
 Universität Hannover
 Welfengarten 1, 30167 Hannover, Germany
 Email: kayser@isah.uni-hannover.de
Kruse, Tanja, Dipl.-Päd.
 Project Area Water, International Womens University *ifu*
 Hinrichsring 26, 30177 Hannover, Germany
 Email: kruse@vifu.de
Kunst, Sabine, Prof. Dr.-Ing. habil. Dr. phil.
 Institut für Siedlungswasserwirtschaft undAbfalltechnik Hannover ISAH,
 Universität Hannover
 Welfengarten 1, 30167 Hannover, Germany
 Email: kunst@isah.uni-hannover.de
Mennerich, Artur, Prof. Dr.-Ing.
 Fachhochschule Nordostniedersachsen
 Herbert-Meyer-Str. 7, 29556 Suderburg, Germany
 Email: a.mennerich@fhnon.de
Metz-Göckel, Sigrid, Prof. Dr.
 Hochschuldidaktisches Zentrum, Universität Dortmund
 Vogelpothsweg 78, 44227 Dortmund, Germany
 Email: smetzgoeckel@hdz.uni-dortmund.de
Pathak, Namrata, Dr.
 Centre for rural Development and Technology, Indian Institute of Technology
 Hauzkhas, New Delhi - 110 016, India
 Email: namratapathak@hotmail.com
Töppe, Andrea, Prof. Dr.-Ing.
 Fachhochschule Nordostniedersachsen
 Herbert-Meyer-Str. 7, 29556 Suderburg, Germany
 Email: toeppe@fhnon.de

Urban, Brigitte, Prof. Dr. rer. nat.
   Fachhochschule Nordostniedersachsen
   Herbert-Meyer-Str. 7, 29556 Suderburg, Germany
   Email: urban@fhnon.de

Wichern, Marc, Dr.-Ing.
   Institut für Siedlungswasserwirtschaft und Abfalltechnik Hannover ISAH,
   Universität Hannover,
   Welfengarten 1, 30167 Hannover, Germany
   Email: wichern@isah.uni-hannover.de

Wittberger, Dolly, Dr.
   Women´s International Network for Sustainability - WINS
   Mandellstrasse 21/10, A-8010 Graz, Austria
   Email: dollyindia@hotmail.com    http://www.wins.at

# 1 The International Women's University – Framework for the Project Area Water

Andrea Burmester, Tanja Kruse

## 1.1 The Future of Higher Education (Aims)

Initial ideas for an International Women's University *ifu*, developed by the 1st Lower Saxony Women's Research Commission, resulted in the plan to organise a research-oriented university for young postgraduate women scientists. The research topics were to be chosen on the basis of their relevance to both current problems and gender studies. Other special features were to be the integration of perspectives on the scientific problems from different countries, interdisciplinary analysis of the topics treated and incorporation of the junior women scientists' own empirical knowledge in addition to scientific knowledge. Teaching, study and research at the International Women's University were to be done in interdisciplinary projects, the topics of the individual projects continuing current scientific, social, ecological and cultural discussions.

Junior women scientists and women lecturers from all over the world were invited to the International Women's University to present their views on, their analyses of and their solutions to scientific, engineering and medical problems: the aim was to initiate an international dialogue without hierarchies.

The main rationale of the interdisciplinary approach was the need to open up new horizons by viewing problems from different angles, based on the insight that the present strict division of disciplines – as reflected in current curricula and the regulations governing these – is inadequate to properly tackle scientific problems. Teaching and research must integrate practical knowledge and skills – of experts, designers, users and clients – enabling them to be incorporated in the overall body of scientific knowledge.

Similarly, *ifu*'s organisers sought to apply methods and media from different scientific, artistic and political domains in the discussion and presentation of the projects. Artists from different spheres were invited to present to *ifu* their ideas on the chosen topics using their various means of expression, such as exhibitions, performances, workshops, etc., with a view to initiating a dialogue with the scientists.

By combining critical gender perspectives on the natural sciences, technology, and medicine with an equally critical artistic approach to the selected topics, it was hoped to stimulate developments in both science and art. The International Women's University was conceived as a research-oriented university. And with only the short summer semester of 2000 available for the event, this meant that only highly qualified junior women scientists could be admitted.

## 1.2 The International Women's University and Intercultural Science

The International Women's University "Technology and Culture" *ifu* began its three-month study period on 15 July 2000, just over three years after the project's inception on 1 July 1997. In the course of its realisation, ideas and emphases changed and the aspect of intercultural research became increasingly prominent, particularly when prospective junior women scientists from the countries of the South began to show more and more interest in the project area Water .

The concept of the International Women's University reflects a new understanding of science based on the following principles: international and intercultural teaching and learning and the adoption of an interdisciplinary approach encompassing not only the natural and engineering sciences but also the humanities and social sciences as well as art. Other important factors are the project's practical and social orientation as well as its integration of the gender perspective in scientific issues.

The controversial topics chosen for the six interdisciplinary project areas opened up new, global perspectives. For the *ifu* pilot semester held in the summer of 2000, an ambitious interdisciplinary teaching and research programme was developed, focusing on six topics of global relevance for the future: BODY – CITY – INFORMATION – MIGRATION – WATER – WORK. It was not only the topics that were of a global nature, the curricula too were designed by groups of leading women scientists that were both international and interdisciplinary in composition. The project involved women researchers from a wide variety of disciplines from every continent in the world.

"The International Women's University is like six international postgraduate study programmes rolled into one" is how *ifu* President Aylâ Neusel described the university reform project. A total of 313 women lecturers and visiting scholars from more than 49 different countries were involved in the intensive teaching and research programmes. The results of the *ifu* study program are impressive: the young women scientists submitted independently produced final papers, including written dissertations, designs, film drafts and Internet presentations. 97% of junior scientists were awarded certification for successful completion of their studies.

## 1.3 Junior Women Scientists

The six project areas, in which teaching and research were conducted in English, were each able to accommodate 150 junior women scientists from all over the world. The minimum requirements for admission were a first degree and an excellent knowledge of English. The help of the German Academic Exchange Service (DAAD) was enlisted to handle the admissions process, from inviting applications world-wide to the eventual selection of junior scientists – a process without precedent in Germany and conducted with the utmost care.

## 1.3.1 Selection Process in the Project Area Water

By the end of 1999, a total of 222 applications had been received for the project area Water, 81% of them from the so-called developing countries, 13% from industrialised countries and 6% from Germany. To process the applications, the project area Water had two women experts, one from Germany, the other from India. Further expertise was provided by the advisory board. Applications were evaluated using special evaluation sheets. For each applicant, an average evaluation was determined on a scale of 1-10 points. An evaluation of 8-10 points led to immediate admission, 7-8 points to admission in the second round.

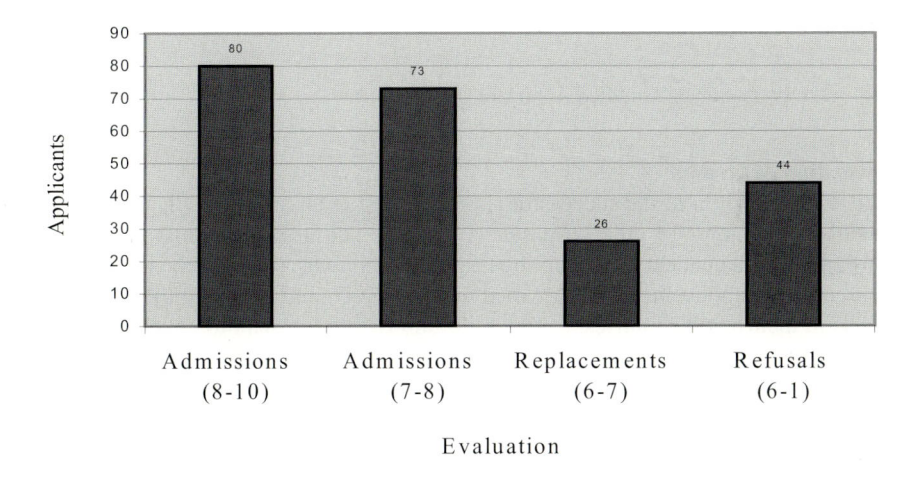

**Fig. 1.1.** Selection of Applicants

At the end of January 2000, 153 applicants from 55 countries were notified of their admission to the project area Water.

According to initial planning, a third of the available places were to be reserved for applicants from Germany, one third for those from other industrialised countries and a third for women from the so-called developing countries. However, a large proportion of the applicants – and the eventually admitted participants – came from the developing countries (cf. Fig. 1.2., the high proportion of admitted applicants from Asia and Africa). The intense interest from these two parts of the world can be attributed to the direct impact of reduced water supplies and water pollution and the consequent threat to the life of those living there.

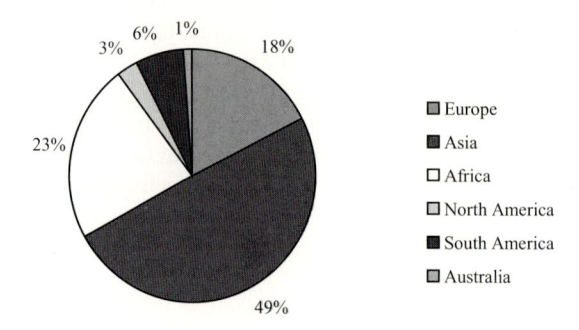

Fig. 1.2. Distribution of Admitted Applicants by Continent

**Awarding of Grants**
The regional origin of the successful applicants led to an unexpectedly large number of applications for grants. About 94% of the admitted applicants made their participation dependent on receiving a grant. This was due largely to the high percentage of applicants from so-called developing countries. Given the low wage levels in their home countries, it was quite impossible for them to cover their own travel expenses and the high cost of living in Germany. Every effort was made, however, to enable these women to participate because of the unique opportunity *ifu* offered them and their respective countries.

Of the junior women scientists participating in the project area water, 85 received grants and 12 partial grants. The other four junior scientists financed their studies themselves.

### 1.3.2 Profile of the Junior Scientists in the Project Area Water

All 101 junior scientists in the project area Water were in possession of a bachelor's or master's degree/diploma or Ph.D./doctorate in the following fields:

- Agricultural Sciences (11)
- Biology (23)
- Chemistry (15)
- Civil Engineering (17)
- Environmental Sciences (11)
- Humanities (6)
- Medical Sciences (3)
- Philology (7)
- Physics (4)
- Other (4)

In the project area Water, the principle of internationality was guaranteed by the fact that junior scientists came from 44 different countries.

**Table 1.1.** Home Countries of Junior Scientists

| Asia | | Europe | | Africa | | S. America | | N. America | |
|---|---|---|---|---|---|---|---|---|---|
| Bangladesh | 1 | Albania | 2 | Egypt | 1 | Brazil | 2 | Canada | 1 |
| China | 1 | Austria | 2 | Ethiopia | 2 | Chile | 1 | USA | 2 |
| India | 13 | Georgia | 1 | Cameroon | 2 | Ecuador | 2 | **Total** | **3** |
| Indonesia | 2 | Germany | 7 | Kenya | 5 | El Salvador | 1 | | |
| Iran | 1 | Luxembourg | 1 | Nigeria | 6 | Cuba | 1 | Australia | |
| Korea | 1 | Madagascar | 2 | S. Africa | 3 | Mexico | 1 | | 0 |
| Malawi | 1 | Rumania | 3 | Tanzania | 1 | **Total** | **8** | | |
| Myanmar | 1 | Slovakia | 1 | **Total** | **20** | | | | |
| Nepal | 1 | Turkey | 1 | | | | | | |
| Philippines | 2 | Turkmenistan | 1 | | | | | | |
| Russia | 6 | UK | 1 | | | | | | |
| Sri Lanka | 1 | Ukraine | 2 | | | | | | |
| Sudan | 5 | **Total** | **24** | | | | | | |
| S. Korea | 1 | | | | | | | | |
| Syria | 2 | | | | | | | | |
| Taiwan | 2 | | | | | | | | |
| Vietnam | 5 | | | | | | | | |
| **Total** | **46** | | | | | | | | |

**101 Junior Scientists**

The average age of junior scientists was 31, making this project area *ifu*'s "youngest".

## 1.3.3  Catering to Participants' Needs: Service Centre

Another element reflecting *ifu*'s reform approach was the Service Centre, an "all-in-one" service facility designed to cater to participants' every need. To this end, study offices were set up at the co-operating universities, providing help and advice to junior scientists, visiting scholars and lecturers on all matters relating to the academic programme and their stay in Germany.

Tasks that in many countries are traditionally attended to a variety of offices or administrative units – e.g. the matriculation and examination office, the student affairs office, the central counselling office and the department administration offices – were taken care of by a single institution, ensuring optimal service.

During the planning phase, it was decided that the Service Centre should also be responsible for the following key areas: arranging accommodation, providing and organising childcare and supervising access to facilities such as libraries, laboratories and computer pools.

On arrival of the junior scientists, visiting scholars and lecturers, the Service Centre took care of all the necessary formalities. The Service Centre was responsible for providing information on the study venues and the curriculum as well as sport and cultural events and planned local excursions. It also provided help in matters relating to healthcare, insurance and dealings with the foreign residents authorities as well as offering social counselling.

Help and advice on academic matters was provided by the Service Centre which offered a general participants counselling service, distributed information material and supervised the allocation of lecture rooms and technical equipment. Besides functioning as a central service facility, the Service Centre was also an interpersonal and intercultural meeting place fostering communication, empowerment and innovation. The key objective was to enhance the quality of the study programme by creating an optimal environment for both junior scientists and faculty staff.

## 1.4 *ifu* as a Platform for Global Dialogue

The teaching and research programme at the International Women's University made extensive use of modern information and communication technologies. *ifu* has its own online multimedia learning and research platform. The *ifu* website (www.vifu.de) offers those who participated in the pilot semester a forum for discussing research issues and exchanging results. The aim is to establish a vibrant and highly diverse world-wide network of women scientists.

*ifu* has generated a great deal of enthusiasm and has had wide repercussions. This is evidenced not least by the extraordinary sense of commitment shown by all those participating in the project. The international framework in which junior scientists were offered a targeted mentoring programme forms a sound basis for global networking between young and already established women scientists. During *ifu*'s three-month pilot phase in the summer of 2000, many women took advantage of the opportunity to establish new research contacts, to draft joint projects or to invite to conferences fellow junior scientists who developed interesting research results during the *ifu* semester.

The International Women's University provided a platform for the exchange of knowledge on a globally unprecedented scale – an exchange in which the voices of women scientists from the southern hemisphere and Eastern Europe were clearly audible. Never before has Western-dominated science been obliged to question its Euro-centric views and assumptions to this extent, and never before has a university made intercultural teaching and learning such a key focus of scientific thought and activity.

With its teaching and research programme, the International Women's University has initiated a global dialogue and laid the foundations for developing a better understanding of "otherness" and for questioning one's own values, interests and scientific perspectives. *ifu* is the start of a journey towards new intellectual horizons. In the age of globalisation, international and intercultural universities and science are an absolute must.

Based on an extensive evaluation of the *ifu* pilot semester, a concept is under development for continuing the International Women's University. There are plans for a one-year international and interdisciplinary master's program that will address globally relevant research issues more intensively and involve representatives from the natural and social sciences and the humanities. The projected programme will be ambitious and interdisciplinary in nature, designed to appeal in particular to young women engineers, computer scientists and economists, and offering an attractive way of enhancing one's skills and qualifications and enjoying targeted mentoring in an international academic environment.

# 2 *ifu* – an Intercultural Innovation in Higher Education ?

Vathsala Aithal

"I thought of myself as being rather progressive: quite up to date on feminist theory, politically committed, quite informed and quite open-minded. But when I saw these women from Africa, when I listened to their perspectives, to their visions, I was thrilled ... I said to myself, wow! But then I thought ... what is happening to me? Of course, I know that there are academics in Africa!?!", says an articulate and self-reflective junior scientist from an industrialised country in the project area Migration. She is talking about her interaction with academics from Africa and is reflecting on the changes that her direct intercultural encounter with women from the "Third World" at the International Women's University *ifu* has initiated in her.

The atmosphere at the closing of *ifu* on 15 October 2000 was highly euphoric. The immediate evaluation of the event by the junior scientists, women lecturers, organisers, visitors and the media was unanimous: in terms of intercultural learning and intercultural communication, *ifu* was a great success. "Just to have been here with so many different women, being exposed to so many approaches from so many different places and to learn from them – all this has been extremely exciting and very worthwhile" – such were the words of many women after spending three months at *ifu*.

It was a process of learning and doing research together, discovering things in common and differences and discussing the scope of sisterhood and solidarity. The intercultural exchange was the most important experience that the junior scientists took home with them. Referring to the energising experience, some said that *ifu* was something like a mini-Beijing at the academic level (Aithal 1996). For the post-*ifu* phase, they made plans to take up studies at the universities of their fellow participants, planned research projects, drafted joint publications, projects at the intersection of arts and science – or, if such a thing is possible given the North-South divide, simply to visit colleagues in their respective countries. A major achievement was the initiation of a research and action network of the project area Water, called WINS Women's International Network for Sustainability. All these activities are being planned across the continents, overcoming several types of boundaries.

## 2.1 The Intercultural *ifu*

Overcoming boundaries – this has been the aim of *ifu* since its inception. It has done so in many ways: by organising a university for women in a male-dominated higher-education system in Germany; by bringing arts and science together; by insisting on the unity of theory and practice. By focusing on these aspects, *ifu* took up great challenges. But the greatest challenge of all was to pursue the idea of interculturality. This internationality and interculturality was ensured by having a quota system for junior scientists. Of the 747 junior scientists, one third came from the "East", one third from "South" and one third from the rest of the world, meaning Europe, North America and Australia. The *ifu* classification also had a economic component, i.e. for countries that do not have a hard currency/economy, the categories "East", "West", "North" and "South" must be seen as political categories, too.

The distribution of the junior scientists over the project areas was quite uneven. The project area Water was outstanding as regards its composition. Almost 80% of the junior scientists came from the South, and 8 out of 11 visiting scholars were from Asia, Africa or Latin America.

Given the disparity and diversity of the junior scientist body, there were fears, anxieties and apprehensions in the preparatory process. Intercultural communication emerged as a factor that had to be dealt with and could not be ignored. The major questions were: How do we deal with this diversity of 1000 women from more than 150 countries? How can we go about it? How can we handle this complex situation? How can we cope with cultural conflicts?

This paper discusses the theoretical prerequisites for intercultural communication. The first part contextualises interculturality within the higher-education, socio-political and feminist debates in Germany. In the second part, I look at some of the categories and linkages of interculturality in the academic context, a site of knowledge production. The aim is to work towards a notion of interculturality as a structuring principle. I do this by unfolding my deliberations at two levels. The theoretical level will be discussed with postcolonial approaches. These seem appropriate because *ifu* is, indeed, a postcolonial context (Spivak 1993). For the practical level, I draw on my experience as lecturer in the project area Water, as a trainer of intercultural communication and as a supporter and observer of the curricular and extracurricular activities of *ifu* junior scientists in the project areas "Body", "Migration" and "Work". Also, my experience as a co-ordinator of the gender area in the Advanced Study Programme "Education and International Developments" at the University of Frankfurt has been extremely useful. The paper is guided by the idea that practical training activities have to be on a sound theoretical footing.

### 2.1.1 Internationalisation or Parochialisation?

The International Women's University was held at a time when a new/old slogan was haunting the intellectual debates in Germany: that of internationalisation and globalisation. For experts in academia as well as in politics, it is clear that the higher-education system is undergoing a deep crisis. The way out of this crisis is sought in internationalisation, largely as it has been put into practice in the Anglo-Saxon system. In the light of globalisation, the emphasis is on the education and training of the labour force for the global market. The issue of nationality seems to be rather irrelevant – capital does not care about borders or skin colour.

But in this context of supposed globalisation, a rather paradoxical process is becoming evident (Aithal 1999): the discourse on internationalisation is accompanied by a parochialisation of higher education. Since the reunification of the two German states in 1990, a process of inward-looking tendencies has been observable. Since then, an increasing number of highly reputed international advanced study programmes have been abolished. In the 1990s, several advanced study programmes dealing with development issues were closed down, and other Third-World-oriented programmes were abandoned in favour of more European programmes. In order to attract international junior scientists, English-language programmes are being introduced. However, it looks as if they are meant to cover up the evident self-isolation higher education in Germany is undergoing. Important questions arise: What is the hidden agenda behind the discourse on internationalisation of higher education? Does the system have the flexibility to meet the emerging challenges? What kind of openings will it have and what kind of options will it offer? Will it help to overcome the old and rigid structures?

In this rhetoric about internationalisation, the structural asymmetry in North-South relations is likely to increase. As far as the participation of junior scientists from countries of the South is concerned, there will be radical changes. Increasingly, it will be the criteria of achievement and efficiency costs that play a role in admissions – and not so much political criteria that enabled young people from politically fragile countries in the "Third World" to pursue their studies at a German university. It is the candidates of financially powerful countries that will find recruitment as junior scientists. The privatisation of the German university system is proceeding on a vast scale and cannot be ignored. An increasing number of programmes will be launched with high tuition fees, which, needless to say, only a very small group of people from the South will be able to afford.

International co-operation is another magic word. International co-operation in higher education is restricted to one region. Co-operation between universities and academic exchange largely focuses on the USA, other European countries, especially Scandinavian countries, Japan and Israel. But does this have any impact on the theoretical perspectives?

It is in this context that *ifu* provides a public space. It took up issues that were dealt with at the Expo 2000: body, city, information, migration, water and work. While at the Expo they were dealt with within the logic of the dominant mode of thinking, *ifu* offered a counter-public space for these issues, by using a feminist perspective. There was also the Summer University that represented a counter-counter-space. But it is with interculturality that *ifu* can create a public space that has an impact on higher education in Germany (Aithal et al 2000).

## 2.1.2 The Socio-Political Context

The year 2000 was the year in which discourses about inclusion and exclusion were at their peak in Germany. When *ifu* was inaugurated at the end of July, the

German version of the "Green Card", which permits restricted entry to non-European IT specialists, was just being introduced. This new move in labour policy was preceded by highly confusing, extremely nationalistic political debates. The new rule itself was seen by a considerable portion of the population as a threat to their national identity. As a reaction, more refugee hostels were set on fire, and right-wing and racist attacks increased. The violence was watched with anxiety by the international community. In the academic field, one world-famous Asian-American scientist invited to hold a chair and head a prestigious institution in Eastern Germany declined the offer at the last minute. The interconnectedness of xenophobia, anti-Semitism, homophobia and discrimination against the disabled and homeless posed new challenges to conceptualising interculturality.

The debate about the "Green Card" in Germany has also had cultural and psychological implications. The general presupposition was and is that the South *is* the recipient – either of aid or of theory. The debate in Germany about the "Green Card" was extremely diffuse and contrasts sharply with the way in which discussions on immigration were previously conducted. Surprisingly, those who had formerly been against a proper immigration law were suddenly endorsing it, and in return for doing so, they demanded that the laws guaranteeing political refugees asylum – legislation unique to Germany – be abolished.

It was also clear at this point that some of the "Green Card" holders would be from India. Obviously, this was difficult to imagine of a country like India – a country considered to be the biggest recipient of German development aid and more than half of whose population live below the poverty line. Naturally, it was hard to believe that India of all countries would be supplying Germany with highly qualified experts. And they would not be coming as beggars, but to boost the economy and to put their knowledge, their know-how to use to lower unemployment in Germany. That was indeed a delicate issue – the headline of one of the leading German newsweeklies, published on Thursdays, asked the question, telling enough in itself: "Are the Germans too stupid for computers?"

It is the perception of the self and others in a socio-political context, the supposed superiority implicit in the reactions that gave way to new directions in the debates on immigration, multiculturalism and interculturality.

### 2.1.3 The Feminist Agenda

*ifu* emerged from discussions among women at higher-education institutions who have also been active in the women's movement. They had been working on this project for more than 12 years. Having undergone a process of institutionalisation, the movement used the wind of publicity that came with the EXPO to advance feminist projects in higher education. This resembles the situation in India, where feminists in academia and in the women's movements use forms of this institutionalisation to create a public space for feminist concerns (IAWS 1995). The fact that internationality played an important role from the very beginning was indeed a reaction to feminist debates that have questioned romantic notions of global sisterhood.

Women's and gender studies may not belong to the elite areas of higher education in Germany, but they do reproduce and consolidate the relations of dominance. It almost seems as if "women" existed in the "Third World" and the agents of theory production lived in the North, as if the institutions of feminism were here. Hence the question of locality must be asked: Where is the site of feminist theory production?

And how is theory produced? Does it help in working against the "sanctioned ignorance" of feminist theories that deny the women of the South their agency? The presence of women, of feminists, of scholars, of junior scientists from countries of the South at *ifu* helped – if only for a short while – to break open this state of imbalance as regards the ownership of knowledge and the "object" of knowing. Their very presence also meant that the women of the South could win control over the content and process of theorising: What and how were they being theorised?

From a feminist perspective, the insecurities of the "Green Card" debate were revealed by the category of "gender" because the above-mentioned magazine had a photograph on the front page. It showed the face of a woman from India, a "Hindu" woman with something like a red dot on her forehead. But the red dot was in fact replaced by the Internet symbol "@". The editorial staff obviously did not see what had happened in this substitution process. What it implied was that, in the group of highly qualified computer experts, there could also be women – a notion quite unimaginable in this country. The dominant view in the North perceives Indian women as a homogeneous group and imposes on them contrary ascription. *The* Indian woman is thought to be traditional, dependent, oppressed, unable to read or write, liable to immolate herself on her husband's funeral pyre, and so on. In Hinduism, the red dot symbolises auspiciousness and the promise of hope and fertility. The replacement of the red dot with the "@" symbol implies different things, at odds with the above-mentioned characteristics. First of all, it implies that the prevalent conceptions about women in India could be misrepresentations; secondly, that at least some of these women might possibly be highly educated, and perhaps have professional qualifications in promising areas; thirdly, that the very fact of their being flexible and mobile, and their professional mobility, could actually bring them to Germany, too, instead of to the USA, the most popular country in this respect; and finally, that the women of the South must considered "agents of hope" for the economic development of Germany, one of the richest countries in the world. Which means development aid in reverse this time – within the structures of domination. It is in this sense that the lack of imagination must stop – namely, by recognising that female experts could hail from the South. And for this shattering of stereotypes *ifu* was a first step.

*ifu* must be seen, then, in the context of these diverse debates in Germany on internationalisation and institutional restructuring, of socio-political debates on inclusion and exclusion and multiculturalism and differences among women. All these debates show how essential it is to have interculturality as a principle structuring *ifu*.

## 2.1.4 Culture as Social Practice

The fact that "culture" became a relevant category for *ifu* is a reflection of a general phenomenon. With increasing globalisation, the category "culture" is gaining importance, too. There are even fears that culture might represent a threat. In his influential book entitled "The Clash of Civilisations and the Remaking of World Order", the US political scientist Samuel P. Huntington (1996) has formulated the prediction that the major dividing lines and the dominant sources of conflict in the future will be primarily cultural – and not ecological or economic. According to Huntington, the clash of civilisations will determine global politics. The breaking lines dividing Christian and Islamic, Confucian, Hindu, American and Japanese, European and African cultures will be the frontlines of the future. Given the present intermix of religions, ethnic groups, local conditions and languages, Huntington assumes that the next world war will be a war of cultures. While the persuasive power of Huntington's arguments cannot be ignored, they must be rejected because of fundamental theoretical flaws. He not only assumes distinct entities, but also assigns them to separate geographical regions, ignoring the irreversible interconnectedness of the world.

In a similar vein, the term "intercultural" can be misleading, too. It seems to refer to the existence of contact or encounter between cultures as a precondition of intercultural learning. But what is culture? In the debates in Germany, there is a tendency to reject "culture" as a category altogether. Firstly, it is rejected for being too monolithic and heavy. Secondly, it is felt to be charged with connotations of "Volk" (people). There is also a tendency to assign "culture" to the private sphere or to reduce it to the techniques of knowing or to patterns of behaviour. Also, it has been criticised for being too static, homogenising and essentialistic.

However, it is possible to think of culture in contrast to society, state or a system. If culture is thought of as social practice, it can be conceptualised in such a way that it is sensitive to power structures (Mecheril 2000). Initially, culture is merely a sign and a sound. It is in the very process of communication that the word attains the power of meaning. In this discursive understanding of culture, there is scope for considering power relations. The "other" is a construct and an imagination of the "self". The "other" is constructed in the discursive process. Consequently, we cannot posit "the migrant" or the "Third World woman"; she is being made that and she emerges in the discursive process.

In this sense, it can be said that *ifu* itself has created a culture. The *ifu* culture that emerged might be characterised as one of mutual respect, understanding, solidarity and sisterhood. But the *ifu* culture existing at the beginning of the courses was certainly very different from the one towards its end. The notion of dynamics, mobility and fluidity of culture is best captured in the words of Gayatri C. Spivak: "Culture alive is always on the run" (Spivak 1999).

This has implications for the understanding of intercultural communication, too. It is impossible to understand the "other" totally. It is necessary to realise that there are limits to understanding the "other". This realisation is crucial to a successful process of intercultural communication.

This concept of culture as social practice prevents us from folklorising and anthropologising the "other". It also helps in analysing the emerging conflicts in communication and interaction. In the context of *ifu*, conflicts among junior scientists can also be understood in terms of the categories of class, rural/urban divide, disciplinary affinity, theory/practice divide. To illustrate this, let us take an example of a role-play that was done during one of the intercultural communication workshops. The visiting scholars were asked to discuss one concrete situation and prepare a role-play on intercultural misunderstanding. The outcome was a scene in which one visiting scholar from the South was trying to keep the dishes she used in the common kitchen separate from those used by her room-mates. The friends wondered why she was doing this, and finally, after discussing it, attributed her behaviour to religious specificity, meaning "culture". However, in the discussion that followed the role-play, it was found that there was another visiting scholar, hailing from Europe, who also disliked sharing her dishes with others and preferred to have her own set of dishes. The visiting scholars came to realise that in intercultural settings there is a tendency to reduce conflicts primarily to culture.

The notion of culture as social practice makes cultural diversity and knowledge systems from other cultures visible without falling into the trap of culturalism. This notion has another aspect: intercultural learning is learning about ourselves – about the images *we* have of the "other", the "migrant", "Third World women". This becomes most evident in the statement referred to at the beginning of this piece. Intercultural learning is about recognising, discovering and unmasking images, stereotypes and prejudices – a process that can be initiated by direct encounter.

## 2.1.5 The Many Differences

*ifu* can be seen as an exercise in dealing with differences. There is an inherent connection between the category of culture and that of difference. Several types of differences – class, language, access to education, but also the geopolitical position, to name but a few – play a role here. These differences lead to exclusionary practices – including participation in *ifu*. The last is a very important one. Women of the so-called South would certainly not have been able to participate in *ifu* by their own means – even if they belong to the elite. Class background and lack of education denies access to higher-education systems as such. If English is spoken, it excludes the speakers of other languages. Even members of dominant language groups like Francophones or Hispanics are at a disadvantage here. It should be possible to imagine an *ifu* at which Wolof is spoken. My vision would be to have an international event at which diversity of languages and multilingualism are taken for granted. But even in this case, it would have to be seen how other mechanisms of exclusion can be eliminated. Also, differences among women and feminist ethnocentrism determine the process of scientific research as it takes place within structures of domination. The reversal of the situation, with women in the South doing research on women in the North, is difficult to put into practice.

At a theoretical level, there are problems with the naming process itself. The process of naming differences presents us with several dilemmas. Let me mention an important one. On the one hand, in order to make difference visible, it is necessary to give it a name, for example "Third World woman". By doing so, women in the Third World are homogenised and reduced to an essence – while at the same time the difference is reinforced and cemented. On the other hand, if you refrain from making a difference visible, you run the risk of levelling out differences. The question is how to recognise difference without essentialising and how to acknowledge differences without levelling them out. The challenge remains to finding ways of respecting and actively acknowledging difference, while at the same time working to abolish this difference. For a strategy for learning and acting in the awareness of difference, a kind of strategic essentialism might prove necessary.

Discussions on feminist theories of difference and political action led to concrete projects and actions at *ifu*. A case in point is the project on biographies across similarity and differences in the project area Work. A concrete action that followed these discussions was the Anti-Racism Rally on 3 October organised by *ifu* junior scientists. This day is the German National Reunification Day. It was renamed Anti-Racism Day by the *ifu* women to mark the rally that took place on that date. With the banners, slogans, songs and leaflets they took along, they did make a difference in the city of Hannover. In addition to the struggle against racism, they extended their solidarity to refugees in Germany. Later on, they joined another rally in Hannover that was protesting against the devastating situation of refugees in the country.

## 2.1.6 Intercultural Training at *ifu*

The discursivity of the notions of culture and difference was the major focus at some of the intercultural communication workshops held at *ifu* while preparing for the main event. There was one workshop for staff members, and all six project areas conducted intercultural training of visiting scholars. This was done on a decentralised basis at the respective universities and in different teams. A major focus of the workshops was self-reflection on anxieties, understanding the process of culturalisation and the processes of reinforcing and cementing stereotypes. Exercises were conducted to develop sensitivity towards difference. Debates also evolved around the ontological difference in ways of thinking and action. Some discussed their visions of the interculturality of *ifu*. Conflict management played a big part here. Exercises were also conducted to help recognise that junior scientists had intercultural competence that needed to be tapped. Group discussions, games, role-play and other forms of theatre were used as methods. All in all, emphasis was put on viewing intercultural communication as enriching rather than trying to neutralise deficiencies.

### 2.1.7 The Interculturality of Knowledge Production

The critiques of the dominant mode of knowledge production in institutions of higher education are manifold. On the one hand, there are the profound critiques of science from a feminist perspective (Fox Keller 1987; Harding 1993). There are also critiques put forward by intellectuals from the South (Lal 2000), feminist critiques in a postcolonial perspective (Minh-ha 1989; Spivak 1999) as well as by migrant intellectuals. The fundamental critique is that knowledge production is largely Western-biased. This is even the case in fields that claim to avoid this bias. As a study aimed at improving the teaching programme at the Technical University of Berlin has shown, even courses on interculturality hardly make any use of references from non-European countries (Basu et al 1998). Knowledge production in non-Western countries finds hardly any recognition in the Western academic community. This phenomenon is best described by the term "Eurocentrism". According to Ernest Jouhy (1995), Eurocentrism is a structure of thought and action that forms the basis for certain claims. The major claim is that the dominant rational-scientific thought and action that emerged in Europe is universal and that it is the only valid norm for assessing other cultures. No doubt each culture has its own norms and values that guide its assessment of cultures other than its own. This phenomenon of ethnocentrism can be found everywhere. However, not all ethnocentrisms are the same. Eurocentrism has a different quality because it has claims to dominance.

The questioning of Eurocentrism has several aspects, the main one regarding its claim to universality. Universalism cannot be considered a given that only needs to be propagated throughout the world. Universalism, as it is understood today, is in itself particular. Universalism must be formulated as a goal, yet to be attained, and attained *together*. And this is where the scope for *ifu* lies: to work towards a mode of thought and action that is shaped by all cultures. Here, understanding universalism emerges within a context, in the in-between and in a process of negotiation (Fuchs 1999). This is well illustrated by the lecture given by Professor Kumkum Sangari in the project area Body. In her talk on the "The Beast and the Bomb", Sangari analysed beauty shows in the context of globalisation and militarisation in India and the commodification of the female body. She put forward a hypothesis on the complexities of the construction of the international woman's body. This prompted reactions from the audience and expressions of opinion by junior scientists from no less than 26 countries on all the continents, endorsing or rejecting her thesis.

In our intercultural exercises, the aim is not only to open up other systems of knowledge. The challenge lies in the methodology as well. Our aim should be to adopt a multiperspective approach to science.

### 2.2   Conclusion

With its structuring principle of interculturality, *ifu* has provided major impulses at the national and international level. At the national level, it has set the agenda for upcoming international courses. However, these remain within the context of neoliberal thinking and traditional scientific understanding. At the international level, plans are already in the making to replicate the *ifu* model of interculturality. Interculturality as a structural principle might be of far-reaching importance not only for *ifu*. It might even turn out to be a matter of survival for Europe. With increasing globalisation, it will be important to learn from Africa and Asia. According to the anthropologist Clifford Geertz (1996), in the process of decolonisation in Africa and Asia, societies have emerged that do not reproduce the idea of the unity of "nation" and "land", as was the case in Europe. Here, history is not repeating itself. Something different is happening on these continents. The heterogeneity of the continents and the violent cartographic divisions in the colonial period have caused new entities to emerge, entities that have solutions to the challenges posed by the age of globalisation. These solutions might be better than the ones that are associated with the European idea of the nation-state and that of the unity of one nation, one language and one culture. Although the political and economic tensions are much greater there than in the West, and although the religious and cultural problems are considerable, these societies function no worse than do the attempts to stabilise the construct of the nation-state. Geertz proposes a change of perspective when analysing the organisation of social, political and cultural processes. According to Geertz, the dominant mode of thinking in the context of modernisation must be abandoned and the new organisational structures that are emerging must be recognised. And he suggests: "We may soon have to recognise that the political restructuring in Africa and Asia is a process that is more important for the change of the US-European conception of social identity than vice versa."

By having a truly multicultural junior scientist body, *ifu* has taken an important step towards the change of conception advocated by Clifford Geertz. It is indeed an intercultural innovation in higher education. In order to counter Eurocentrism and the process of the monoculture of the mind, this must be complemented by even greater interculturality of faculty and curriculum (Doro 2000). This will enable the decentralisation of science and initiate the epistemological shift that is indispensable for the survival of the world.

## References

Aithal V (1999) Gender and Higher Education in the Context of International Developments. Paper presented at the Euroconference - Gender, Higher Education and Development, 24-26 April at the Centre for the Cross-Cultural Research on Women in Oxford (unpublished)

Aithal V (ed) (1996) Vielfalt als Stärke, Beijing '95. Texte von Frauen aus dem Süden zur 4. UN-Weltfrauenkonferenz [Diversity seen as strength, Beijing '95. Texts by Southern women on the 4th UN Conference on Women], epd-Entwicklungspolitik, Materialien II/96, p 84

Aithal V (et al) (2000) 100 days of feminism. texte zur kunst [texts on art] June 38: 88-97

Basu S (ed) (1998) Eurozentrismus in der Lehre [Eurocentrism in the Teaching Programme]. Abschlußbericht des Studienreformprojekts am Fachbereich Erziehungs- und Unterrichtswissenschaften der TU Berlin

Doro (2000) When "East" meets the "South" and the "North" sets the Agenda. diskurs 12: 27-31

Feldhaus A (1993) Water and Womanhood, Religious Meanings of Rivers in Maharashtra. Oxford University Press, New York

Fox Keller E (1987) The Gender/Science System, Or Is Sex to Gender as Nature is to Science? Hypatia 3: 71-85

Fuchs M (1999) Kampf um Differenz, Repräsentation, Subjektivität und soziale Bewegungen, Das Beispiel Indien [Struggle for Difference, Representation, Subjectivity and Social Movements, The Example of India]. Suhrkamp, Frankfurt am Main

Geertz C (1996) Die Welt in Stücken, Kultur und Politik am Ende des 20. Jahrhunderts [The World in Pieces. Culture and Politics at the End of the 20$^{th}$ Century]. Passagen Verlag, Vienna

Harding S (1991) Whose Science? Whose Knowledge? Thinking from Women's Lives. Cornell University Press, Ithaca

Huntington S (1996) The Clash of Civilizations and the Remaking of World Order. Simon and Schuster, New York

Indian Association for Women's Studies (IAWS) (1995) Remaking Society for Women, Visions – Past and Present. Background Volume for the Conference, New Delhi

Jouhy E (1985) Bleiche Herrschaft – Dunkle Kulturen [Pale Dominance – Dark Cultures], IKO Verlag, Frankfurt am Main

Lal V (ed) (2000) Dissenting Knowledges, Open Futures, The Multiple Selves and Strange Destinations of Ashis Nandy. Oxford University Press, New Delhi

Mecheril P (2000) Culture, Statement at the Intercultural Communication Workshop for *ifu* Staff on 10.2.00 in Hannover (unpublished)

Minh-ha TT (1989) Women, Native, Other, Writing Postcoloniality and Feminism. Indiana University Press, Bloomington

Spivak GC (1999) A Critique of Reason, Towards a History of the Vanishing Present. Harvard University Press, Cambridge

Spivak GC (1993) Outside in the Teaching Machine. Routledge, New York

# 3 The Project Area Water

Sabine Kunst, Andrea Burmester, Tanja Kruse

## 3.1 Water is Life – Background Information

Water is scarce! Only a few years ago, water was regarded – at least in many regions of the world – as a inexhaustible resource. This might, at first sight, appear to be a correct assumption. In actual fact, though, freshwater accounts for only 2.5% of earth's total water resources – and even this small amount is unevenly distributed across the world and constantly declining. Today, there are already some two billion people without access to clean drinking water, and experts are agreed that this number will continue to grow as the world's population increases. The water available to these people – whose provision normally involves considerable effort – is in most cases highly contaminated. Supplying the world's population with sufficiently clean drinking water is – besides providing adequate food supplies – currently regarded as the major problem facing mankind, a problem that wars are likely be waged over.

The urgency of this issue is felt most acutely in the countries of the so-called Third World. Here, the ever-growing divide between different sections of the population also applies to the provision of clean drinking water. Brackish seawater finds its way into depleneshed groundwater channels, rendering precious drinking-water resources useless. Sinking groundwater levels cause entire regions to subside and buildings to topple. Monsoons trigger off huge tidal waves accompanied by landslides; water gushes unused into the ocean, carrying with it tons of fertile soil, and thousands of people lose their livelihood or even their lives.

Both the surface waters and the groundwater are polluted by industrial or domestic wastewater that is often discharged untreated into the landscape; by contaminated leachate from uncontrolled waste dumps; and by the frequently improper use of pesticides and fertilisers. This results in bacterial contamination, sedimentation and increasing heavy-metal and nitrate pollution. This growing pollutant load not only exacerbates the problem of water shortages, it also leads to a rapid increase in water-borne diseases, particularly in conurbation and their mushrooming slum districts (favela syndrome). And utilisation of the existing scarce water resources is far from effective: leaks in water-supply networks, kilometre-long irrigation channels and irresponsible behaviour on the part of many industrial and private users result in ever-increasing losses. This wasting of precious water can only be countered by education and information campaigns, by promoting more efficient methods of irrigation, distribution and purification, and by stringent watermanagement measures on the part of the authorities responsible.

Pursuing an innovative policy of sustainable water management can help to save water, distribute the scare supplies more evenly and ensure the replenishment of depleted resources. "Management", as the term is used here, should include recording data on water supplies (quantity, quality and location) and supervising planning and exploration, distribution, quality control and conservation measures. If all elements of the hydrological cycle and all levels of utilisation are taken into account, one could speak of "integrated water-resource management". Since the way soil is utilised has an impact on the water balance, what is needed is an integrated concept of land and water utilisation. This in turn requires planning on the basis of water and river catchment areas, often transcending administrative or even national boundaries. (BMZ 1999) It becomes apparent that water and soil management are clearly connected. The correlation between excessive depletion of water resources and soil salinification, desertification and increasing erosion is as serious as the dependence of people on both water and soil: those in possession of land, soil and water resources are also in possession of power – and this is not only the case in so-called Third World countries. Without access to these resources, people are totally dependent on the goodwill of such "owners".

The importance – or rather the necessity – of regulating water supplies by introducing water charges that cover not only supply but also disposal costs has been under discussion for several years now. Many experts emphatically demand such a move. Critics of this policy, which is already being pursued in a number of countries, point to its social drawbacks. The poor sections of the population would then no longer be properly supplied with water. It would mean clean water for the "rich", dirty water for the "poor". However, studies by the Federal Ministry for Economic Co-operation and Development have shown that subsidising water, as is common practice today, actually benefits the middle classes and the "rich" who – unlike the "poor" – are connected to the water-supply network anyway. The sums in question are in fact needed for the construction, operation and maintenance of the water-supply network and for the disposal of wastewater. According to estimates by the World Health Organisation (WHO), 40-60% of all water-supply equipment in rural regions is not in working order. Socially acceptable solutions make provision for supplying the "poor" with subsistence amounts of water at a low rate ("lifeline tariff"), with increasing prices for consumption exceeding this amount.

The United Nations Development Programme (UNDP) has elaborated a concept that takes into account social, ecological and economic aspects in an effort to ensure sustainable development in the future. "Current water utilisation must be geared to available resources, without impairing the options of future generations." (BMZ 1999)

An international transfer of knowledge – as an alternative to conventional foreign aid – based on the principle of "helping people to help themselves" and accompanied by scientific exchanges is an important step towards meeting basic needs and hence ensuring peace. Unlike conventional foreign-aid programmes, this concept makes it possible to avoid the mistakes made in imposing Western standards, knowledge and technology on developing countries. For instance, failure to consider cultural aspects when planning and conducting foreign-aid projects can prevent such projects from getting off the ground. If projects designed to safeguard the resource water are viewed as an international transfer of knowledge, local conditions such as region-specific user perspectives or cultural obstacles can be discussed and given proper consideration. People in the so-called developing countries, who stand to benefit most from such a transfer of knowledge, work as project partners. Instead of decisions being taken over their heads, they have a say throughout the decision process. Projects are planned and executed by local people sharing the same cultural background. It would, however, be arrogant to assume that only so-called Third World countries can benefit from this approach. The industrialised countries themselves are not exactly awash with ideas on how resource management can be translated into truly sustainable development. This means that international projects, which often deal with urgent issues of water management, can and will yield new, energy-optimised solutions based on decentralised organisation of the water supply system. Such solutions will provide important input for the discussion on how sustainable resource management can be achieved in the future. The benefits these projects offer may not make a great impression on the industrialised nations, but their cultural and scientific impact should not be underestimated.

## 3.2 Feminist Perspectives at the Action Level

In many parts of the world, it is women who provide their families with water and who thus have the most intensive contact with this resource day after day. This is particularly true of regions where water is scarce and its provision involves covering long distances. The enormous effort involved in fetching water each day is one of the reasons for the growing awareness and perception of the need to conserve this resource. Women have recognised its vital role in maintaining their families' health. There has been a growing understanding of the need to use water sparingly and of the options available for conserving and processing it. Water availability, water quality, potability and simple storage and processing methods – these are some of the categories in which women think in their daily dealings with water. Based on women's water-related responsibilities and their ability to give birth, Vandana Shiva (among others) has developed the thesis that women have a special, closer relationship with nature which makes them "more capable of preserving nature".

In this context, Vandana Shiva, a quantum physicist who was awarded the Alternative Nobel Prize for Ecology in 1993, argues that sustainable development should mean, among other things, the preservation or restoration of a large variety of species, i.e. "abundance". She sees women, as they go about their work, as "safeguards" of biological diversity, giving proper consideration to ecological and sustainable relations – in contrast to the predominantly linear, purposeful approach generally ascribed to men. Others (e.g. Agarwal 1994; Quisumbing et al 1995; Khan-Tirmizi 2000) go even further, arguing that gender inequality has a strong material basis. This argument is based on acknowledgement of the fact that independent ownership and the control of resources and means of production provide a potential source of income and wealth. Economic independence, in turn, is an important element in determining one's bargaining position, both within households and within communities. Agarwal (1994) argues that the links between gender subordination and property, which in many cases also means access to resources such as water, must be sought not only in the distribution between households, but also within households. She also shows that it is not just access to resources or ownership of property that determines one's relative position of power, but also control over resources or property. Agarwal's final argument, then, is that the links between gender inequality and access to resources and property do not simply apply to private situations but are equally valid for common property and access control. Access and control crucially hinge on having a voice in the institutions that manage the resources. Over the past decade, many mainstream scholars and policy-makers have pinned their hopes on the capacity of user communities to manage natural resources. The question now is whether current trends towards more decentralised forms of resource management will open the door to efforts to establish women's access to and control of these resources.

This view, focusing as it does on questions of resource management, neglects the issue of how to accept and integrate cultural differences, beliefs, norms and social and economic conflicts. And another crucial factor here is women's evolving role as experts in the field of resource management. Here there is scope for developing individual concepts that run counter to the mainstream. In recent years, the foundations were laid for national and international networks of experts, which it was hoped would be strengthened by the work done during and after the *ifu* semester These hopes were fulfilled. In Germany, there are several acknowledged papers and research projects by social scientists dealing with feminist perspectives in the engineering and natural sciences (e.g. Schultz 2000; Schultz a. Weller 1995). These are, however, of a strongly theoretical nature and thus of little help to practitioners in determining how to proceed with the development of eco-feminist work approaches.

The project area Water, then, as part of the International Women's University, faced the task to exploring new ways of integrating feminist perspectives into the practical work of engineers and natural scientists. What is needed here are innovative and sustainable, i.e. resource-saving, concepts that will help to considerably improve the overall situation with regard to water quality, availability and supply. By bringing together scientists from all over the world – women with highly diverse scientific backgrounds, but who all shared the desire to exchange ideas and to learn from each other – *ifu* provided a cornerstone the development of such concepts.

## 3.3   The Concept of the Project Area Water

The task the project area Water has set itself is to exploit and develop the potential for knowledge transfer provided by the research results of *ifu*'s three-month residency phase in the summer of 2000. The aim in bringing together women lecturers, visiting scholars and junior scientists from diverse scientific and cultural backgrounds was to generate impulses for organisational, ecological and social action and promote a "globalisation of thinking". Discussions integrated socio-scientific, cultural and political concerns with technical, scientific, ecological and economic questions. The ultimate goal was to ensure a practice-oriented and interdisciplinary transfer of knowledge by establishing teaching/learning networks which would facilitate the dissemination and implementation of project results.

The international composition of the project area's junior scientist body and faculty would, it was hoped, encourage participants to take a critical view of their own culture, values and practice. By tapping and combining the knowledge and experience of numerous scientific experts and practitioners, it would be possible to develop concepts and approaches for the sustainable management and just distribution of water and soil. This was why, even in the early stages of curriculum design, the project area Water placed such emphasis on close co-operation with experts from Asia, South America and Europe. This ultimately enabled a wide and highly heterogeneous range of topics and interests to be covered. Conditions in the junior scientists' home countries with respect to water management, development technology and the political situation differed considerably. The aim was to combine the know-how associated with traditional irrigation technologies, drinking-water recovery, rainwater harvesting and wastewater treatment with knowledge relating to gender perspectives, hygiene, health, social and political conditions in order to develop ideas on how to influence future planning decisions in the environmental sphere.

For many years now, there has been a consensus among experts that many regions of the world will suffer from water shortages in the coming century. Supplying the world's population with water is regarded as *the* overriding problem. Developing sound solutions to this problem means combining the knowledge and experience of as many experts as possible and pursuing a policy of gender equality in terms of gender mainstreaming. By initiating a transnational and interdisciplinary dialogue, *ifu* encouraged an intercultural and intergenerational exchange of knowledge that constitutes an important step in the search for such solutions.

The lecturers in the project area Water came from a wide range of professional and scientific backgrounds: structural engineering, biology, medicine, city sociology, area planning, political science, philosophy, social sciences and art. Experience and knowledge were exchanged and discussed, and with them essential information about the junior scientists' cultural backgrounds. As is evident from the project reports (cf. chapters 4–10), this exchange generated highly interesting, innovative ideas on how to develop new and sustainable integrated water-management concepts – integrated in the sense that they take into account the requirements of both water and soil management.

## 3.4 Curriculum of the Project Area Water

### 3.4.1 Knowledge Transfer

Junior scientists (visiting scholars and women lecturers) were able to deepen and enhance their knowledge during the first part of the semester – the theoretical knowledge transfer phase – in interdisciplinary and intercultural lectures, workshops and plenum discussions. The practice- and action-oriented project phase in the second part of the semester provided junior scientists with the opportunity to apply and further develop their acquired knowledge and skills in the scientific and cultural context of their respective countries. The international composition of the junior scientist body encouraged participants to take a critical view of their own culture, values and practices. It helped make them aware of the manifold social, economic, political, and especially cultural differences in their way of looking at problems (including scientific problems) as well as in the concrete approaches adopted to solve these problems. The idea was that these differences should be subjected to critical reflection. By encountering fellow women scientists from such a wide variety of backgrounds, by working together and holding lengthy discussions with them, junior scientists were able to broaden their horizons, extend their knowledge and even change their way of thinking. This provided a basis for developing new approaches to improving the quality of life.

Water was discussed as a political issue, from a feminist perspective, by reference to examples from selected regions of the world. Discussions combined social, cultural and political considerations with technical, scientific, ecological and economic issues. The teaching programme in the first phase gave a comprehensive treatment of the topics "Water and Culture", "Water and Gender", "Water Quality and Health", "Water Availability and Utilisation", "Wastewater Treatment and Watershed Management" and "Drinking-Water Processing".

Approximately one week was dedicated to each of the above key topics. The daily schedule was as follows: 2 lectures, each lasting 1.5 hours, in the morning; small-group workshops in the afternoon.

### 3.4.2 Practical Projects

The scientific work consisted principally in conducting problem-centred projects addressing controversial issues from a practice-oriented perspective.

We summarise here the topics treated and aims pursued in these projects. Detailed results are presented in subsequent chapters.

**Project: Rural Development with Special Emphasis on Women, Water and Environment (see chapter 4 for details)**
This project is concerned with the application of sociological principles and community planning to the development of rural areas. Based on concrete examples from India, special emphasis is given to the concerns of women, water and the

environment. Innovative development concepts resulting from the intercultural exchange are presented. Other key issues addressed are the integration of the village community in planning processes, the importance of empowerment through women's pressure groups and the development of strategies for conflict avoidance and management. Inter-disciplinary aspects play a prominent role here.

**Project: Water Treatment and Rainwater Harvesting (see chapter 5 for details)**
A comprehensive overview is given of the wide range of methods used for processing drinking water, from simple UV disinfection to industrial methods. The focus is on application options for rainwater harvesting. The differences in water-quality requirements for different applications are discussed. Based on the large amount of material collected during excursions, planning concepts are developed, incorporating a gender perspective, for facilities designed to collect and utilise rainwater for different purposes in different countries.

**Project: Wastewater Treatment (see chapter 6 for details)**
An overview is given of the current status of wastewater treatment in Europe, in particular Germany. The fundamental elements of wastewater treatment plants (mechanical and biological treatment stages) are explained in textbook terms, and typical biological treatment plants are presented on the basis of the evaluated excursion material. The processes at work in biological treatment plants are explained. This is followed by a brief description of how activated sludge and biochemical processes can be recorded. On a more abstract level, there then follows an explanation of how the biological processes can be illustrated using modelling approaches. This in turn provides the basis for the dimensioning of wastewater treatment plants. Finally, it is attempted to determine which elements of WWTPs developed in Europe can be used in countries that have a different infrastructure and climate and in some cases have no concept of wastewater treatment.

**Project: Decentralised Wastewater Treatment, Wastewater Treatment in Rural Areas (see chapter 7 for details)**
Chapter 7 gives a comprehensive overview of the state of the art with respect to the different decentralised wastewater treatment options: lagoon plants, soil filters, constructed wetlands and combinations of lagoon plants and soil filters. The way these different wastewater treatment plants function is discussed, and it is determined which types are suitable for which applications. This is followed by a discussion of the importance of gender perspectives for decentralised wastewater disposal concepts in rural areas. Finally, ideas for practically implementing these concepts are presented.

**Project: SIRDO Integrated Recycling System for Organic Waste (see chapter 8 for details)**
The connection between faecal waste and disease has been known since the 18$^{th}$ century. Today, lack of hygiene is still responsible for the epidemic-level spread of water-borne diseases in many areas of the world. In the course of the last century, many so-called developing countries have adopted the water closet and sewage disposal systems developed in the industrialised countries (and thus inherited

the problems inherent in such systems). Different approaches have also been adopted, though – approaches better adapted to the specific local situation (e.g. the availability of water). An example of such an approach is the SIRDO system developed in Mexico. For these innovative combined wastewater- and waste-treatment concepts, the critical variables were examined to determine their influence on the composting process. The aspects of user participation and women's empowerment as a result of marketing systems were systematically discussed.

**Project: River Development Planning (see chapter 9 for details)**
The River Stahlbach is an approx. 7km-long tributary of the Hardau in the River Elbe catchment area, near the town of Suderburg, Germany. A river development plan makes provision for the passage for fish and invertebrates and takes into account the hydraulic state and condition of the bed structure, the utilisation of the river meadows and future planned river-structure measures. On this basis, an exemplary catalogue of renaturalisation measures was drawn up, designed to restore the Stahlbach to a natural creek from its spring to its mouth over a period of three decades.

**Project: Water and Soil: Towards Sustainable Land Use (see chapter 10 for details)**
The connections between soil utilisation and groundwater quality are explained, taking into account soil properties and trickling capacity. A wide range of field- and laboratory-research methods are illustrated by reference to a study of the arable land, meadows and fallow areas along the River Elbe, including determination of physical, chemical and biological properties of the soil. The evaluation of project results is followed by an examination of the future prospects for sustainable soil utilisation taking into account the aspect of gender mainstreaming.

## References

Agarwal B (1994) A Field of One's Own, Gender and Land Rights in South Asia. Cambridge University Press, Cambridge, UK
BMZ Bundesministerium für wirtschaftliche Zusammenarbeit und Entwicklung (1999) Wasser, Konflikte lösen, Zukunft gestalten. Berlin
Khan-Tirmizi M (2000) Gendered Inclusion and Marginalization in Water Managament. *ifu*-Reader, Hannover
Quisumbing AR, Brown LR, Sims Feldstein H, Haddad L, Peña C (1995) Women, The Key to Food Security. International Food Policy Research Institute, Food Policy Report, Washington DC
Schultz I (2000) Ein Blick auf die deutsche Debatte. Hintergrundtext für den Niedersächsischen Forschungsverband für Frauen und Geschlechterforschung in den Naturwissenschaften, Technik und Medizin, Vortrag, 4/2000, Hannover, Schriften des ISOE Frankfurt, Forschung im Rahmen des Themas Gender and Environment, http://www.isoe.de/ftp/2001
Schultz I, Weller I (ed) (1995) Gender & Environment, Ökologie und die Gestaltungsmacht der Frauen. Frankfurt/Main

# 4 Rural Development with Special Emphasis on Women, Water and Environment

Leelamma Devasia

## 4.1 An Experiment in the Creation of Knowledge, Skills and Attitude

More than three-fourths of the population in the developing countries live in rural areas. The economy, occupation, socio-cultural background and problems of this segment of population are vastly different from that of the urban population (Augustine 1989). The majority of the developing countries with their colonial history have made a sustainable effort to develop their rural areas (Devasia a. Devasia 1994). In fact, the notion of rural development has been deeply ingrained in the minds of the political leaders, planners, intellectuals and professionals of these countries while working on the wide spectrum of national development. Thus, in most of these countries, rural development has become a movement for national emancipation. It takes the form of strategies, policies and programmes carried out in spheres such as agriculture, forestry, fishing, rural crafts, rural industries and building up social and economic infrastructure. The ultimate aim here is to further utilisation of available physical and human resources to ensure higher income and better living conditions for the rural population. But for tradition-bound South Asia, rural population development must be viewed as quality of life, something very different from macroeconomic hypothesis (Sapru 1989). It must set a trend for the technologies, organisation and values of that particular society. India alone has 57,800 villages. These villages provide food grains to a population of some 1,000 million, raw materials for industry and are, in fact, the support system and lifeline of one-sixth of the world's population.

In this context, rural development for women (Devasia 1988, Dube 1986) means:

1. Multiplying women's opportunities by increasing knowledge and skills as well as extending opportunities to participate both constructively and presumably in the activities of their culture (Dietrich 1992).
2. Progressively providing more effective means for rural women to resolve as peacefully as possible the conflicts and injustices that invariably arise as technological and cultural changes take place.
3. Maintaining an optimum balance between rural women's opportunities for freely chosen self-expression and the corporate needs of the culture in which they live (Agnihotri 1995).

4. Increasingly putting all potentially available farmland to the most effective use without damaging the earth's ecological systems.

There were junior scientists in this international project partnership. The home countries were: Austria, Bangladesh, Bolivia, Brazil, Cameroon, Canada, Greece, India, Indonesia, Luxembourg, Malawi, Nepal, South Africa, Sri Lanka, Vietnam. All junior scientists had a bachelor's or master's degree/diploma or Ph.D./doctorate in one of the following disciplines: Agriculture, Biology, Engineering, Psychology, Philosophy, Philology.

The project was an interdisciplinary endeavour to redefine the concept of rural development and the application of relevant technology and methodology to suit each participant's reality, with women, water and the environment as the focal points. In fact, the fundamental object was to generate knowledge, promote a positive attitude and develop skills enabling the junior scientists to work with rural women in diverse situations in different parts of the globe. The approach was participatory. To begin with, the attempt was made to develop concepts that are relevant to the project. This helped the junior scientists to define and redefine their understanding of development. Most of the junior scientists shared the experience of enormous difficulties encountered while working with rural women. They believed that they possessed the required know-how, but applying their knowledge in practice was the most difficult task, which they later on realised was due to the lack of skill and professionalism. They had to learn, unlearn and re-educate themselves, because they wanted to involve rural women in their countries in development.

The project had the following objectives:

1. To develop the junior scientists' knowledge about working with rural women so as to attain scientific knowledge appropriate to their own milieu.
2. To acquire skills relevant to working with rural women and apply these skills in order to initiate rural women's development.
3. To inculcate a proper attitude in the junior scientists so that they would be ready to work with rural women, even though the problems they face are seemingly insurmountable and the social situations they encounter are chaotic.
4. To do practical work either in the form of an action or research project so that the knowledge, skills and attitude thus developed are tested and verified.

The present paper is based on the author's experience at the International Women's University *ifu*, as a 'Professor in the project area Water. The input is mainly from the author's work with rural women in Vidarbha, Maharashtra, India, and the project "Rural Development With Special Emphasis on Women, Water and the Environment" conducted at *ifu* from August 1999 to October 2000.

## 4.2 Feminisation of Water Management – an Indian Concept

### 4.2.1 India – the Land and People

India is one of the world's oldest civilisations with a rich cultural heritage. It covers an area of 3,166,414 sq. km, about 2.42 per cent earth's surface, extending from the snow-clad Himalayan mountains in the North to the tropical rain forests in the South. It is bounded by the Bay of Bengal in the East, Arabian Sea in the West and Indian Ocean in the South. India features a number of mountains and lakes which make it a distinct geographical entity. The mainland has four regions: the great mountain zone, the plains of Indus and Ganges rivers, the desert region and the southern peninsula.

Approximately 16 per cent of the world's population lives on the Indian subcontinent. The country comprises 26 states the boundaries of which mainly follow language lines. Indian society is blend of different cultures, castes and backgrounds. Democracy is the practised form of government.

The country is criss-crossed by a network of rivers, which are the main sources of water. There are snowed and rained rivers. The northern rivers are perennial and snowed and have a reasonable flow throughout the year. The southern rivers are rained and a large number of streams are non-perennial. The coastal streams are short in length and have limited catchment areas.

The climate can be broadly described as tropical monsoon type, with the following seasons: winter (Jan-Feb), summer (March-May), rainy (June-Sept) and post-monsoon, also known as north-east monsoon (Oct-Dec). The south-west monsoon brings most of the country's rainfall.

**Water Policy and Safe Drinking Water**
With water as a resource under tremendous pressure from population growth, rapid urbanisation, industrialisation and environmental degradation, long-term planning for water resources is a felt need. Government water requirements are met through major, medium-size and minor irrigation projects tapping groundwater and by rainwater harvesting. The Central Ground Water Board (C.G.W.B.) is the national umbrella organisation for the management and development of the country's groundwater.

Water is a free gift of nature and India has plenty of it. As mentioned in the introduction, the subcontinent is surrounded by oceans and has an extensive river system. But providing safe drinking water to millions of people across India would appear to be an impossible task. The oceans, rivers and perennial monsoons seem unable to quench the thirst of the India's masses, especially those living in the rural areas.

**Water and Health**
Water intended for human consumption should be not only "safe" but also "wholesome". Safe water is water that cannot harm the consumer, even when ingested over prolonged periods. Water may be safe, but if it has an unpleasant

taste or appearance, it may drive the consumer to other less safe sources. Drinking water should therefore be not only completely safe, but also agreeable to use and wholesome. Such water may be termed "acceptable" or "potable". Hence safe and wholesome water is defined as water that is

1. free from pathogenic agents
2. free from harmful chemical substances
3. with a pleasant taste
4. usable for domestic purposes

Water is said to be contaminated if it contains infective and parasitic agents, poisonous chemical substances, industrial or other wastes or sewage. The term polluted water is synonymous with contaminated water. Pollution and contamination are the result of human activity.

### 4.2.2 Rural Women's Participation in Water Management in Maharashtra State

It is through a participatory research that women in the rural areas of Maharashtra, India were galvanised into action on the key issues of safe drinking water and better basic health-care facilities. The women participating in this endeavour are from 20 villages in the Vidarbha region. The women of these 20 villages came together, raised and discussed issues, found some answers and organised concrete measures to improve their lives and living conditions. In the process, their growing awareness and understanding encouraged them to shed their apathy and traditional vulnerability. While development remained for them a distant dream, they realised that women's vulnerability is reinforced by traditional structures of male authority that limits women's access to and control over the productive resources of land and labour, impose a gendered division of labour and restrict women's physical mobility. They had concrete experience of this with regard to the provision of safe drinking water for household use and the availability of better health facilities.

Vidarbha is one of the largest but least-developed regions of Maharashtra. It comprises 10 districts and accounts for about one quarter of the state's total population. About 82% of the population of Vidarbha live in rural areas, women making up about 49% of the population. For religious and socio-cultural reasons, women are not treated as equals in decision making processes in the rural areas of Vidarbha. Decisions on matters relating to the home, local self-government and developmental programmes are mainly taken by men, which means that women are consistently denied the right to make decisions about themselves in all spheres. Moreover, they are directly or indirectly denied access to various kinds of resources. These factors contribute to the subservient position of women in the rural society of Vidarbha.

With a group of social workers and activists from Community Action for Development (CAD), an NGO (Non-Governmental-Organisation) based in Nagpur, the researcher has, since 1990, been mainly in contact with the women of 10 remote villages in the Nagpur District. The chief purpose of such contact was to learn about the status of women in rural areas, their problems, their potential, the processes of social change taking place among them, their contribution to sustainable development and possibilities for making them literate. The villages in which the researcher worked were Pardi, Ashti, Yerla, Bodala, Chicholi, Fetri, Mahurzari, Walni, Khandala and Boregaon, all in the Nagpur District of Maharashtra State. In 1995, with the help of the rural women, the researcher initiated a participatory research study on the acquisition of safe drinking water and water management by women in these villages.

The rapport established by the researcher with the rural women greatly facilitated the participatory research. The main objectives of the research were:

1. to become acquainted with the sources of safe drinking water in the 10 selected arid villages in Vidarbha;
2. to elucidate the nature and characteristics of rural women's participation in the quest for safe drinking water;
3. to find out which mechanisms facilitate the participation of women in the acquisition of safe drinking water, especially in the summer months;
4. to understand the dynamics of better water management so that all households may share available water according to their needs and to their satisfaction;
5. to explore the possibilities of attaining sustainable rural development through effective participation and better water management.

These objectives were jointly developed by 890 women from these villages, who were also members of the local mahila mandals (women's groups), with the help of the CAD social workers. The main research issues were based on the following questions:

1. How and why do rural women participate in efforts to procure safe drinking water?
2. What is the significance of drinking water for the families of these rural communities?
3. How and why is safe drinking water related to better health?
4. How could sufficient acquisition of safe drinking water solve certain physical, emotional, monetary and other resource problems and create assets for the wellbeing of the family?
5. How does better water management lead to sustainable development?

In certain villages, the men were antagonistic towards the idea of women acquiring a greater say in their day-to-day life. The men thought that women must be kept occupied from early morning to late at night with household responsibilities such as looking after children, cooking, cleaning, washing, attending to other members of the family and animals and lending a helping hand on the farms. If running water were provided, what, they asked, would the women do with the time gained from not having to fetch water. They also expressed the fear that if women had more time and space for themselves, they might elude men's control, and that might disturb the rural social fabric. Hence, the men insisted, there was no real need for such facilities and participatory efforts to maintain the traditions of the rural community.

The members of the mahila mandal realised that most of the men in these villages were very conservative and they could expect little but the worst form of oppression from them. Yet, because of the consistent motivation of certain women leaders, the mahila mandals in all the villages were able to survive and draw conclusions from their discussions and dialogues.

## Participation of Women for Safe Drinking Water

Through the mahila mandal meetings, the women realised that participatory work for the acquisition of safe drinking water was possible in their rural communities.

There were three reasons for this:

1. Village women have a great capacity for initiative if they are free to act.
2. They can work to attain a better standard of living if they are exposed to development in nearby and similar communities and if they realise that water is the key to development.
3. In comparison with men, village women are more practical and willing to experiment, especially in the procurement of water and better health care.

The participatory efforts for water management by these rural women became self-supporting in villages such as Boregaon and Yerla. Women contributed a certain amount to cover incidental expenses for organising meetings, demonstrations, etc. The mandals of Boregaon, Mahurzari, Chicholi and Walni wanted to be registered in order to acquire legal status. The mahila mandal members were ready to contribute to cover expenses and many women came forward and offered to become official members of the mandals. When the news spread to other villages, the women there tried to replicate what had happened elsewhere. Legal status meant more responsibility and more power. It also helped the women to administer their own affairs in a democratic manner and gave them a feeling of equality and a sense of belonging. The registration of the mahila mandals helped them to assume responsibility in a participatory atmosphere. The women were responsible for policymaking, planning and the implementation of the mahila mandals' various programs, relating to the main objectives: acquisition of safe drinking water and better health-care facilities. The formation of people's organisations with grassroots-level participation helped them to understand their needs in the larger context of shouldering more community responsibilities. It was also a new learning process through which they were able to generate new knowledge about organisation, management, mobilising resources and developing negotiation skills for dealing with other people and the bureaucracy.

Initially, the research team tried to stimulate the women by asking critical questions about the availability of safe drinking water, e.g. drinking-water sources, number of taps, wells, tube-wells, farm wells, the distance the women have to cover to fetch water and the hours spent doing so. Questions such as who was responsible for providing safe drinking water, how much water was needed and to what other uses the water they collected was put, were asked to provoke ideas and discussions among the women. At the mahila mandal meetings, they discussed at length their villages' water sources.

**Fig. 4.1.** Tube-Well in India

During the initial stages of the participatory research, the situation with regard to safe drinking water in all the villages could be summed up as follows:

1. Most of the public/community wells were dug by the panchayat (village self-government) many years ago. There was no regular cleaning of the water in the wells. People and animals used the water for drinking and bathing. Women washed their clothes near these wells. As the wells were common property, no one was seriously interested in their upkeep; children played around them and threw stones, old shoes and other rubbish into them. During the summer months, the water level was depleted considerably and some wells even dried up.
2. There were some tube-wells in the villages which had been installed by the panchayat. The authorities in the panchayat determined the location and number of tube-wells needed for each village, without taking into account the needs and convenience of the villagers. In all these villages, the water table of the tube-well was found to be very low. Very often, the tube-well equipment failed and for its repair the women were dependent on the mechanics working for the panchayat. The ensuing red tape delayed the repair work for months or even years.
3. All the private wells belonged to rich landowners. As owning a well was very expensive, water became a rare commodity for the poor women, especially in the summer months.

4. Farm wells were owned by rich farmers and were intended for protective irrigation. Most of them were situated on farms, at quite a distance from the village. Normally, the water available in the farm wells was not potable, but during the summer months these wells were the villagers' only source of water. A fair number of the wells were not always available to everyone because of caste and social stratification.

Through discussions and dialogue, quite a number of the women in these 10 villages realised that the water they used for drinking and cooking was large unsafe and contaminated. The members of the mahila mandal realised that when water was a scarce commodity, most women ignored questions of quality, being more concerned with the quantity they could acquire. The group discussions helped the women to recognise that all 10 villages had serious water problems. During the initial discussions with women to generate more knowledge about themselves and their villages, it was perceived that women's awareness about the issue of safe drinking water was low, except among the women of Fetri and Boregaon. The contaminated water from different sources created a number of problems, especially for children. Fetching water from farm wells situated 2-3 km away also caused physical discomfort, compounded by mental tension due to the antagonistic attitude of the well owners. Exposure to temperatures of up to 47°C during the summer months caused cramps, dehydration and was in some cases fatal. At their informal meetings in different villages, the women discussed the problem of safe drinking water again and again.

In almost all families, the housewife and the girl children had the responsibility for fetching water. An average family needs at least 250-300 l of water per day. Using an earthen pot or brass vessel, a women can carry 5-8 l of water, which means that she or one of the girl children has to walk 35-40 times to the water source to fetch the minimum quantity of water required by the household. By sharing their knowledge, most of the women realised the enormous nature of their problem: they have to get up at around 4.30 a.m. and fetch water until 6.30-7.00 a.m., repeating the same routine in the evening. This means that approximately three to four hours are spent each day merely on fetching water. An experience also shared by the women was the fact that on most days they had barely enough water for their own personal use, e.g. for bathing or washing their clothes. Their personal water requirements were of the lowest priority within the family, a fact the group discussions and dialogues helped them to make them aware of.

Lack of sufficient safe drinking water was exclusively the women's problem in most of the villages. The mahila mandals helped them to realise that there were ways of solving this problem, if they pooled ideas, planned and worked together. This growing awareness gave them hope. They were also encouraged to think in terms of alternatives and their viability. In general, these rural women understood the extent of the problem and the dilemma they were in, and this knowledge afforded them profound insights into the nature of the water issue.

As they became increasingly aware that the non-availability of safe drinking water was a serious problem in their village, the women were concerned and disturbed. Merely understanding the problem helped them to see that there were ways of solving it by themselves. In addition, they learned that the problem of safe drinking water had various dimensions, relating not only to the family but also to the community and its infrastructure. Women felt the need to have a sufficient supply of water. They also realised that water must be free from impurities to keep them in good health. In all 10 villages, the people used the same water for drinking, cooking, washing, cleaning and bathing both themselves and animals.

Prior to the present participatory research, if they saw any problem at all, it was their effort to procure a sufficient quantity of water. However, as the research raised their level of awareness, they began to understand that the quality of the water constituted an equally – or even more – important issue.

## Water Management by Women

A pertinent issue in relation to water management was determining whether women alone should be responsible for providing water. Through dialogues and discussion, they came to the conclusion that everyone in a family was equally responsible for procuring safe drinking water. This realisation provoked innumerable conflicts, both within themselves and within the mahila mandals, in relation to the family and the community. A fair number of women in the mahila mandals thought it was women's sole responsibility to provide water for all the household's needs. But a few women suggested that men could/should help women to fetch water, which would leave women more time for other household chores and looking after their children.

**Fig. 4.2.** Women Preparing Food in an Rural Area in India

In the early stages of the participatory research, none of the women thought they were entitled to have some time for themselves. Most felt that they lived only for others. Gradually, as the research progressed, it dawned on some women that they too were individuals who needed rest, leisure, privacy and time for themselves. This was indeed a revelation for many of them. Almost all the women were accustomed to get up at 4.30 a.m. and retire for the night at 11.00 p.m. or midnight. When they totted up their working hours, they found that they worked for nearly 16-18 hours per day.

At the start of the participatory research, when discussing the problem of safe drinking water, the women stated that there was no need to involve men in their discussions, because they thought that women alone were responsible for providing safe drinking water and they could solve the problem themselves. But as the discussions advanced, the women began to realise that the problem was tied up with the structure of their society and that everyone must have a say in solving the it. The women in Walni and Yerla began to invite the menfolk to the meetings. Some came out of curiosity, others from a sense of insecurity because they feared their womenfolk were taking matters into their own hands. But the presence of men at the meetings enabled the community to take a holistic view of the problem of safe drinking water. Through such face-to-face interaction, men had an opportunity to understand the difficulties and problems experienced by the women in procuring water for the household. On many such occasions, participation became a complex task because meetings were attended by individuals with widely differing values and expectations, which, however, ensured the democratic nature of the process. As it progressed, the participatory process sought to accommodate everyone's needs, while at the same time aiming for social justice. The endeavour became multidimensional, with the simultaneous goals of justice, freedom and equality. This helped the women and men to understand not only themselves, but also each other and the community in general. Thus, the participatory research initiated a close interaction among the villagers and was instrumental in shaping views on the wide range of issues involved in the problem of providing safe drinking water. Women directly influenced and controlled the participatory research process in terms of decision-making, implementing decisions and evaluating results, because they generated their own knowledge.

From the outset, the research process brought women together in small informal groups. Gradually, such groups became stable and focused and evolved into mahila mandals. Later on, more and more women started participating in the mandals. They began to collect information relating to themselves, their children, their menfolk, their household, animals, farms, cultivation and marketing of products, ownership and use of money, education and health care of themselves, their family and their community. The women were motivated to collect an abundance of information on themselves and their surroundings, which helped them to realise that there were two types of ownership in the villages, individual and collective, and they acknowledged that both are essential.

The former enhanced self-respect and self-support, whereas the latter encouraged community participation and fellowship. A rural community with no collective ownership was, they felt, a poor community because it discouraged people from coming together, participating and developing.

Problem-solving questions were the technique used by the women at the mandal meetings. By gathering information, the women came to realise that they possessed sufficient knowledge and that this could help them to understand the problem in all its different dimensions. They also realised that the problem must have a solution or that information was available that could help solve the problems. These problem-solving questions related to safe drinking water, the water sources and collection of water. Such questions also helped the women to reflect on their standard of living, social justice and equality. By posing such questions, they were able to develop individual and collective knowledge on the various issues of very real relevance to rural women.

Early on in the participatory research process, the women had a number of fears. They were even afraid to meet, discuss and share their views. They doubted themselves and their capabilities. But the group interaction guided by the CAD team made them feel comfortable and confident and inspired them to continue working for a solution to the lack of safe drinking water. The women discovered that they could be equal to men and that they too could be the advocates of rural development. At their meetings, they succeeded in creating an informal and casual atmosphere by singing together and sharing their experiences with the group. This helped to remove whatever psychological blocks there may have been and alleviate their fears.

The mahila mandal invited some prominent women of the village, including some of the women representatives of the panchayat, to join them. After overcoming their initial inhibitions, they became members of the mandals, boosting the self-confidence of the mahila mandal members. The women realised that the panchayat was a key institution in achieving their goal of procuring sufficient safe drinking water, and having the women members of the panchayat on their side strengthened their determination and optimism. They were in a position to take decisions conducive to an improvement in the women's situation.

The women of Pardi, Ashti, Chicholi and Bodela even organised 'morchas' (protest marches) to convince and pressurise the panchayats about the need and urgency of launching safe drinking-water schemes in their villages. The women of Pardi and Ashti organised street plays, demonstrating this need and exposing the apathy of the panchayats. Street plays, protests and a constant struggle for water created community awareness. The panchayat of Ashti was quick to respond.

A special meeting was convened to take swift action on submission of a Drinking Water Scheme proposal to the Zilla Parishad (district-level self-government). Other villages such as Bodala, Mahurzari and Yerla tried to replicate the work done in Ashti. The problem of safe drinking water became an important talking point in all the villages. This helped to boost the women's self-confidence and promoted mutual trust. It also generated support and solidarity among people who shared the same views. The group meetings organised by the mahila mandals, especially in Fetri, Boregaon and Khandala, offered the opportunity for women to interview other women face to face, and this paved the way for building rapport between them. The educated women recorded the information collected in such interviews. Notes and records helped them at a later stage to prepare policy decisions, to plan and implement the programmes and ensure proper evaluation of the actions undertaken.

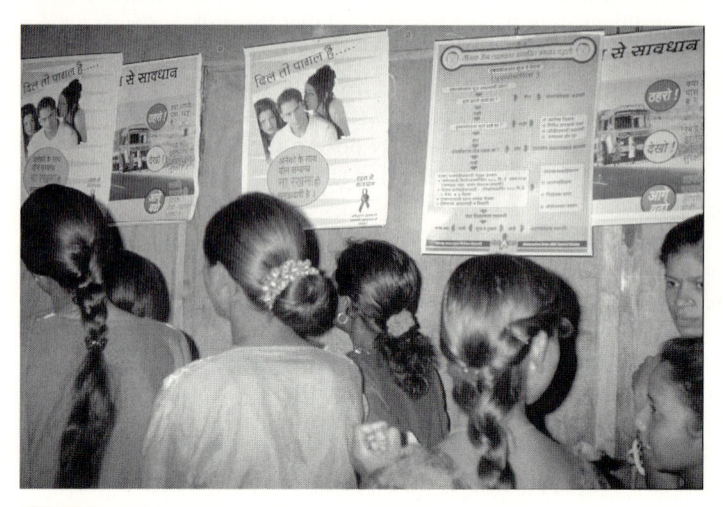

**Fig. 4.3.** Group Meeting to Discuss the Water-Supply

In some of the villages, the panchayat set about repairing the malfunctioning tube-wells and reviving the running-water supply scheme. A couple of the panchayats took the initiative in preparing new water-supply schemes for submission to the Zilla Parishad. It was also decided in most of the panchayats that whenever tube-wells were to be installed, the panchayat would consult the mahila mandal members as to the exact place of installation. It was decided to provide a concrete platform surrounding each tube-well, including cemented drainage. The mahila mandal members also decided that the villagers should be educated and disciplined about keeping the immediate vicinity of the water source clean. Some of the women from the mahila mandal volunteered to keep an eye on non-conforming villagers.

Men and women blamed each other for the repeated breakdown of the tube-wells, but ultimately it was the women who suffered most. The women thought that if young people were trained, they would be able to repair the tube-wells when they were out of order. Many women also expressed their willingness to undergo such training. The women insisted that the panchayat assume a supervisory role here.

The participatory research was used as an effective method to enable the oppressed women of these 10 villages to begin taking control of the social, economic and political forces affecting their lives. Safe drinking water was one of the prime concerns of the women. Based on the knowledge created by examining their own environment – comprising other women, men, young people, children, animals, farmland, water sources, hills and forests – the women also increased their awareness of the various technologies available. Such knowledge was extremely valuable because it was based on the women's life experience. The women took an active part in the participatory research because they wanted to learn about themselves and their villages and set about solving their problems, particularly those that affected their everyday life.

The new knowledge created was based on the life of the community or the physical environment or both. It was used to improve the living conditions of women in particular and the rural population in general. The participatory research enabled the women to bring about radical changes in the panchayat with regard to the supply of safe drinking water. The women compelled the panchayat to take certain decisions of a substantive, meaningful and developmental nature.

In the participatory research process, women became the owners of the research. They generated new knowledge and disseminated it, which helped them to realise that they could often do the same kind of work that the so-called experts did. They showed intelligence in their decision-making and were analytical in their approach. They were able to develop new perceptions and put into practice methods and techniques that were appropriate to a rural setting. In a way, they made effective use of this new knowledge for their own self-education. By sharing their experience and knowledge, women who were not the members of the mahila mandals also benefited. In Khandala, Yerla and Bodala, the women made a point of sharing this information with others, calling women's meetings on a regularly fortnightly basis. Such meetings created links between the women and the establishment. This helped to communicate to society at large the needs of the village, especially in relation to safe drinking water. It was felt by the research team that such meetings were absolutely essential for solving the social and financial problems involved and for generating political awareness.

## The Results of the Women's Endeavours

The efforts to secure safe drinking water and ensure proper water management had an all-pervading effect. The younger generation of women were able to assume the leadership here as the problem of safe drinking water and better health care directly affected their lives. This was a major shift in power because the older women were in most cases comparatively free from family obligations and thus found time for community responsibilities in these rural settings. But as a result of the research, they encouraged the younger women to attend the meetings. The CAD as an activist group was able to foster the creation of mahila mandals in all the villages, a process that went quite smoothly.

It was, however, extremely difficult to generate an atmosphere in which women could analyse their own lives and motivate themselves to achieve specific goals such as the acquisition of safe drinking water and proper water management. The participatory research succeeded in making a lot of positive changes, especially with regard to the issue of water management by rural women. The women became aware of the economic dimensions of their needs, such as the cost-benefit analysis of running-water schemes, tube-wells and dug wells. When they actually did the calculations, they were amazed at what they had learned and the results they had obtained. The women also analysed the political dimensions of the panchayat, the Zilla Parishad and the prevalent decision-making processes in those institutions. This enabled them to assess their own ability to respond to the problems they faced in their villages.

Nearly all the members of the mahila mandals were keen to develop long-term strategies for solving the issue of safe drinking water in their respective villages. In the process, they also evolved the ability to fix priorities and the vision to have appropriate and tangible goals. In Pardi, Chicholi and Boregaon, women believed that their problems could be solved by digging a couple of tube-wells in different corners of the village. In other villages, the women wanted to have community wells. But none of them realised the implications of groundwater depletion. In order to maintain the level of the groundwater, they learned through detailed discussions that they had to develop social forestry and explore the possibility of setting up rainwater-harvesting programmes.

Developing a running-water supply scheme connecting all 10 villages was also considered a long-term viable solution to the problem of securing safe drinking water. The mahila mandals therefore strongly petitioned the different panchayats to explore the possibility of developing a special water-supply scheme for all 10 villages as they are situated in fairly close proximity. With the help of experts, they drew up a plan that also included the effective harvesting of rainwater in a couple of ponds. The decision was taken to deepen all the village wells and ponds and the move was implemented with the help of the menfolk. Within six months, 17 community wells in eight of the villages had been deepened, and with the help of the panchayat pipelines had been laid to supply drinking water to two of the villages. Sludge and waste material were cleared from an existing dry pond to enable the collection of rainwater. A couple of women volunteers from each village attended a one-week vocational training camp organised by the CAD on plumbing, repairing tube-wells and carrying out minor electrical work.

All the tasks were performed on the basis of "Shramdan", i.e. Mahatma Gandhi's concept of free labour. The participatory process of water management also helped to break down barriers of caste by providing an opportunity for women from all castes to sit down together as equals and discuss their problems. In six villages, the 'low-caste' women were allowed to draw water from private wells owned by the 'high caste'.

The members of the mahila mandals took serious measures to keep all the wells and ponds clean. The children were given special advice and guidance in this regard. For the first time in the recent history of these villages, there was sufficient drinking water in seven of the villages during the summer months of 1997. There is now more greenery in all the villages thanks to rainwater harvesting. The women hope that eventually every family in all the villages will have their own individual water supply and a water tap in their kitchen.

### 4.2.3 An Alternative Vision Planned and Directed by Rural Women

The rural women's efforts in the management of safe drinking water have provided a number of insights and lessons. First, the process of obtaining safe drinking water has definite class-gender micro-implications. It is the women of the poor rural households who are most adversely affected by the problems of water and health. Second, the adverse class-gender effects of the problem of acquiring safe drinking water are manifested in the erosion of the survival and knowledge systems of poor and low-caste rural women. The nature and impact of the acquisition and management of safe drinking water are rooted interactively in ideology on the one hand (in notions about development, knowledge generation, ownership of knowledge) and in status, freedom, social justice and women's power on the other. Implicit in the acquisition of safe drinking water is the attempt to carve out a space for an alternative existence based on equality rather than dominance, on co-operation rather than competition, between women, and between women and men. Indeed, what is implicitly being called into question is the existing development paradigm which is gender-biased and devoid of sensitivity to women's issues.

The adoption of a realistic alternative is, of course, no easy matter. An alternative might be transformational rather than welfare-related, indigenous rather than alien. This would necessitate complex and interrelated changes within the rural community itself: for instance, in the context of safe drinking water, the inculcation of the concept of 'Pani Panchayat', and rainwater harvesting leading to radical changes and development, bringing more prosperity to women and the villages. A change in the decision-making process would imply a shift from the present top-down approach to one that ensures the broad-based democratic participation of disadvantaged groups, especially the women in rural areas.

What is needed is a radical shift from the vertical or bureaucratic planning of water management to a community-based women-centred water-management system. This would require decentralised planning and control and institutional arrangements that ensure the involvement of rural women and the sharing of the associated benefits. Similarly, the continued use and growth of local knowledge and innovations on safe drinking water and water management must be encouraged. There should be a flow of knowledge from rural women to academics and experts, enabling village women to become researchers, teachers, consultants and scientists in the India of the 21$^{st}$ century.

Taken together, safe drinking water and water management highlight the need to re-examine, and the possibility of throwing new light on, many long-standing issues relating to development, equality, redistribution of natural wealth and institutional change. That these concerns preclude easy policy solutions underlines the deep entrenchment (both ideological and material) of interests in existing structures and models of development prevalent in India. And this fact also underlines the critical importance of grassroots political organisations of women as a necessary platform enabling their voices to be heard. Most of all, it stresses the need for a shared alternative vision, planned and directed by rural women, that can channel dispersed rivulets of resistance into a creative tumultuous flow.

## 4.3 An Interdisciplinary and International Approach to Rural Development within *ifu*

### 4.3.1 Women and Rural Development

Rural women make an essential contribution to the socio-economic milieu of every country (D'Souza 1975). Currently, women produce 50 per cent of world's food supply, account for 60 per cent of the workforce and make up to 30 per cent of the official labour force but receive only 10 per cent of the world's salary (Desai a. Patel 1985). In the rural economy, the role of men in capital formation has been recognised, stressed and studied, but women's role, though substantial, has been ignored (Devasia a. Devasia 1990). It is a common experience in South Asian countries that men migrate to urban areas as a result of industrialisation and modernisation, leaving the burden of agriculture to their womenfolk (Parasuraman 1995). And in thousands of villages, women become the major agricultural producers, especially at the subsistence level, without having entitlement rights (Chatterjee 1990). Though they constitute half of the rural population, they become vulnerable in the tradition-bound, male-dominated rural set-up.

In the *ifu* programme, junior scientists from Malawi, Nepal and Bangladesh reported that in their countries men frequently migrate, establish another family in their place of work and visit their native village and family only occasionally or not at all, but without ever relinquishing their claim to property or giving these rights to their wife and children in the village, thus leaving them in the lurch. Women's productive role is not recognised where such a patriarchal system ensures their subjugation, but their labour plays a key role in keeping themselves and their families alive. In most cases, they wait endlessly, and in vain, for both the man and money. In India, Pakistan, Bangladesh, Indonesia and South Africa, women are important economic agents, particularly at the subsistence level, and their income in the poverty household is essential for survival. In Thailand, Vietnam and many of the African countries, the work contribution of rural women is much higher than that of urban women. When there is a paradigm shift in economic development due to liberalisation – such as happened in Ethiopia, South Africa, Burma, India, Afghanistan and Iran – the subsistence economy of women shifted to the global economy of capitalism and the resultant failure of traditional farming methods, catastrophic famine, death and doom for women and children became a common phenomenon.

In developing countries, rural women are an underprivileged group without access to economic resources, power, authority or decision-making processes (D'Souza 1980). About 90 per cent of the female workforce are in the unorganised sector where social benefits and economic returns are very low (Devasia a. Devasia 1991). A large number of women depend on daily wages earned in agricultural work and their economic contribution is invisible. The societies in which they live are not ready to accept any change in women's role, work positions, status and so forth. Illiteracy and poverty marginalise the participation of women in gainful employment (Gangrade 1988). Not only is lack of education an obstacle to conceptualising and articulating their problems at a higher level, it also means dependence on men. Owing to cultural factors, women and their female offspring in developing countries are denied education, are underfed and malnourished but are forced to play a major role in performing farming and household chores (Rehana 1988). A large number of rural women depend on their female offspring for support, and both spend their earnings on the welfare of the family. Women are directly or indirectly denied access to various types of resources and these factors contribute to their subservient position in marginalised rural societies (Devasia a. Devasia 1999). They are implicitly discouraged from participating in development programmes specifically designed for them by not being allowed to attend meetings and by having restrictions imposed on them by their husbands and in-laws. Decisions affecting them are made by male family members (Ghosh 1994). Bina Agarwal (1992) reports that during a discussion with men and women in one of the villages in India the men categorically said: "Women are unable to handle money because they are illiterate and intellectually inferior to men". Our experience regarding women's participation around the world is that at the initial stage of a struggle women participate much more actively, but when the organisation becomes established men take over and women give in. The majority of rural women work fulltime on farms and in the household, with no time left for education, health care or self-preservation, let alone participating in meetings (Acosta a. Bose 1990). For them, it is a never-ending struggle from morning till night merely to secure the survival of their families (Gothoskar, 1992). Rural development therefore involves a fairer deal for rural women, with water and the environment as key factors. The strategies and programmes offered need to be more women-oriented. Education, health care, skill development, promotion of equality, social justice and protection of the environment are some of the essential ingredients of rural women's development (Guy 1987). The challenges of rural women's development include not only their financial involvement but also motivational and socialisational aspects. Traditional rural structures oppress women; they attain their due status only if they are allowed to participate. To achieve this, existing structures, prevailing prejudices and poor organisational facilities should be eradicated (Devasia 1988).

Rural women spend most of their time working in the service of others, e.g. managing the household, looking after the children or working in the fields to ensure the family's survival. Their development is possible by bringing about changes in the existing structures and priorities characteristic of each culture (Gupta 1986). The realisation of equal rights for women by changing social attitudes and practices will make a profound impact on the emancipation of rural women. In India, Bangladesh, Pakistan, Nepal, Thailand, Vietnam and most African countries, planning for rural women has been done by men, without any consideration being given to the nature, characteristics and depth of the problem. This leads to women being treated as beneficiaries of, and not as contributors to, progress. Such a one-sided approach and the belief that men are agents of change, the family breadwinners, the mainstay of the human race itself should be seen as myths. A reorientation in planning and policymaking at the global level is essential. Let women plan the programmes according to their needs and problems, and then the world will see how efficiently they participate in them (Devasia 1998). For women, development is only possible if they acquire the capacity to resist subjugation and oppression. Women's organisations are therefore very much needed to empower women to resist victimisation and human-rights violations. Rural women must be enabled to build grass-roots organisations. Their struggle is for survival, especially in the developing countries.

### 4.3.2 Rural Women - Water and the Environment

Water is one of the scarce commodities rural women spend their life securing (Devasia 1998). Scarcity of water is a major cause of concern in most developing countries. Studies conducted in India and elsewhere show that rural women spend on average 4-6 hours per day in fetching, providing and managing water, only at the household level (Devasia a. Devasia 1999). The socio-economic consequences of the water problem in rural areas are highly significant. It leads to violence against women who are unable to fetch water and to conflict among communities over entitlement and distribution. It also causes displacement as villagers are forced to move to other parts where there is apparently a surfeit of water. And, even worse, it causes migration to cities, creating uninhabitable slums. *Ifu* junior scientists reported instances of men taking more than one wife, to fetch more water for the household.

Massive water depletion is a cause for concern. Water scarcity world-wide is leading to the large-scale displacement of agricultural labourers from villages. A report by the School of Public Health, John Hopkins University, warned that unless corrective measures are taken by the year 2025 one in every three persons will live in countries that are short of water. It is widely recognised that water must be treated as a scarce commodity and managed carefully and prudently.

Since 1980, when the United Nations declared a decade of water and sanitation, the governments of South Asian countries have been trying to tackle water scarcity in various ways. India alone has to deal with 117,000 "habitations" which have no access to safe drinking water. There exists a direct correlation between the welfare of rural women and the availability of water because in rural households it is the women who fetch, manage and determine the quality and quantity of water. It is primarily their working day that is lengthened by the depletion of water. Every aspect of the life of rural women depends on water, and thus conflicting demands are made on the availability of water. Women are, in fact, those most affected by the continuous degradation of land, which results in scarcity of water, fuel, wood, fodder and other resources, essentials for subsistence. The increasing number of rural women participating in the ecological movement is directly related to family survival.

The women's movement and the ecological movement both stand for egalitarian and non-hierarchical principles (Barret 1988). Rural women from developing countries depend on nature for their own sustenance and that of their families and their societies. The destruction of nature therefore means the destruction of women's means for "staying alive" (Bose 1993). In the case of the village women in Asia and Africa, essential items like food, fuel and fodder medicinal herbs are gathered from village commons. They are used for personal needs or sold. Village commons are the basic source of sustenance for the rural poor. The health of forests depends on the availability of water and soil. Environmental degradation is manifested in dwindling forests, deteriorating soil and increasingly depleted water levels. Studies have shown that rural women are the group most affected by the loss of food and earnings through environmental degradation. Rural women who depend on the forest for subsistence and survival suddenly face starvation and displacement due to the destruction of forests. Their health and well-being are placed at tremendous risk by the disappearance of village commons and forests (Unni 1992). Displaced women find it extremely difficult to survive on unknown sources of income and to live with unfamiliar neighbours who may not be able to provide social support. Displacement due to mega-projects cuts them off from lifelong social relations and mutual support practices.

Living close to nature and enjoying nature's bounty equips rural women with a wealth of indigenous knowledge about nutrition and medicinal herbs. Valuable knowledge about roots that can be consumed during drought seasons enable them to stay alive during critical periods, like the farmers in Bolivia with their extensive knowledge of potato varieties. However, such indigenous knowledge and wisdom has no place in existing development models. The increasing participation of women in environment movements all over the world is a clear sign of their realisation that it is their life and existence that is at stake owing to the ruthless attack on nature (Chatterjee 1990). In developing countries, agriculture is the only industry in which the majority of women are employed en masse. Owing to patriarchal attitudes and the system of land rights, women's social status has deteriorated and their rights have been denied them (Tinker 1990). As a result of generations of indoctrination and misconceptions about their weakness and backwardness, it is very difficult and an excruciatingly slow process to get rural women out of their cocoons. This is an observation made by many *ifu* junior scientists. A Sri Lankan participant reported that women in Sri Lanka do not hesitate to play a leading role on the war front, but feel safe playing a supporting role in the social and economic sectors. Women's development should therefore have a dimension of quality, quality depending on the ability to conceptualise from a democratic perspective. The philosophy of development itself is based on respecting the worth and dignity of all human beings, on social justice, equality and the belief in the people's capacity and desire to change and progress (Dacis 1994). It can only take place in one's own environment.

In this context, *ifu* junior scientists defined the following elements of women's development:

- Meeting the needs of women in all sectors
- Women living in an atmosphere of love and harmony with the family, society and nature
- Improvement in their living conditions

- Change from the negative situation to a positive one
- Better education and facilities to improve women's lifestyle
- Economic independence, social change and equal opportunities for political participation
- Improvement of women's social well-being
- Equal opportunities for achieving economic, social and professional growth with equality for all and no discrimination
- Opportunities for spiritual advancement
- Eradication of women's illiteracy and poverty through self-help
- Availability of food, clothing, shelter, information and purchasing power
- Well-being of every individual family and the community as a whole

Based on this broader view of women's development, the model of rural women's development is defined by *ifu* junior scientists as follows:

- Agricultural sector enhanced by land reform, equal rights, supply of inputs, soil conservation, water management, fisheries, poultry, dairy farming, post-harvest technology and the exploitation of new and alternative forms of energy, eco-management and conservation
- Women-oriented programmes focusing on rural health education and family welfare
- The strategies advocated are long-term, based on felt needs and incorporating scientific participatory planning
- Social initiative and administrative reforms
- Simple technology relevant and suited to the local situation
- Tapping of traditional and rural wisdom
- Moving at women's own pace
- Women's participation at all levels

*ifu* junior scientists had a clear view of rural women's development, which was more or less based on the background of each individual participant. The emancipation of rural women, their economic independence and equality of opportunity constituted the key to such a development perspective. Junior scientists also felt that rural women's skill development was imperative for their equality and progress.

## 4.3.3  Skill Development

Community development is not a new phenomenon in the South Asian countries: in these countries planning and policy formulations to ameliorate rural poverty have been in existence for half a century (Devasia 1986). The time-tested methods and strategies include charity relief and welfare as well as government efforts coupled with the activities of NGOs. Various modules have been tried to uplift the rural population, but the problems of underdevelopment, especially that of rural women, still lingers throughout Asian, African and Latin American countries. The fruits of development have not reached women, especially those in rural areas.

The realisation on the part of development workers and technocrats that women must be enabled to question their situation, analyse the system on whose fringes they are compelled to stay, attack the root cause of their poverty and injustice and try to improve their situation through their own ideas and actions. In fact, the mainstream development strategies pursued in developing countries have excluded women from reflection, planning, consultation and involvement. Given this situation, the *ifu* project was planned and developed with an inbuilt element of skill development designed to prepare junior scientists for working with rural women in their own countries. Working with rural women is an extremely complex process because, for any urban-educated professional, entering into their life stream requires a great deal of skill, dedication, motivation and self-discipline, and this proved true for most *ifu* junior scientists. Constant self-analysis, updating of communication skills and reflection were necessary. Self-analysis not only strengthens professionals but also acts as a constant indicator in assessing one's own strengths and weaknesses from time to time and modifying and redirecting one's approach. The *ifu* junior scientists welcomed such a process.

To promote personal development and create a sense of togetherness among junior scientists, each morning session would begin with meditation, yoga practice, group-dynamic sessions, songs, and reflections initiated by the junior scientists. Through discussion, workshops, seminars role-playing and dialogue, junior scientists realised that the involvement of rural women could be made possible only through education and organisation in homogeneous groups to improve their bargaining capacity and receiving mechanisms. The pressing need here was to get professionals ready to work with these women, to equip them with social skills, awareness and organisational competence. Group discussions and the sharing of experience from Mexico to Chile, Cameroon to South Africa, Bangladesh to Vietnam revealed that purely economic and technological solutions have not only aggravated poverty and increased the marginalisation and inequality of rural women but have also made them victims of oppression and subjugation. What was needed, then, in the junior scientists' view, was the social development of rural women. The junior scientists realised, however, that though highly educated themselves, they lacked proper orientation and the skills required to actively engage in the helping process or to establish authentic communication channels with the rural women. This was because many of them were not familiar with the vision of and approach to women's development involving a process of participatory reflection and action. Many were not told that they must be sensitive change agents identifying with the people, alive to women's needs and aware of the power relationships that determine the course of development at the macro- and micro-levels in their country. What they needed was a firm belief in the potential of rural women and a deep-seated hope that the women themselves would be the architects of their own destiny and of social transformation.

This awareness and experience helped junior scientists to understand their own native reality. Problems were outlined, questions raised for discussion and research. People remember best what they actually discover and do for themselves.

The junior scientists were allowed to try out their solutions, draw general conclusions from particular cases and come up with their own answers. The personal and group reflection experience helped junior scientists to analyse events and situations better, to draw the line between what is subjective and what is objective, between what is and what should be, by dividing them into groups to share their ideas and then creating a genuine listening climate in which each participant had the chance to share her views and experience with the entire group, thus affirming the wisdom of each individual and treating them as experts with a contribution to make. This method should be replicated when working with rural women in their own countries.

The way one presents something and the method one uses are, after all, as strong as a political statement. They reflect respect for individuals, the importance of community, democracy and are closely related to the way one organises meetings, projects, decision-making procedures and the monitoring and evaluation of programmes. The idea is to perceive things as they are and the junior scientists as well as the rural women to assume responsibility for shaping their lives, their community and their environment.

Working with rural women implies using a number of skills in the process of development. It is only by practice that such skills are acquired. Instead of theorising, the junior scientists did skill-building exercises like listening, conducting meetings, writing and submitting reports, analysing a situation. These exercises led to new assumptions, hypotheses, theories and to originality in methods and approaches. The junior scientists discovered knowledge by themselves in situations of group dialogue. This "dialogical" discovery process encouraged a stronger personal conviction about the principles and orientations that evolved and provided the junior scientists with models for their work with rural women when they returned to their own countries.

Education is not a matter of learning more, but of how to behave differently. Both the project co-ordinator and the *ifu* junior scientists felt that, by coming together and critically examining the reality of their existence and the society in which they live, they would cease to be mere "historical persons" and would decide and act and help rural women to decide and act for themselves in order to transform their world. The project Water, then, was an introduction to a method for developing critical awareness, the fundamental politicisation of women, their rights, equality, empowerment, potential and a critical consciousness of their milieu and the world. The assumptions underlying each development model, its micro- and macro-significance were discussed in detail by the junior scientists. In a gradual and consistent endeavour, the junior scientists set about learning skills and techniques that could be practised across "borders" while at the same time becoming partners in social change and the development of women. This partnership was strengthened by personal commitment, deeper motivation and concerted efforts of the co-ordinator and her colleagues.

Together, they realised that women needed a global vision of development, but with hard work at the micro-level. The unreflected importance of some development concepts would certainly not stimulate rural women to come up with their own solutions. It was important that an indigenous development model, sustainable and participatory, should be evolved for rural women. The junior scientists realised that the values underpinning the system, such as ultra-individualism, excessive competition, rivalry and profit motive, male domination, exploitation, destruction of the environment, consumerism and false prestige, tended to be unfavourable to women. The *ifu* junior scientists felt that for women to develop and shape their communities and societies, they needed to try to integrate the values of sharing, co-operation, openness to learning from others, objectivity of judgement, equality, justice, respect for the environment and simple living.

Junior scientists from Bolivia, Brazil, Mexico, Indonesia, Bangladesh and South Africa were eager to take part in role-playing and street theatre to depict certain themes such as women's meetings, environmental protection, conflict resolution, water management by women which had a profound impact on other junior scientists because such techniques could be demonstrated easily and meaningfully in most rural situations. The exercises enabled junior scientists to draw valuable conclusions and led them to realise that such methods could be used to convey powerful ideas such as:

- Developing a competitive spirit among rural women
- Helping them to take the initiative in different situations
- Showing how rural women in traditional and orthodox societies can develop leadership qualities
- Explaining to rural women that some people can continue to work as individuals despite being in a group
- Showing that in every situation and in all societies women take time to relate to others in the group
- Illustrating rural women's inherent capacity for conflict resolution
- Demonstrating that water management can be practised, even in groups of illiterate women
- Showing that user groups can participate effectively and democratically

### 4.3.4 Exposure to Different Realities, Field Trips and Excursions

According to Freire (1973), exposure to a different reality enables one to acquire new knowledge, to relate one's own reality to that of others, to compare, adopt and understand common aspects from other cultures. Visits to some rural areas of Germany such as Strachau, Konau and Hitzacker enabled junior scientists to experience this for themselves (see chapter 7 for details). These trips helped them to realise that in the industrialised countries as well there are villages that in many respects resemble villages in developing countries. In developing countries, poor employment opportunities drive people from rural to urban areas. In Europe, too, pluralistic societies exhibit a great deal of social tension, like that in India, Pakistan, Indonesia and South Africa. Villages in Germany are less populous, enjoy a high standard of living, allow greater participation of people in the community's affairs and demonstrate the professional nature of women's work.

From a technological point of view, after visiting the different villages in Germany, the junior scientists advocated local technology, simple and adaptable to the needs of rural women in their own countries. Since the project focused on water, water processing and the low-cost water purification methods adopted by the German villages for water treatment became an important concern of the junior scientists.

On some of the trips, the junior scientists were exposed to laboratory demonstrations of low-cost water purification methods such as:

- Filtration of faeces-contaminated water using sand filters (with and without a grid)
- Irradiation with ultraviolet light
- Incubation of moringa seeds as a coagulant
- Use of sunlight, lime (bleaching powder) and boiling water to inactivate organisms

Such methods were discussed in detail and were found to be adaptable to or currently in use in most developing countries. The junior scientists felt that rainwater-harvesting techniques were one of the cheapest available methods for water conservation and environmental protection. South Asian countries have a comparatively large amount of rainfall, and yet these are the countries where the water table is being depleted day by day. The rainwater runs off to the oceans causing floods and soil erosion. The majority of the population depend on rainwater for domestic and agriculture use. In the junior scientists' view, rainwater harvesting is the best remedy for water scarcity in these countries. Studies and experiments in Indian communities have shown that this technique has helped many villages to overcome water scarcity. Internationally, the Kronsberg rainwater harvest in Germany provides valuable experience in rainwater (rooftop) harvesting technology (see chapter 5 for details).

## 4.4 The *ifu* Experiment as a Beginning of a New Endeavour

The project provided experience and lessons that could be used for rural development, especially in developing countries. The junior scientists realised that the world is experiencing an ecological and developmental crises, both of which are interacting and reinforcing one another as people show scant respect for the environment. In the name of development and greater material comfort, forests have been denuded and rivers polluted. There is a systematic degradation of the ecosphere and an appropriation of community assets. Drought and floods are common phenomena in many countries. Deforestation causes soil erosion, silting of rivers, lakes and tanks, often causing flooding. Deprived of tree coverage, the earth does not allow the rainwater to sink, and thus there is no way of replenishing the groundwater that is being rapidly used up for agriculture and industry, causing desertification. The commercialisation of plantations with a heavy concentration on monoculture is leading to the destruction of the polyculture of the age-old forests and the irrecoverable loss of certain plant genes. Added to all this is the indiscriminate use of chemical fertilisers and pesticides, causing health hazards especially to rural women the majority of whom work in the fields.

The massive deforestation is resulting in the extinction of wildlife and in gene erosion. The valuable genetic material found in forest canopies must be preserved. It is the fruit of thousands of years of natural selection. The strategy for rural development in developing countries have been largely dictated by the donor agencies, the junior scientists believe. The imposition of development models on developing countries on the assumption that these countries are unable to think for themselves and lack the resources to try out their own solutions is baseless, they observed. When it is in vogue to pursue an integrated rural development approach, donors steer the recipient local organisations in that direction. When "feminine" concepts are defined as the prime target in donor countries, all of a sudden huge amounts of money are poured into women's issues. When donor countries get fed up with these models, they try to jump on the bandwagon of so-called "appropriate technology", and suddenly mud-walled homes, manual pumps, solar heating and windmills appear in their development parlance.

Based on these insights, the *ifu* junior scientists sketched out the following strategies to facilitate women's participation:

- Rural women's opinion is to be valued in developmental issues, especially issues relating to water and the environment.
- Rural women need to be empowered (as individuals and as groups) through education and awareness building.
- Issues affecting women's daily lives should be chosen in order to attract them to rural development, water management and participation.
- Women's representation is imperative in all community-level decision-making;
- Gender equality is needed in the socialisation and education of children.

Other junior scientists thought that

- gender equality should be enforced by national integration
- women's representation in politics is absolutely essential
- international networking with and for women is important for their development
- the formation of pressure groups, activism/media presence is required to highlight women's issues
- capacity building among rural women needs to be encouraged
- advocacy for and by women should be initiated in rural settings

The work of the group culminated in the construction of the model of an ideal "Women and Eco-friendly Global Village". Though hypothetical, it demonstrated the ideas, skills and aspirations of the junior scientists, the main features of which were:

### Water
Provision of clean and plentiful water within easy access of women. The water sources were surface, ground and rain.

### Landscape
Sufficient forest cover with areas designated as protected community grazing land for animals; streams and lakes for harvesting rainwater.

### Waste Management
Composting systems for organic waste, landfills for other waste, and a reed bed for wastewater treatment.

### Biogas Plant Using Animal Dung

### Sanitation
Provision of individual and community toilets (wet or dry as the case may be); preference for dry hygiene like the SIRDO.

### Infrastructure and Social Facilities
Community centre, school, market, health centre, religious centre, park, open space and public transport.

### Energy Source
Biogas, wind energy, solar and water mills.

*ifu* provided an excellent opportunity for encouraging a democratic process of decision-making, collective thinking, creation of new knowledge, skill development and participatory efforts among women from all over the world by combining technology with women's issues. It was a great step forward in women's development, in conceptualising their issues and familiarising all women with the process. It was strongly felt during the *ifu* experiment that the vicious circle of oppression, underdevelopment and inequality could be broken by women enabling women to establish their own democratic organisations in which they can come together and fight injustices. Such organisations are vehicles for uniting women all over the world by stressing specific women's issues, especially the establishment and strengthening of rural women's rights. Thus *ifu* played a role in promoting scientific understanding of rural women's issues. The *ifu* experiment stressed women's role in society and attempted to see women as equal partners in development and sustainability of society. It stressed that a political system needs to support rural women and their initiatives and ensure a process of growth and a method for establishing rural institutions. Ultimately, it was realised that rural women's development is not a crash programme to be implemented. It must be pursued at women's own pace, taking into account their needs, aspirations, potential and limitations.

For this, political will and administrative vigour are needed. Rural women can now no longer be dismissed as a backward, powerless, illiterate, apathetic and change-resistant group of people, but must be recognised as committed individuals aspiring to grow and develop. The test of rural women's survival is their assumption of their rightful place and the return to them of the fruits of their hard work.

This may call for a change not only in the thinking process but also in finding the appropriate strategy for achieving this developmental goal. The *ifu* experiment made a good start here. It began as a stream, but will grow into a magnificent river quenching the thirst of millions on different continents. The *ifu* junior scientists are the catalysts of this gigantic endeavour.

## References

Acosta-Beleti E, Bose C (1990) From Structural Subordination to Empowerment. Gender & Society, Vol. 4, No. 3

Agarwal B (1992) The Gender and Environment Debate: Lessons from India. Feminist Studies 18 (1)

Aggarwal A , Arora D R (1989) Women in Rural Society. Allahabad, Vora Publishers and Distributors

Agnihotri I (1995) Evolving a Woman's Agenda: Report from Beijing. Economic and Political Weekly 30 (5): 3195-3198

Augustine J S (1989) Strategies for Third World Development. Sage Publications, New Delhi

Augustine PA (1991) Social Equality in Indian Society: The Elusive God. Concept Publishing Co, New Delhi

Bardhan K (1990) Women's Work in Relation to Family Strategies in South and South-East Asia. Samya Shakti, 4-5: 83-120

Barret M (1988) Women's Oppression Today: The Marxist/ Feminist Encounter. Verso, London

Basu A M (1993) Women's Role and Gender Gap in Health and Survival. Economic and Political Weekly 26 (43): 2350-2362

Brown L D and Rajesh T (1983) Ideology and Politica Economy in Inquiry, Action Research and Participatory Research. Journal of Applied Behavioural Science 19 (3): 277-294

Chatterjee M (1990) Indian Women, Their Health and Economic Productivity. World Bank Discussion Paper, The World Bank, Washington D.C.

Chaudhari M (1993) Indian Women's Movement, Reform and Revival. Radiant Publications, New Delhi

Dacis L V (ed.) (1994) Building on Women's Strengthes: A Social Work Agenda for the Twentieth Century. Hawroth Press, New York

Desai N, Vibhuti P (1985) Indian Women: Change and Challenge in International Decade. Popular Prakashan, Bombay

Devasia L (1986) Co-operatives and Rural Development. Kurukshetra 34 (6): 12-21

Devasia L (1988) Developing Rural India: The Co-operative Strategy. Dattsons, Nagpur

Devasia L, Devasia VV (ed) (1990) Women in India: Equality, Social Justice and Development. Indian Social Institute, New Delhi

Devasia L, Devasia VV (ed) (1991) The Girl Child in India. Ashish Publishing House, New Delhi

Devasia L, Devasia VV (1994) Empowering Women for Sustainable Development. Ashish Publishing House, New Delhi

Devasia L (1998) Safe Drinking Water and its Acquisition: Rural Women's Participation in Water Management in Maharashtra, India. International Journal of Water Resources Development 14 (4): 537-547

Devasia L, Devasia VV (1998) Women Social Justice and Human Rights. APH Publishing Corporation, New Delhi

Devasia L, Devasia VV (1999) Women Participatory Research and Development. Dattsons, Nagpur

Dietrich G (1992) Reflection on the Women's Movement in India: Religion Ecology and Development. Horizon Books, New Delhi

D'Souza A (1975) Women in Contemporary India: Traditional Images and Changing Roles. Indian Social Institute, New Delhi

D'Souza S (1980) The Data Base for Studies on Women: Sex Biases in National Data System. In: D'Souza A (ed) Women in Contemaporary India and South Asia. Manohar Publications, New Delhi

Dube L (1986) Introduction. In: Dube L (ed) Visibility and Power: Essays on Women in Society and Development. Oxford University Press, Bombay

Freire P (1973) Pedagogy of the Oppressed. Continuum, New York

Gankrade KD (1988) Sex Discrimination: A Critique. Social Change 18 (3): 63-70

Ghosh J (1994) Gender Concerns in Macro Economic Policy. Economie and Political Weekly 29 (18)

Gothoskar S (1992) Struggles of Women at Work. Vikas Publication House, Delhi

Gupta AK (1986) Women and Society: The Development Perspective. Criterion Publications, Delhi

Guy G (1987) An Annotated Guide to Global Development Capacity Building for Effective Social Change. Resources for Development and Democracy, Maryland

Parasuraman S (1995) Development Projects and Displacement: Impact on Families. The Indian Journal of Social Work 56 (2): 195-210

Rehana G (ed) (1988) Women in Rural Society: A Reader. Sage Publications, New Delhi

Sapru R K (1989) Women and Development. Ashish Publishing House, New Delhi

Tinker I (ed) (1990) Persistent Inequalities: Women and World Development. Oxford University, Oxford

Unni J (1992) Women's Participation in Agriculture. Oxford and IBH Publishing Company, New Delhi

Vyas A, Singh S (1993) Women's Studies in India. Sage Publications, New Delhi

Whyte W F (ed) (1991) Participatory Action Research. Sage Publications, California

Young K, Harris O (1982) The Subordination of Women in Cross-Cultural Perspective. In: Evans M (ed) The Women Question. London

# 5 Water Treatment and Rainwater Harvesting

Namrata Pathak

## 5.1 Overview

Water is a valuable resource that is essential in everyday life, domestically and industrially. As such, it is now considered to be the blue gold of the $21^{st}$ century. However, increased human demands and usage tend to degrade water quality, which could be devastating to the natural environment and consequently to human health. With the ever-growing human population, the problem of availability of adequate supplies of safe drinking water is worsening. According to UNICEF, 80% of diseases and death among children is caused by unsafe water. Water-borne diseases such as diarrhoea, dysentery, cholera, amoebic infection contribute to nearly 4 million child deaths every year. The use of unsafe drinking water and poor sanitation is part and parcel of the daily existence for more than one billion people world-wide.

The problem is particularly serious in most of Africa, South Asia, parts of south-east Asia and South America. Until recently, projects to improve water and sanitation were planned on grand scale with the promise that everyone would benefit. In practice, this is hardly true, since the better off and those with political influence were able to claim the benefits of improved supplies and the poorest were left without access, especially in rural communities. With this in mind, the emphasis of new projects has shifted from grandiose nation-wide plans and instead focused on the needs of local communities. Therefore, it is very important that affordable, efficient and reliable water-treatment methods that can be accepted and used by local communities be established, which in turn could bring about a change in the lifestyle of the people.

The water-quality standards recommended for drinking water by the WHO and other similar bodies are stringent. The maximum permissible limits of various physico-chemical parameters defining water quality (WQ) are given in Table 5.1, as prescribed by the WHO, respectively. Each country, while generally following international standards, may allow for slight deviations depending on the agroclimate, availability of water resources, socio-cultural practices, water conservation rates, etc.

**Table 5.1.** WHO Guidelines for Drinking Water Constituent (Microbiological and Biological Quality)

| Organism | Unit | Value | Remarks |
|---|---|---|---|
| **I. Microbiological Quality** | | | |
| *A. Piped water supplies* | | | |
| *A1. Treated water entering the distribution system* | | | |
| Faecal coliforms | Number/100ml | 0 | Turbidity 1 NTU: for disinfection with chlorine pH preferably 8.0: free chlorine residual 0.2-0.5mg/litre following 30 minutes contact |
| *A2. Untreated water entering the distribution system* | | | |
| Faecal coliforms | Number/100ml | 0 | In 98% of samples examined throughout the year in the case of large supplies when sufficient samples are examined |
| Coliform organisms | Number/100ml | 0 | |
| Coliform organisms | Number/100ml | 3 | In an occasional sample, but not in consecutive samples |
| *A3. Water in the distribution system* | | | |
| Faecal coliforms | Number/100ml | 0 | In 95% samples examined throughout the year in the case of large supplies when sufficient samples are examined. |
| Coliform organisms | Number/100ml | 0 | |
| Coliform organisms | Number/100ml | 3 | In an occasional sample, but not in consecutive samples |
| *B. Unpiped water supplies* | | | |
| Faecal coliforms | Number/100ml | 0 | Should not occur repeatedly if occurrence is frequent and if sanitary protection cannot be improved an alternative source must be found if possible. |
| Coliform organisms | Number/100ml | 0 | |
| *C. Bottled drinking water* | | | |
| Faecal coliforms | Number/100ml | 0 | Source should be free from faecal contamination |
| Coliform organisms | Number/100ml | 0 | |
| *D. Emergency water supplies* | | | |
| Faecal coliforms | Number/100ml | 0 | Advise public to boil water in case of failure to meet guideline values |
| Coliform organisms | Number/100ml | 0 | |
| Enteric viruses | | | |
| **II. Biological Quality** | | | |
| Protozoa (pathogenic) | | Do | Do |
| Helminths (pathog.) | | Do | |
| Free living organisms (algae others) | | | |

Reference: "Control of Brackishness in drinking water" published by Rajiv Gandhi National Mission for Drinking Water Mission & Central Salt and Marine Chemicals Research Institute, Bhavnagar 364002, September 1997: 13

Water quality is a very important issue. The major contaminants can be classified as biological and non-biological. The diseases connected with some of these are listed in Table 5.2.

**Table 5.2.** Water-Based Diseases Transmission and Preventive Strategies

|   | Transmission | Examples | Preventive strategies |
|---|---|---|---|
| 1 | Disease is transmitted by ingestion | -Diarrhoea (cholera)<br>-Enteric fevers<br>-Hepatitis A | -Improve quality of drinking water<br>-Prevent casual use of other unimproved sources<br>-Improve sanitation |
| 2 | Transmission is reduced with an increase in water quantity: Infections of the intestinal tract; Skin or eye infections; Infections caused by lice or mites | -Diarrhoea (amoebic dysentery)<br>-Trachoma<br>-Scabies | -Increase water quantity<br>-Improve accessibility and reliability of domestic water supply<br>-Improve hygiene<br>-Improve sanitation |
| 3 | Pathogen spends part of its life cycle in an animal which is water-based. The pathogen is transmitted by ingestion or by penetration of the skin. | -Guinea worm<br>-Schistosomiasis | -Decrease need for contact with infected water<br>-Control vector host populations<br>-Improve quality of the water<br>Improve sanitation |
| 4 | Spread by insects that breed or bite near water | -Malaria<br>-River blindness | -Improve surface-water management<br>-Destroy insects' breeding sites<br>-Decrease need to visit breeding sites of insects<br>-Use mosquito netting, use insecticides |

1: water-borne and water-washed; 2: water-washed and water scarce; 3: water-based; 4: Insect-vector Waterlines. "Technical Brief No. 52: Water: Quality or quantity", Vol. 15, No. 4, April 1997

**Objectives of the Study**

The focus was on the use of simple and appropriate small-scale technologies and disinfection methods for polluted water, which after treatment could be used at the household level. The treatment methods used were: physical (heating, UV radiation and slow sand-filtration), chemical (treatment with aluminium sulphate and calcium hypochlorite) and biological (use of *Moringa oleifera* seeds), respectively.

Studying the efficacy of the treatment methods based on microbiological testing for total coliform (TC), faecal coliform (FC) and heterotrophic plate count (HPC). Overviewing Rainwater Harvesting Technology (RWH) with suitable case studies from India and Germany. Undertaking visits to various rainwater harvesting sites (eg. Hannover /EXPO grounds, Kronsberg and Darmstadt).

Investigating the quality of rainwater samples, collected from various sites during excursions, in the laboratory for cations and heavy metals. Drafting of project proposals on Rainwater Harvesting by the junior scientists with reference to the climatic variations (rainfall pattern/temperature/water demand) in their respective countries.

Surveying traditional water-processing practices in Germany, along with excursions to different waterworks and institutions responsible for the water supply in the country.

## 5.2 Water Disinfection Methods

### 5.2.1 Physical and Chemical Methods

There are a number of factors affecting the method of choice for disinfecting polluted water. Some of these are: 1) efficacy against water-borne diseases, 2) accuracy with which the process can be monitored and controlled, 3) aesthetic quality of the drinking water, 4) the availability of technology for adoption on the scale required. In short, disinfection is the elimination of pathogens that are responsible for water-borne diseases. However, in addition to potential pathogens, raw water may contain contaminants that can interfere with the disinfection process or maybe undesirable in the finished product. These include inorganic and organic contaminants, particulate and other organisms. These may 'demand' a capacity to react with and consume the disinfectant, thereby necessitating a higher dosage. Preliminary treatments like filtration, coagulation, oxidation, sedimentation and flotation can therefore be carried out to obtain a clear water sample.

Coagulation processes are conducted to promote the aggregation of small particles into larger particles, which can be subsequently removed by sedimentation or filtration. This treatment differs from chemical precipitation in the way it destabilises stable particulate suspension in water. Clay- and silt-based turbidity and natural organic matter are some of the materials commonly removed by coagulation. Alum and iron (III) salts are the most common treatment chemicals, while synthetic organic polymers are sometimes also used alone or in combination with the chemicals [1]. These methods do not kill pathogens, but they reduce their level besides removing particles that could shield the pathogens from chemical or thermal destruction.

Microfilters are small-scale filters designed to remove cysts, suspended solids, protozoa and, in some cases bacteria from water. The filter material is ceramic or fiber element that can be cleaned to restore performance as the units are used.

Reverse osmosis forces water under pressure through a membrane that is impermeable to most contaminants. The most common use is aboard boats to desalinate salt water. The water produced by reverse osmosis, like distilled water, will be close to pure water, so that mineral intake is suggested to compensate for loss of the normal mineral content of water. [7]

Sedimentation and flotation are solid-liquid gravity processes. Sedimentation promotes the settling of solid particles to the bottom, these being subsequently removed. The method is generally used in combination with coagulation and flocculation to remove floc particles. In contrast, flotation introduces gas bubbles into the water which attach to solid particles and create bubble-solid agglomerates that float on the top of the water column and are then removed.

Filtration is the most commonly used process in most water-treatment facilities for the removal of suspended particulate matter like clay and silt, colloidal and precipitated organic matter, metal salt precipitates and micro-organisms. Consequently, it can also be considered a disinfection process. In potable-water treatment, granular media filters are most commonly used. The material could be: sand, crushed anthracite, coal, garnet or granular activated carbon.

Slow sand-filtration functions by passing water slowly through a bed of sand. Pathogens and turbidity are removed by natural die-off, biological action and filtering. Here, it is important that water should have as low a turbidity as possible.

Activated charcoal filters water by adsorption. Chemicals and some heavy metals are attracted to the surface of the charcoal and attached to it. This method can be used in conjunction with chemical treatment. The chemicals could kill the pathogens and the charcoal will remove the treatment chemicals.

Disinfection has proved to be one of the multiple barriers for ensuring the production of microbiologically safe water quality. It can be performed in any of the following ways: using free and combined chlorine, chlorine dioxide, ozone, ultraviolet (UV) irradiation and other miscellaneous disinfectants like potassium permanganate, iodine, heat, or extremes in pH.

However, the performance of each is dependent on the type of disinfectant, dosage, pH, temperature, presence of interfering substances, feed-water microbiological quality and contact time.

Chlorination is the most widely used method because of its convenience and satisfactory performance. It is a strong oxidising disinfectant that has been in use for the past 60 years. The active forms of chlorine are:

$Cl_2 > HOCl > OCl > NHCl_2 > NH_2Cl > R-NHCl$ (where R- alkyl group)

The following points are considered in chlorination: time of contact between bacteria and bactericidal agent, temperature of water (the action of chlorine decreases at low temperatures), pH of water (the action of chlorine decreases at higher pH). The action of chlorine in *E. Coli* is the leakage of cytoplasmic materials in the order of protein, RNA and DNA. However, the presence of by-products of chlorination invites the application of alternative disinfectants. Ordinary household bleach contains 5.25% sodium hypochlorite and can be used to purify water. However, the chlorine content is unstable during long storage. Calcium hypochlorite has nearly 70% available chlorine. It is normally used to superchlorinate water to ensure the death of pathogens, hydrogen peroxide being subsequently added to remove excess chlorine. Chlorine dioxide is more powerful than chlorine but causes adverse taste and odour problems. Also, its by-products like chlorate and chlorite have been found to cause harmful effects in animals.

Ozone is the most powerful disinfectant, but it is highly unstable, so secondary disinfectants are still required. It is produced by exposing air or oxygen to a high-voltage electric arc.

UV light in the 240-280nm region is an effective agent for killing bacteria. The mechanism of action is that the inactivation of UV light acts through direct absorption of UV by micro-organisms, causing a molecular rearrangement of one or more of the biochemical components that are essential to the organisms' functioning. The major sites are the purine and pyrimidine components of the nucleoproteins, causing the reversible formation of pyrimidine hydrates and pyrimidine dimers. It also causes a break in the bonding structures [5]. However, the efficacy of the treatment is very much dependent on the turbidity of the water sample.

Heat treatment, in particular boiling, is one guaranteed way of purifying water. Water is normally boiled for 5 or 10 minutes, plus an extra minute for every 1,000 feet of elevation. At below-boiling temperatures, contact time can be increased. African aid agencies figure it takes 1kg of wood to boil 1 litre of water. The cost of boiling water often consumes up to 25% of a family's income (Safe Water Systems, Honolulu, Hawaii). In some places, solar pasteurisers have been employed.

Solar photo-oxidative disinfection is a form of water treatment whereby, if sufficient dissolved oxygen is available, sunlight will cause the temporary formation of reactive forms of oxygen such as hydrogen peroxide and oxygen-free radicals. This treatment has been shown to dramatically reduce the level of faecal coliform. Exposure to 4.5 hours of bright sunlight has been shown to cause a thousandfold reduction in faecal coliforms in laboratory tests. The activity occurs when the temperature is above 50°C.

Micronutrients are substances that are needed by the body in small amounts because they cannot be synthesised by the body. This means they must be provided by the diet. Most minerals are micronutrients and are essential for the body to maintain its normal functions. However, except for iron, zinc, chromium, selenium, manganese, molybdenum and copper, all trace minerals are toxic at high levels.

The presence of metal ions in drinking water can be caused by a number of factors. Its presence can be good or harmful depending upon the type of metal ion. Its quantitative determination can be done using spectroscopic techniques.

Lead is generally not present in water, but it is typically introduced by corrosive water. Leaching lead from household plumbing, lead pipes, etc. contributes only very little contamination. It accumulates in our body – once it gets in, it stays in. In reality, our systems can eliminate some lead. Adults can get rid of the lead they consume in their drinking water if the level is 15μg/l or lower.

Copper is not present either in the source water, but rather enters the water as corrosive water flows through copper plumbing. The best way to eliminate it is to make the water less corrosive. Iron (Fe)and manganese (Mn) become a problem only when they are oxidised. In their reduced state (+2), Fe and Mn stay in solution and do not cause discoloration. The oxidised (+3) state, however, is insoluble. If Fe and Mn are already oxidised, we need some kind of filter.

When we test water for hardness, we are measuring the divalent ions in the water. These ions include calcium and magnesium. These are essential minerals needed by our body. But unfortunately, hard water has some unpleasant properties. The hardness comes out of solution when the water is heated. Hard water also keeps soap from lathering and makes it harder to rinse out.

### 5.2.2 Biological Method

For the biological treatment of the water sample, the seeds of *Moringa oleifera* were used. One of the most common vernacular names for *M.oleifera* is the horseradish tree. This arose from the use of the root by Europeans in India as a substitute for horseradish. Such a practice would not now be recommended as the root has been shown to contain 0.105% alkaloids, especially moriginine, and a bacteriocide, spirochin, both of which can prove fatal following ingestion. The leaves are widely used, particularly in India, the Philippines, Hawaii and parts of Africa, as a highly nutritious vegetable supplement. Analyses of the leaf composition have revealed them to have significant quantities of vitamins A, B and C, calcium, iron and protein. The leaves are considered to offer great potential for those who are nutritionally at risk and may be regarded as a protein and calcium supplement. The flowers, which must be cooked, are consumed either mixed with other foods or fried in butter. They have been shown to be rich in potassium and calcium. The seeds are utilised in some regions of India either as a green 'pea', in their immature state, or fried, in their mature state, possessing a peanut-like flavour.

### 5.2.3  Bacterial Contamination

To ensure that the drinking water satisfies public-health requirements, it is necessary that samples be examined at regular intervals for indicator micro-organisms. Pathogenic organisms are difficult to monitor as the methodology involved is time-consuming and highly intensive.

Any indicator bacterium selected should be closely associated with the source of the pathogen. It should be able to provide an accurate estimate of the number of pathogens present at levels which pose a health risk. It should have same survival characteristics of the most resistant pathogen and be accurately measurable using simple methods. It should occur in greater numbers than the intestinal pathogens concerned and should be more resistant to disinfectants and natural processes than the pathogens.

The recommended bacterial indicator is the coliform group of organisms. The coliform group of organisms are not pathogenic, they are aerobic, facultatively anaerobic, gram negative, non-sporulating, rod-shaped bacteria, which ferment lactose with gas formation and aldehyde/acid within 48 hours at 35°C inculation. Coliform are capable of growth in the presence of bile salts or other surface-active agents that are cytochrome oxidase negative. Coliform belongs to the family Enterbacteriaceae, which includes the genera *Escherichia, Klebsiella, Salmonella, Shigella, Aerobacter, Scrratia, Citrobacter* and *Proteus*.

The coliform group (total coliform) are almost always present in large number in the faeces of humans and other warm-blooded animals. They can be detected even after dilution. However, unlike many others that are not thermotolerant, *E coli* groups grow well at 44.5°C, i.e. they can ferment lactose and also mannitol at elevated temperatures with the production of acid and gas and form tryptophan.

The presence of faecal coliform cannot definitely identify human contamination of water systems, and this has led to the introduction of supplementary indicator organisms. Another important component of human microflora is faecal streptococci, a group that includes all streptococci found in the intestines of warm-blooded animals. The group 'D' streptococci known as Enterococci includes *Streptococcus bovis*, Streptococcus equines and *Streptococcus fecalis*. Faecal

Streptococcus (FS) reducing *Clostridia* are useful in determining the origin of faecal pollution and also assessing the efficiency of water treatment processes.

The human population has a much higher ratio of FC-FS, usually in excess of 4.0, whereas other warm-blooded animals have an FC-FS ratio less than 0.7. This kind of ratio can indicate whether a contaminant in a water sample is of human origin or not.

To test the bacteriological quality of water after disinfection, microbiological analysis can be performed on the treated water sample. Routine tests like Heterotrophic Plate Count (HPC), Total Coliform (TC), Faecal Coliform (FC) can be used as an indication of the sanitary quality of water. HPC quantifies the viable aerobic bacteria in water samples. These bacteria do not represent the total number of micro-organisms in water, but rather those that are able to form visible colonies in plate count agar at 35°C for 48 hours. In contrast, Total Coliform organisms refer to a group of gram-negative aerobic and facultatively anaerobic, non-spore-forming, rod-shaped bacteria that ferment lactose at 35°C in 24-48 hours. The colonies can be identified as dark red colonies with a greenish-gold sheen. They are considered to be nonpathogenic under normal conditions. The faecal organisms refer to the thermotolerant forms of the total coliform group which ferment lactose at 44.5°C in 18-24 hours. Within the group, *E coli* and *Klebsiella* species are the organisms of interest because, when present, they indicate that recent faecal contamination has occurred with the possibility of accompanying enteric pathogens.

## 5.3 Rainwater Harvesting – Two Scenarios

**Indian Scenario**
At the dawn of the 21st century, India among other countries in the world is facing a water crisis caused by growth of the human population, industrialisation and urbanisation. This is compounded by ubiquitous groundwater contamination by pathogens and chemicals (fluoride, arsenic, iron, nitrites, etc.), improper water management and inadequate implementation of water legislation. Water is no longer considered a cheap and unlimited resource. Poor sanitation and unsafe drinking water are the cause of a large number of water-borne diseases like cholera, typhoid, hepatitis, etc. Since freshwater resources are limited in quantity and fast becoming elusive, the only alternative is to conserve every drop of rainwater by appropriate and economical water-harvesting techniques.

In situ collection of rainwater is felt to be desirable as the water contamination is low. Also, expensive transportation costs can be eliminated. Thus it is very important to appreciate the potential of rainwater harvesting for mitigating the problem of water.

Rainwater can be captured from the roofs of buildings in residential areas, courtyards, playgrounds, hill slopes, places of worship, institutions, etc. A rainwater-harvesting system consists of the following subsystems: catchment area (roof), conveyance system (guttering, downspouts and piping), filtration (screen), storage (cistern), disinfection (filtration, chlorination, etc.) and distribution. A rainwater-harvesting system may be designed both at the domestic and community level. It can serve the following purposes: (i) providing drinking water, (ii) increasing groundwater recharge, (iii) reducing groundwater contamination, (iv) reducing stormwater discharge, urban floods and overloading of sewage-treatment plants, (v) reducing seawater ingress in coastal areas, (vi) reducing soil erosion. Water harvesting involves the community and gives it ownership and responsibility over water projects. Thus water is used judiciously instead of being squandered by the elite's.

India is one of the wettest countries in the world with an average annual rainfall of 1,170mm. 80% of the country receives more than 80% of the rain (as liquid rain) during a three-month period. The rainfall varies from 100mm in the deserts of western India (Jaisalmer) to 15,000mm in the high-rainfall hills in the northeast (Cherapunji). Theoretically, the potential of water harvesting for meeting household needs is enormous. Rain captured from 1-2% of India's land can provide the population of 950 million with as much as 100 liters of water per person per day.

**German Scenario**
On average, each square metre of Germany receives 837 litres of rainwater annually. However, rainfall varies considerably from place to place, the annual amount of precipitation in European towns and cities ranging from 500 to 2,500 litres per square metre. Most of this rainwater either evaporates or flows off into the groundwater before being used by humans. In dry regions suffering from climatic water shortages, the rain sometimes evaporates completely before it can enrich the water supply. Another problem often encountered is that conurbations with high industrial and domestic water needs do not have sufficient groundwater resources. The damage resulting from sinking water levels and the fact that in many areas the drinking water has already become nonpotable make it convenient to use rainwater. Areas of rainwater usage may be flushing lavatories (requirement 18-30 litre per person per day), washing machines (requ. 16-40 l/person/day), for cleaning (requ. 3-5 l/person/day) and last but not least for watering the garden (requ. 60 l/person/day). [3]

In private buildings where no efforts have been made to save water, up to half the water used can be replaced by rainwater. In public buildings such as schools or offices, as much as 80% of the water requirements can be met by rainwater. The following factors have a direct influence on the quantity of usable rainwater resources, and thus the amount of drinking water that can be saved:

- Precipitation amounts and weather conditions
- The location, incline, alignment and type of the collecting surface
- The size of the collecting surface
- Storage-tank size
- Service-water requirements

**Rainwater Usage at a Glance**
The arguments for integrating rainwater into water consumption may be summarised as follows:

Water can be saved, thus eliminating the need to develop previously untapped water resources and expand the long-distance water-supply network, as well as reducing expenditure for sewage treatment.

Rainwater is initially retained in storage tanks, reducing the risk of sewage systems' being flooded during heavy storms.

Water pollution is reduced thanks to the decrease in detergent usage (less detergent is required as rainwater is soft).

Awareness of the need to save water is raised among consumers, thus tackling the root of the problem.

## 5.4   Description of the Project

**Project Junior Scientists**
The junior scientists of this *ifu*-Project were of different nationalities and ages (between 22 and 40), with different educational qualifications and experience. The project was a unique experiment in international and interdisciplinary co-operation. The junior scientists came together by choice to achieve similar goals and objectives. They came from the Philippines, from Albania, Nigeria, Tanzania, from USA, Cameron and India. Their special qualifications varies from chemistry, biology, public health, geography to agriculture.

**Treatment with UV Radiation**
Suitably diluted water samples (cf. chapter 5.6) were exposed to UV radiation and withdrawn at time intervals of 3.0 seconds, 6.0 seconds, 1.2 minutes, 5.0 minutes and 20 minutes, respectively. The efficacy of the method was examined by determining the microbiological quality of the UV-treated samples .

**Heating at Elevated Temperatures**
The water samples were heated at temperatures 55°, 65° and 75°C and were removed after 10, 15, 20 and 25 minutes, respectively. The samples were diluted 10 times and tested with regard to their microbiological quality.

**Slow Sand-Filtration**
The slow sand-filter was constructed such that the bottom layer was composed of stones and gravel with a diameter >1.5mm, the middle layer of coarse sand with a diameter of 0.9-1.5mm and the uppermost layer of fine sand with a diameter 0.3-0.5mm. Sample 2 (Rainwater from the EXPO lake) was passed through the filter and withdrawn at time intervals of 15, 30, 45 and 60 minutes. It was later subjected to microbiological analysis to check the efficacy of this technique.

**Reaction with Calcium Hypochlorite, Aluminium Sulphate and Slow Sand-Filtration**
1mg/litre calcium hypochlorite and 30mg/litre aluminium sulphate were separately used for treating water samples. The treated samples were then passed through a slow sand-filter and collected at different time intervals (15, 30, 45, 60 minutes). The efficacy of the method was examined by subjecting the treated samples to microbiological-quality tests.

**Treatment with *Moringa Oelifera* Seeds**
The effect of *Moringa oleifera* seeds on the water samples was observed. *Moringa* seeds were powdered and weighed. 2.0mg of the seed powder was added to 100ml of the water samples and placed on a shaker for constant stirring. Water samples were withdrawn at intervals of 5, 10, 15 and 20 minutes and filtered using a sterile filter and a double-layered muslin cloth. The filtered samples (diluted 10 and 100 times) were then checked for microbiological activity .

**Analysis**
The efficacy of the physical, chemical and biological treatments was determined by testing the microbiological and chemical quality of the treated samples and comparing them with the blank and control (distilled water and untreated) samples.
   Biological Quality of Water Samples: Microbiological analysis was done for faecal coliform (FC), total coliform (TC) and heterotrophic bacteria (HPC). [6]
   Chemical Quality of Water Samples: Besides testing for microbiological activity, a few samples from the physical, chemical and biological treatment were also tested for heavy-metal content, along with rainwater samples collected during the excursions to various sites, using inductively coupled plasma-atomic emission spectrometry (ICP-AES).

### 5.4.1 Presentation of Excursions

The junior scientists examined different water-processing methods at household and community levels for four weeks, focusing on the techniques and theory of water treatment. Simultaneously, the principles and techniques of rainwater harvesting were overviewed along with case studies from Germany and India. This generated enthusiasm for developing plans to harvest rainwater for the community in order to help relieve women of the burden of carrying water long distances. To help better understand the technology of rainwater harvesting and its application, excursions were made to the following sites: Kronsberg and the EXPO at Hannover, Darmstadt and Wiesbaden. A three-day excursion was also undertaken to waterworks in northern Germany, providing hands-on experience of various techniques for water management and processing for community use.

## Rainwater Harvesting: Case Studies

### Kronsberg

The visit to Kronsberg was intended to familiarise junior scientists with a rainwater-harvesting system that was not just designed for collecting rainwater to counter water scarcity or as a supplement, but as a seminatural management system. It was initiated as part of the Kronsberg development programme for ecological optimisation. The principle of rainwater harvesting applied here was to try and follow the natural system, with most of the rainwater runoff being infiltrated back into the ground rather than being lost as runoff down slopes, which in turn would have led to soil erosion and aggressive damage to the topsoil and the environment.

The system was an example of one of the decentralised systems for the retention and infiltration of rainwater using soakaways. The soakaways were created along the sides of the roads, as furrows or open gullies covered with grass and layered with pebbles to ensure that the water that eventually enters the underground storage basin is first filtered. This was one way of ensuring that the groundwater level was maintained and filtered. Also, the underground storage basin in the housing area was connected to the drainage pipes around the houses. Each of these pipes had a restricted flow-outlet shaft and small holes enabling the rainwater to be gradually released into the surrounding retention areas.

Apart from the underground storage basin, there were also two open water sources, created as hollows filled with different layers and sizes of sand and gravel, situated downhill between the housing blocks at strategic points to collect rainwater. These ponds were not only created for aesthetic purposes, to provide the residents with a lake in their neighbourhood, but also served to replenish the groundwater by slow infiltration into the groundwater basin.

The Kronsberg project demonstrated the living conditions in terraced houses with large and small apartments for families of different income levels. It was meant to create and attract a socially balanced residential population. The apartments on the top floor had green roofs which were used for collecting rainwater. The idea was to utilise some of the water from these green roofs to water the gardens before it was eventually infiltrated into the ground.

### EXPO

The EXPO 2000 World Exposition at Hannover was a global event attracting many thousands of people. One of the major concerns was to ensure proper toilet facilities at the site. To this end, numerous structural changes were made, both underground and on the surface.

Students from the University of Hannover assumed responsibility for providing water for the toilets and also for greening the exposition site. This led to the creation of three retention ponds at the far end of the site. The first pond covered an area of 2,600m$^3$, the second and third ponds each covering 50,000m$^3$. Reeds were planted to reduce the rate of evaporation and ensure more surface retention, the ponds being an open-surface feature. The ponds had a retention volume of 12,200m$^3$ and a storage capacity of 4,000m$^3$. Rainwater from these ponds was processed by flocculation, slow sand-filtration and UV treatment being used for disinfection. The water was then stored in tanks of 44m³ capacity and was eventually utilised for flushing thirty of the seventy toilets installed at the exposition site.

**Darmstadt**

The visit to the 'Fachvereinigung für Betriebs- und Regenwassernutzung e.V. (FBR)' offered first-hand experience of the rainwater-harvesting plant at the Technical University of Darmstadt. FBR is a private firm handling the treatment and supply of rainwater for flushing toilets, for laboratory water needs and for watering lawns on the university campus. It has a catchment area of 500,000m² and a cistern volume of 300m³. Rainwater is treated using two-stage flocculation, and sand-filtration is used for disinfecting the water. An average of 70,000m³ of water/year are processed for further use. The firm plans to extend the plant to increase its capacity to 100,000m³ water/year.

The public transportation depot at Darmstadt was also visited. Here, the water obtained by rainwater harvesting and wastewater recycling is used for cleaning buses. The rainwater is collected from the roof, with a total area of 6,000m², and stored in a tank of 80m³ capacity. After the buses have been cleaned, the wastewater is collected and recycled again.

**Wiesbaden**

In Wiesbaden, the market garden was visited to study water-conservation techniques in use there to supplement the water supply to plants during periods of low or no rainfall and to reduce the salinity of the groundwater in that area. The high salinity is due to salt being used on the roads and railway tracks in the wintertime to facilitate vehicular movement.

The rainwater is stored at the site in the form of a lake. Since the water collected here is used solely for plant growing, the bottom of the collection pond is lined with polythene sheeting to prevent contact with the saline groundwater. Before being used, the water is allowed to sediment and is then filtered. The plants' water requirements are carefully monitored using a sensor which indicates the level of water in each section of the garden. This results in an efficient water-management system for individual plants.

## *Excursions to Water-Treatment Plants Serving the Community*

The aim of these excursions was to familiarise junior scientists with the ways in which the groundwater is procured, treated and supplied for service use in Germany. There were many similarities between the various waterworks, e.g. their obtaining water from bore wells, the concept of protection zones and management by the community members. But they also differed in a number of respects, e.g. in terms of charges, maintenance, disinfection methods, etc.

**Uelzen: Three Waterworks**
- Stadensen II is run by the WBV-Uelzen/ SVO Water Association of Uelzen
- Bad Bevensen is run by the WBV-Uelzen/ SVO Water Association of Uelzen
- Stadtwald waterworks is run by the Water Association of Stadtwerke, Uelzen

All three stations procure water from bore wells, use multimedia pressure filters and perform aeration using oxygen gas.

At the Uelzen/SVO-Uelzen waterworks, water is not subjected to any form of disinfection because it does not contain any detectable harmful chemicals. Oxidation of ferrous ions in the groundwater is the only treatment carried out before the water is supplied for use. Also, the station is operated using a multimedia filter (anthracite/gravel and pressure filter with diameters of 3.50m each). The waterworks at Stadtwald has eight filters with Magno-Dol quartz sand of 1-2mm, quartz gravel of 3.12-5.6mm and hydrolith-Mn for manganese removal. Treatment of water at this station involved the removal of iron, manganese and carbon dioxide.

**Hannover: Two Waterworks**
- Wettmar waterworks is run by the Water Association Wasserverband NORD, Hannover.
- The waterworks at Fuhrberg is run by the Water Association Stadtwerke, Hannover.

The Wettmar waterworks obtains water from bore wells. The treatment involves the removal of iron by filtration with gravel, of manganese by the introduction of dolomite and potassium permanganate in three filters, and the removal of carbon dioxide by filtration using two filters with dolomite followed by aeration in the tank. In all, the waterworks uses eight filters.

At Fuhrberg waterworks, the groundwater had the following water quality: iron 16mg/l, manganese 1-2mg/l, SI = negative, humates (humic acid). The water is obtained from five horizontal filter wells with 8-10 horizontal filter pipes, all within a depth of 30m. It is pumped using three subaqueous pumps at a rate of 400-600m$^3$/h. The treatment includes four cascades of aeration with hydrogen peroxide ($H_2O_2$) as the oxidising agent and aluminium salts as flocculants. Potassium permanganate ($KMnO_4$) is used for oxidation and limewater for adjusting the pH. The water is chemically fed in a four-chamber flocculator at a rate of 900-1,100m$^3$/h. Also, the system operates with 9-chamber open, double sand-filters with 864m$^2$ available surface and a filtration rate of 4.5m$^3$/h. Sodium hydroxide is used for pH adjustment and $P_2O_5$ to provide protection against corrosive action.

The waterworks at Peine is unique in the sense that there were long-term differences of opinion between the farmers and the relevant authority with respect to the protection zones demarcated by the waterworks. This waterworks is run as a closed system with a 1,200km distribution network. It also operates from six wells with waters of different qualities. Some of the wells, for example, provide water with a very high iron content. Water treatment in this plant is designed to remove iron and manganese and achieve carbon balance. Four of its filters are used for iron treatment. The filters work for six to seven hours before they are automatically switched off.

The excursions gave junior scientists the opportunity to learn about the principles of rainwater harvesting and its application at both the individual and community level and provided hands-on experience of techniques used at different water-treatment stations.

## 5.4.2 Results

### *Microbiological Analysis of Treated Water Samples*

**Samples**
1. Rainwater samples taken from the EXPO lake (before filtration)
2. Rainwater samples taken from the EXPO lake (after filtration)
3. Rainwater samples taken from the EXPO lake (before soil infiltration)
4. Rainwater samples taken from the EXPO lake (after soil infiltration)
5. Rainwater samples taken from the Kronsberg settlement area in Hannover
6. Rainwater (control), i.e. Sample 1
7. Wastewater taken from the Suderburg wastewater-treatment plant (control), i.e. Sample 2
8. Wastewater treated with *Moringa* seeds
9. Wastewater treated with *Moringa* seeds and passed through slow sand-filter
10. Wastewater passed through slow sand-filter
11. Wastewater treated with calcium hypochlorite and passed through slow sand-filter

**UV Radiation**
Table 5.3 shows the effect of UV radiation on water Sample 2, respectively. It was observed that the UV exposure of the water samples for about 10 minutes was effective in arresting microbial growth.

**Table 5.3** Treatment by UV Radiation on Sample 2

| Time | Dilution | HPC Heterotrophic Plate Count | TC Total Coliform | FC Faecal Coliform |
|---|---|---|---|---|
| 0 min | - | 41+377 small | 77 | 5+1 big mass |
|  | 10(-1) | 31 | 2 | - |
| 3 sec | - | 299 | 22+1 mass | 1+1 big mass |
|  | 10(-1) | 17 | - | - |
| 6 sec | - | 198 | 15+1 mass | 30 |
|  | 10(-1) | 12 | 2+2 mass | - |
| 1.2 min | - | 43+ many | 9+1 mass | 20 |
|  | 10(-1) | 17 | - | - |
| 5 min | - | 39 big | Mass | 16 |
|  | 10(-1) | 16 | - | - |
| 20 min | - | 15 | Mass | 12 |
|  | 10(-1) | - | - | - |

**Heating at Elevated Temperatures at Different Time Intervals**
From examinations it was concluded that an exposure for 10 minutes at 55°C was enough to kill the bacteria present in the sample. It was also seen that as the temperature and exposure time increased, the colony count declined.

## Slow Sand-Filtration set-up with and without Aluminium Grid
Note
- The water sample collected from the Mensa pond (Sample 1) did not show any bacterial growth in the control. Thus, it was not used for further treatment and testing for the microbial count.
- The water sample from the effluent of the sewage plant was inoculated with the influent to ensure that enough bacteria were present in the sample (control). It was then treated with different methods and tested for the microbial count.

## Reaction with CaClO and Al$_2$(SO$_4$)$_3$ and Use of Slow Sand-Filtration

**Table 5.4.** Treatment of Calcium Hypochlorite and Aluminium Sulphate on Sample 2

| Treatment | Dilution and Timing | Heterotrophic Plate Count | Total Coliform | Faecal Coliform |
|---|---|---|---|---|
| CaClO Treatment | 10 times diluted sample | | | |
| | 15 min | w.p. dense | m.c | ~1/4 m.c dense |
| | 30 min | 203+23 b.c | m.c | ~ 1/3 m.c |
| | 45 min | <200+10 b.c | m.c dense | 22 |
| | 20 times diluted sample | | | |
| | 15 min | 103+9 b.c | ~50 | 1 |
| | 30 min | 98+5 b.c | m.c whole plate | 4 |
| | 45 min | 1 b.c | m.c 1/ 8 plate | 0 |
| | 60 min | 7 | m.c scattered | 0 |
| Al$_2$(SO$_4$) Treatm. | 10 times diluted sample | | | |
| | 15min | w.p. dense | > 200 | 1/3 plate m.c dense |
| | 30 min | w.p. less dense | m.c w.p scattered | 1/2 m.c dense |
| | 45 min | w. p. dense | m.c 3/4 plate dense | 2/3 m.c dense |
| | 60 min | w.p. | m.c 3/4 plate | 1/2 plate m.c |
| | 20 times diluted sample | | | |
| | 15 min | 78+11 b.c | 15 | 0 |
| | 45 min | 112+19 b.c | m.c 1/4 plate | 0 |
| | 60 min | w.p uniform & big | 4 | 0 |
| Blank (distilled water) | | - | 0 | 0 |
| Control | 10 times diluted sample | w.p dense | | 0 |
| Control | 20 times diluted sample | w.p less dense | | |

Where: w.p = whole plate, m.c = mass of colony, d = dense, b.c = big colony

From Table 5.4 it was observed that in Sample 2, during treatment with calcium hypochlorite at 45 minutes incubation, there was a decrease in the number of bacterial count. However, during treatment with aluminium sulphate, there was an increase in the number of colonies as time passed.

Note
The chemical treatment (especially with aluminium sulphate) did not yield the expected results. Actually, one would need to repeat the experiment with different doses, dilutions, incubation times, etc. to reach conclusions as to its effect on the bacterial growth.

## Biological Treatment

The Faecal Coliform FC petri dish for the untreated sample with a concentration of 0.1 did show a small uncountable dot that could be colony growth. The sample after five minutes at a concentration of 0.01 was difficult to count in both the TC and FC cases. In the TC sample (Total Coliform), four countable colonies and a clump of colonies existed. The FC petri dish had seven countable colonies and many smaller specks. Without the use of a microscope, it was impossible to determine which of the smaller specks were actually colonies. Moreover, the fifteen-minute sample at a concentration of 0.1 contained one huge colony and several smaller uncountable ones in the FC case. The sterile distilled water was not treated with the *Moringa* seed powder and, as expected, yielded no colony growth.

With the exception of the distilled water, the results were different from those expected. The *Moringa* seeds were expected to remove, not add, contaminants to the water. The untreated pond water with a concentration of 0.1 should have contained the most colonies of all samples, and the 0.01 concentration sample after twenty minutes should have contained the fewest colonies.

In the FC case, the complete opposite was observed. The untreated pond water contained one colony and the treated twenty-minute sample thirteen. The TC case does not behave as expected either. The fifteen-minute 0.1 concentration sample shows the highest number of colonies. These results would lead to the conclusion that the *Moringa* seeds do not kill the total and faecal coliforms present in nonpotable water, but add the bacteria to the water.

However, the seeds were old and this may have impaired their effectiveness. Moreover, the optimal number of colonies per filter is between 20 and 200. None of the samples were within the optimal range, so our data cannot be considered reliable. Furthermore, the controls, zero-minute samples, contain almost no colony growth, demonstrating that the pond water is not contaminated enough for the purposes of our study.

## *Chemical Quality of Water Samples*

### Determination of the Metal Concentration by ICP-AES

The analysis was performed to determine the quality and quantity of the metal and cations present in the water samples and to check the effect of various treatment methods on the amount of metals present in the treated samples. The technique used (ICP-AES ) was capable of analysing 72 metals simultaneously, but the researchers focused only on the metals for which standard values for the maximum allowed limits in drinking water are given.

**Table 5.5.** Amounts of Na, K, Ca, Mg (ppm) in Water Samples after Undergoing Different Treatments

| Cations | 1   | 2     | 3     | 4         | 5     | 6         | 7         |
|---------|-----|-------|-------|-----------|-------|-----------|-----------|
| Ca      | 400 | 43,37 | 44,53 | 76,39     | 34,28 | 36,32     | 15,17     |
| Mg      | 50  | 3,908 | 3,835 | 7,027     | 1,583 | 3,520     | 1,776     |
| K       | 12  | 4,523 | 4,482 | 5,999     | 0,898 | 2,418     | 2,315     |
| Na      | 150 | 17,97 | 17,40 | saturated | 4,473 | saturated | saturated |

Where: 1) Drinking water standards; 2) Rain water control (Sample 1); 3) Wastewater from treatment plant, control (Sample 2); 4) Sample treated with *Moringa* seeds; 5) Sample treated with *Moringa* seeds and also sand-filtered; 6) Slow sand-filtered sample, after 60 minutes; 7) Calcium hypochlorite reaction and slow sand-filtered sample, after 60 minutes

Table 5.5 shows the amounts of Na, K, Ca, Mg in water samples after different treatment methods. All the samples had values below the maximum allowed limit for the given cation, except for sodium which was shown to be saturated. At the same time, all the samples exhibited amounts that were still below the daily recommended dietary allowance. Moreover, it was observed that filtration of the water sample appeared to reduce the calcium, magnesium and potassium concentrations in the water samples.

The amounts of various heavy metals in water samples after different treatment methods were analysed. Nearly all the samples had a metal ion concentration below the maximum allowed limit. However, it was noticed that the strontium concentration (0.069-0.337ppm) in all the samples was relatively high compared with the standard value. On the other hand, it was observed that those samples that underwent slow sand-filtration exhibited a higher concentration of most of the metals in comparison to the control.

Also the amount of various heavy metals in rainwater samples, indicates that almost all the samples had a metal concentration below the maximum allowed limit, except for iron at the EXPO lake, and zinc from all water samples. It was also observed that the strontium concentration (0.158-0.424ppm) was way above the allowed concentration. Moreover, it was observable that filtration of the samples did not change the concentration of metals significantly.

**Table 5.6.** Showing the Effect of Soil Infiltration on Amounts of Na, K, Ca, Mg (ppm) in Rainwater Samples

| Cations | 1   | 2     | 3     | 4     | 5     | 6         |
|---------|-----|-------|-------|-------|-------|-----------|
| Ca      | 400 | 43,37 | 44,53 | 43,17 | 28,65 | 48,40     |
| Mg      | 50  | 3,908 | 3,835 | 3,887 | 3,429 | 10,46     |
| K       | 12  | 4,523 | 4,482 | 4,243 | 2,638 | 11,34     |
| Na      | 150 | 17,97 | 17,40 | 17,79 | 16,28 | saturated |

Where: 1) Drinking water standards; 2) EXPO site sample before filtration; 3) EXPO site sample after filtration; 4) EXPO lake sample before filtration; 5) EXPO lake sample after filtration; 6) Kronsberg lake

Table 5.6 shows the effect of soil infiltration on the amounts of Na, K, Ca and Mg in the rainwater samples. It was observed that all cations were below the maximum allowed limit, except for that of sodium in Kronsberg lake. This might be attributable to the fact that these samples were mainly rainwater samples, and rainwater is known to lack minerals and ions.

## *Conclusions and Recommendations*

### Biological Disinfection

There was definitely a reduction in the bacterial count when the combination of treatment with *Moringa* seeds and sand filtration was used. The experiment required a continued assessment of the treatment technique with a greater number of water samples before any conclusion as to the effectiveness of the treatment could be arrived at. Other factors that determined the efficiency of the treatment method included the shelf life of the seeds. The seeds that were used in the experiments had been stored over a very long period of time. It is possible that they had partially or fully lost their biological activity and were thus not very effective.

### Physical Disinfection

Exposure to Ultraviolet Radiation: The treatment was effective on exposure to UV radiation for a duration of 10 minutes. The literature review for this method stated that it is effective for treating water in smaller quantities, that it requires highly skilled personnel to manage and operate the plant, is not easily available at the community or household level, is capital-intensive and therefore not economically viable. However, it could form part of a centralised treatment plant for larger communities.

Boiling: For both water samples, heating for at least 10 minutes at 55°C was sufficient to kill the bacteria, as could be seen from the results of the bacterial count. Slow sand-filtration can also be considered a good method because no bacterial growth was seen in the samples.

Of the above methods of treatment, boiling and slow sand-filtration can be considered the best because of their availability, affordability, simplicity and cost-effectiveness.

### Chemical and Physical Treatment

A combination of chemical treatment and slow sand-filtration was able to reduce but not completely remove the bacteria present in the sample. It is recommended that a microbiological analysis of a sample that is only chemically treated be done to check the effect of chemicals on the sample.

### Heavy-Metal Analysis

Based on the spectroscopic analysis, the amount of cations Ca, Mg, Na and K in rainwater was found to be below the maximum allowed limit. At the same time, the values were in fact below the required daily allowance. If rainwater is to be used as a potable water source, it is recommended that mineral supplements be added to it.

Further trials for metal analysis should be performed to check the effect of filtration on the removal/increase of the metal content of the water sample.

***Scope and Limitations of the Study***

- The experiments were performed for a duration of four weeks only.
- For all three methods of water disinfection – physical, chemical and biological – multiple tests should be performed to check for microbiological contamination of the treated water sample and to ensure the accuracy and precision of the results.
- The project focused on the use of simple and appropriate technologies for small-scale treatment and disinfection methods. It was intended to use these technologies to bring the polluted water up to drinking-water standards, ready for use at the household level.
- The water-treatment methods used were: physical (heating, UV radiation and slow sand-filtration), chemical (reaction with calcium hypochlorite and aluminium sulphate) and biological (use of Moringa oleifera seeds).
- The efficacy of the treatments was determined by microbiological quality testing for Heterotrophic Plate Count (HPC), Total Coliform (TC) and Faecal Coliform (FC) in the treated water samples. The metal content of the water samples was also determined using atomic emission spectroscopy, with inductively coupled plasma (ICP) as the excitation source.
- The junior scientists were encouraged to develop rainwater-harvesting project proposals for their own countries, taking into account water demand, climatic variations, rainfall patterns, etc. To mitigate the problems associated with the nonavailability of clean water, rainwater harvesting was found to be one of the suitable technologies.
- The junior scientists were able to gain first-hand experience of the water-processing methods practised in Germany.

## 5.5 Rainwater Harvesting Project Plan Developed by the Women Junior Scientists

The project activities were not confined to laboratories and lecture halls. Junior scientists were motivated and encouraged to ponder on the water-related problems faced in their own countries. They took back home with them seven innovative project proposals for rainwater harvesting, along with traditional and small-scale water-treatment methods for water purification.

### 5.5.1 Rainwater Harvesting for Household Consumption
(Philippines by Angelica R. Martinez)

Rainwater harvesting has been in use in many parts of the world where conventional water systems are lacking. It is an old technique, but the technology can be simple or complex depending on the demands. It cannot be considered the perfect solution to household-water problems, but given its flexibility and adaptability, it can be applied to a wide variety of conditions. For planning purposes, cost, climate, technology, hydrology and social and political issues would play a role as regards the choice of the water-supply scheme.

Rainwater is considered to be the softest water available, with zero hardness. It contains no dissolved minerals or salts. The use of such soft water avoids mineral deposits and corrosion damage to water-using equipment, so it not only extends the life of appliances but also prevents scaling in pipes.

This proposal is concerned with possible rainwater use in a six-member household for domestic applications such as washing clothes, toilet flushing, gardening and cleaning cars. Since it is not meant for direct human consumption, water treatment is unnecessary. However, the proposed use of harvested rainwater requires large quantities of water.

Nowadays, people in Manila are settling away from the city, because of pollution. Subdivisions and villages catering mainly to middle-income families are sprouting within the suburbs. This feasibility study is designed for a household located in the suburbs of Manila/Philippines, in a subdivision 16km away from the city centre. The main water supply is from the groundwater, which is provided by the homeowners' association. However, the presence of large amounts of iron is evident from the rust colour of the tap water. Acquiring water for domestic use is, then, a real problem for the households. The water could make clothes turn yellow if used for washing, besides causing stains in toilet bowls and marks on car paintwork. Since the average rainfall in Manila is around 2,000mm per year, harvesting rainwater could be a practical solution to the above-mentioned problems. However, a number of parameters must be considered before designing a domestic rainwater-harvesting system.

There are six components in a rainwater system:

1. Roof or catchment area
2. Gutters and downspouts
3. Roof washers
4. Storage tank
5. Piping system
6. Water treatment

The most expensive component in any rainwater-harvesting system is the storage tank, so careful planning should be done to avoid costly oversising of this tank. The first flush of a rain shower carries all the dirt, debris and bird droppings from the roof, so the downspouts can be designed to divert or remove the first flush of rain from the storage system. If the water is to be used for washing clothes, gardening or washing cars, treatment beyond sediment removal is not required.

One must also consider the following factors when designing a domestic rainwater-harvesting system (DRWH):

- local rainfall data and weather patterns
- roof or other collection area
- runoff coefficient which varies depending on the roof material and slope
- user number and consumption rates

Bearing these things in mind, one could now design a DRWH system. There are a number of approaches that can be used for the design process, one of these being the 'supply side approach'.

## Supply Side Approach

In areas where there is uneven rainfall, this method can be used to calculate the size of the tank.

Table 5.7. Calculations

| | |
|---|---|
| Consumption per capita per day | 40 litres (clothes washing, flushing the toilet, gardening, washing the car) |
| Number of people: | 6 |
| Daily consumption: | 240 litres or 0.24 cubic metres, (7.2 cubic metres/month) |
| Runoff coefficient (for new corrugated roof): | 0.90 |
| Roof area: | 100m² |
| Average annual rainfall: | 2,050.5mm/year |
| Annual available water: | 184.5 cubic metres |
| Daily available water: | (184.5/365) = 0.506 cubic metres or 505.6 litres |

Since the available water greatly exceeds demand, one can build a tank with a capacity of approximately 8.0 cubic metres to meet the monthly requirement. The storage tank should be located close to the point of collection. It could be made of concrete or plastic depending on the availability of materials and the cost involved.

### 5.5.2 Case Study on Rainwater Harvesting
(Albania by Gentiana Haxhillazi)

Rainwater-harvesting systems are not new to Albania. Rainwater-catchment systems have, for many years, been successfully utilised by people all over the country. In the past, rainwater was collected from many types of surfaces especially for agricultural purposes. The system is designed to collect rainwater from the roofs of buildings in a residential area. My aim is to find out how useful a rainwater-harvesting system would be for my city, Tirana. The project can be implemented at an individual household or community level by members of a self-help group with similar interests and objectives.

## Planning Groundwork

Rainwater collected for use in public and private buildings must satisfy the following conditions:

- The use of rainwater must not present a risk to health or hygiene
- The rainwater must not be used if it contains any solids (e.g. sand)

Other quality requirements for service water can be derived from the purpose for which it is to be used. I propose the use of rainwater for washing machines and for watering the garden. Here, the quality of water is not so important.

## Rainwater System Components

The design of rainwater systems depends on factors such as local conditions and the required quality of the water. However, all systems have the following components in common:

- The roof catchment system - the collecting area: It is generally used as the collecting surface.
- Rainwater pipe: The actual rainwater system begins with the diversion of the water from the rainwater pipe into the storage tank.
- The storage tanks - collecting rainwater: The job of the storage tank to which the rainwater pipe is connected is to make rainwater available, even during dry periods.
- Filters - purifying rainwater are installed in the system to improve the quality of the water: They can be located at various points before or after the storage tank.
- Consumer's taps: The final link in the chain are the consumer's taps.

## Example of How this System will Work in an Apartment Block:

- S: surface of the apartment block's roof = $450m^2$
- N: number of persons living in the apartment block = 45 persons
- D: domestic use such as washing cars, watering garden, etc = 40 l capital/day
- F: % of the rain entering the tank = 70%

## Tirana

Located at about 41.33°N 19.70°E. Height above sea level approx. 89m/292 feet. Average rainfall data is given in Table 5.8.

**Table 5.8.** Average Rainfall in Tirana/Albania

| Jan | Feb | Mar | Apr | May | Jun | Jul | Aug | Sep | Oct | Nov | Dec | Year |
|---|---|---|---|---|---|---|---|---|---|---|---|---|
| 143.7 | 133.7 | 116.0 | 92.5 | 94.6 | 66.8 | 34.4 | 39.2 | 69.0 | 99.3 | 170.6 | 148.2 | 1210.2 |

Source: Global Historical Climatology Network. 302, months between 1951-1990

**Calculating Household Water Demand**
To ensure availability throughout the year, it is important to make sure that

- the tank is full of water at the beginning of the dry season and
- the volume of water drawn per day does not exceed the calculated demand.

The volume of a water tank is estimated in relation to a family's demand, the supply capacity of the roof area and the rainfall.

## Calculating the Volume of the Tank

The required volume of a tank can be estimated as follows:

First, estimate the water demand for the dry season by

(1) determining the number of persons in the household
(2) determining the number of litres of water consumed per person per day and
(3) determining the number of days without rain in the longest dry period of the year.

Multiply the figures obtained in (1), (2) and (3). The result is the required volume of water to be stored in the tank.

The demand is (following the above steps):

- 45 · 40 =1800 l/d
- 1,800 l for 1 day
- 54,000 l for 30 days

The demand is therefore 54m$^3$/month.

To calculate the amount of rainwater that can be harvested, we can also follow the above steps: e.g. for January the demand is:

143.7mm · 70% · 450m² = 45,27m$^3$/January, and so on for the other months.

## Results

All the estimated values exceed the estimated demand (which is 54m$^3$/month). Thus we can see from the results that there is no difference between the values calculated for each month and the estimated demand. As a result, we can say that the volume of the storage tank must not exceed 54m$^3$.

## Conclusion

In this case it is not necessary to build the rainwater-harvesting system as there is high rainfall throughout the year. Nevertheless, it could be used for agricultural purposes such as irrigation, etc.

### 5.5.3 Rainwater Harvesting - A Proposal for Secondary Schools
(Tanzania by Eng. Immaculata Nshange Raphael)

The proposal is made for three day secondary schools currently under construction in the context of a self-help project. The objective is to provide clean and safe drinking water to three different secondary schools in the district. The community is carrying out the construction work and the District Council will be supporting the project by providing supervision, supplying roofing materials (aluminium/galvanised iron sheets) and offering consultancy.

The proposal will be duplicated for other schools, too. Technical data on population and average rainfall are given in Table 5.9, respectively.

**Table 5.9.** Average Rainfall in Moshi/India

| Jan | Feb | Mar | Apr | May | Jun | Jul | Aug | Sep | Oct | Nov | Dec | Year |
|---|---|---|---|---|---|---|---|---|---|---|---|---|
| 38.4 | 45.5 | 117.2 | 328.3 | 179.6 | 34.6 | 20.8 | 17.2 | 14.8 | 32.8 | 67.3 | 52.6 | 955.6 |

Source: www.worldclimate.com

**Catchment Area and Material**
600 square metres of galvanised and/or aluminium iron roof sheets.

- Assumed losses: 25%
- Assumed dry months: 4 (from June to September)

**Water Demand**
- Total population to be served: 400 person
- Average daily requirement for cooking and drinking: 10 l/person/day
- Total daily requirement per school: 400 people · 10 l/day = 4m³/day
- Monthly demand: 4m³/day · 30 days = 120m³/month
- Total demand for the dry months: 120m³/month · 4 months = 480m³

**Possible Quantity of Water to be Collected**
Q= 700mm/a · 600m² = 420m³; Deficit: 480 - 420 = 60m³

**Proposals and recommendations**
- Provide three above-ground concrete tanks of 100m³ each.
- All tanks must be interconnected and each inlet equipped with a 'first-flush mechanism'.
- Only one tank is to be provided with an outlet and a disinfection mechanism (chlorination).
- A dead-end, simple pipe network to be laid, connecting the kitchen, laboratory and cafeteria to the storage tank.

### 5.5.4 Rainwater Harvesting (USA by Margaret Fredricks)

Appleton, Wisconsin, USA is my home town. Although I do not currently reside at home, my parent's place is my permanent address. Therefore, I selected their home for the purpose of this project. They are the only two people living in the house at present. Currently, the water from our roof runs either onto the lawn, where it is absorbed, or down the driveway, into the street. The water from the street flows through the gutters into the nearby Fox River. Flooding is not a problem for the river due to a deep river basin and the numerous dams which control the river's flow.

The average American uses approximately three times as much water per day as the average German or about 300 l per day. My parents would need 219m³ per year if they were to no longer use the city's drinking water supply and exist entirely on rain water. Table 5.10 shows the average monthly rainfall in Appleton for the years from 1931 to 1995. Table 5.11 shows the average monthly high temperature, and Table 5.12 contains the average monthly low temperature. Both the high and the low temperatures are averages from 1961 to 1990. The summer months get the most rainfall. The rainfall for the months of January, February, March, November and December cannot be included in the calculations for how much rain water can be harvested since the average low temperature is below freezing. This precipitation falls in the form of snow, and therefore cannot be harvested.

**Table 5.10.** Average Rainfall in Appleton/USA

| Jan | Feb | Mar | Apr | May | Jun | Jul | Aug | Sep | Oct | Nov | Dec |
|---|---|---|---|---|---|---|---|---|---|---|---|
| 29.8 | 29.0 | 48.0 | 70.4 | 80.2 | 96.8 | 81.7 | 86.9 | 88.5 | 55.7 | 54.3 | 35.4 |

**Table 5.11.** Average High Temperature in Appleton/USA

| Jan | Feb | Mar | Apr | May | Jun | Jul | Aug | Sep | Oct | Nov | Dec |
|---|---|---|---|---|---|---|---|---|---|---|---|
| -4.5 | -1.9 | 4.2 | 14.5 | 20.0 | 25.0 | 27.7 | 26.1 | 21.2 | 14.5 | 5.9 | -1.8 |

**Table 5.12.** Average Low Temperature in Appleton/USA

| Jan | Feb | Mar | Apr | May | Jun | Jul | Aug | Sep | Oct | Nov | Dec |
|---|---|---|---|---|---|---|---|---|---|---|---|
| -13.7 | -11.5 | -5.2 | 1.6 | 7.9 | 13.4 | 16.6 | 15.3 | 10.8 | 4.8 | -2.1 | -4.0 |

Source:www.worldclimate.com

A total of 563.2mm or 0.5632m of rain falls during the seven warm months. Assuming the roof of the house is 100m$^2$, 56.32m$^3$ of water could be collected. About ten percent will evaporate from the roof. Thus, only about 50.7m$^3$ of water can be collected. This is only about 23% of the yearly water consumed by my parents. The most likely application of such rainwater harvesting would be for watering the garden and lawn. Since enough rain generally falls during the summer to keep the grass green, the lawn is watered only a couple of times during the entire summer. A garden requires about 2 l/m$^2$ of water. The garden is only about 20m$^2$, so it needs about 40 l of water. The lawn is about 500m$^2$. For the occasional watering of the lawn, 1000 l of water are required. A storage tank of about 1.5m$^3$ should be sufficient to water the garden on the days when it does not rain and for the couple of times the grass would need to be watered.

The storage tank should be emptied each fall, so as to prevent freezing and cracking of the tank. No water would be collected during the winter months. Overflow can either be directed into the city sewer system or connected to the street gutter system that delivers rainwater to the river.

## 5.5.5 Promotion of Rainwater Harvesting in the Arid Area
(Cameroon by Michele Denise Akamba Ava, Case Study by Maroua Salak)

### Situation of the Area

The project will be implemented in one quarter of Maroua Salak, located in the arid region of Cameroon at about 10.45°N and 14.20°E. It is a pilot project which in the future has to be adopted at different levels, and the technique will be disseminated in different parts of North Cameroon where every year many people suffer from long and severe dry seasons. The area is characterised by:

- a dry period of 8-10 months
- Annual rainfall is 814.5mm/annum
- 80% of the population are not connected to the water-distribution system, and those who are connected suffer from water scarcity and insufficient water quality
- No drinking water is available in the schools
- The only water source for domestic use is a small river situated at a considerable distance, and this also serves livestock
- Water collection is a task performed by women and young girls only
- Numerous diseases spread during the dry seasons

In order to provide water to the population of this region throughout the year, it is necessary to develop a sustainable system of rainwater harvesting, rainwater being the under-utilised resource in the region.

### Objectives

- To provide drinking water to the schools and households.
- To promote and disseminate the technique of the rainwater harvesting in the region.
- To improve the standard of living of the population.
- To reduce the workload of women and young girls in terms of carrying water for long distances.

### Materials

The roof of the primary school will be used because of its quality and location. The use of local materials, e.g. bamboo and bricks, must be encouraged for reasons of cost and accessibility. The different components involved in a simple rainwater-harvesting system are: a catchment area, gutter, downpipe, filter, tank and an outlet to collect water.

## Calculations

Catchment area: 30m length · 15m width = 450m², the loss due to splashing over the roof can been estimated at 20%
Water demand:

- Uses: 2 litres per day per schoolchild and 15 litres per day per household
- Daily demand: 105 schoolchildren and teachers · 2 litres + 30 households · 15 litres = 660 litres/day
- Monthly demand: 660 · 30 days = 19.8m³/month

**Table 5.13.** Average Rainfall Data – mm, effectiveness of 80%, Deficit (m³)

| Jan | Feb | Mar | Apr | May | Jun | Jul | Aug | Sep | Oct | Nov | Dec | Year |
|---|---|---|---|---|---|---|---|---|---|---|---|---|
| 0 | 0 | 4.5 | 17.5 | 70 | 111 | 197 | 240 | 148 | 29.3 | 0 | 0 | 814 |
| 0 | 0 | 16.2 | 63 | 252 | 401 | 684 | 864 | 531 | 104 | 0 | 0 | |
| 19.8 | 19.8 | 3.6 | | | | | | | | 19.8 | 19.8 | 63 |

Source: www.worldclimate.com

Table 5.13 shows rainfall per month; this is the quantity that can be collected from the roof per month (20% losses), the deficit being registered every month on the basis of monthly requirements. The annual demand is about 63m³. To meet this demand, we need a storage tank of about 65m³. This tank must be full at the beginning of the rainy season.

## Impact

The school will be a demonstration unit; after some time people will have to implement it themselves at the household level. The project will involve men, women and children. The maintenance and the management will be done by the communities themselves. The water could be used for: vegetable growing, agroforestry and for overall house improvement.

### 5.5.6 Rainwater Harvesting Draft Plan for a Vegetable Garden (India by Nandini Sankarampadi, Sanjulata Prasad)

Water is a very important component of our daily lives. It is used for several purposes at the community level including agriculture/irrigation, large-scale or kitchen gardens, cooking, washing, cleaning, personal hygiene, etc.

Per capita availability of renewable freshwater resources in India fell from around 6,000 cubic metres per year in 1947 to about 2,300 cubic metres in 1997. It is estimated that by 2017 India will be 'water-stressed' and per capita water availability will be as low as 1,600 cubic metres. It is believed that better demand-side management and supply enhancement will help overcome this potential problem on the demand side, and that demand could be reduced by as much as 50% of current consumption without compromising on output or lifestyles. [4]

The dependence on water resources for agriculture and minor irrigation has always been on the groundwater. With excessive groundwater utilisation, the groundwater resources are rapidly depleted, leading to a scarcity of available water. Harvesting rainwater can help minimise groundwater depletion.

## *Objective*

The proposed rainwater-harvesting scheme for a medium-sized vegetable garden is designed mainly to supplement the water supply from the existing groundwater sources.

### For the Proposed Project it is Assumed that:
- There are 10 households and the total number of all persons is 50.
- It is a colony of similar houses, each of 100 square metres.
- The total average per capita water consumption per day is taken to be 50 litres: 5 for cooking, 10 for washing, 25 for personal hygiene, 2 for drinking and 8 for other activities.
- The per capita wastewater release is taken to be 40 litres. Of this, 15 litres per day, could be treated and utilised for the vegetable garden to supplement the available groundwater supply.

### Details of Proposed Vegetable Garden
Area: 200m²
Annual water requirement: 1,000m³ per year
Annual water availability from the households (wastewater): 15 litres · 50 persons · 365 days = 273.750m³

### Rainfall Data
- total water available from roof runoff = 635 m³/year (as shown in Table 5.14)
- Assuming only 70% of the runoff can be harvested, due to evaporation and other processes
- Annual wastewater available after natural filtration = 82.125m³
- Assuming 30% losses due to infiltration during treatment
- total water available for the vegetable garden = 718.055m³

**Table 5.14.** Average Rainfall Data (mm) and Available Runoff from Households (m³)

| Jan | Feb | Mar | Apr | May | Jun | Jul | Aug | Sep | Oct | Nov | Dec |
|---|---|---|---|---|---|---|---|---|---|---|---|
| 4.9 | 5.9 | 11.5 | 38.7 | 114.4 | 78.0 | 105.8 | 137.6 | 174.9 | 156.8 | 62.5 | 16.7 |
| 3.4 | 4.13 | 3.1 | 27.1 | 80.1 | 54.6 | 74.6 | 96.3 | 122.4 | 109.8 | 43.8 | 11.7 |

Source: www.worldclimate.com

## *Conclusion*

The rainfall is quite high for 6 months, i.e. above 5m³. Of the total requirement of 1,000m³, almost 71% of the demand is met using two storage tanks: one containing the natural sand and soil layers to treat the wastewater, with a capacity of 30m³, and another with a capacity of 60m³ to store the harvested rainwater.

### 5.5.7 Rainwater Harvesting Plan for Loyola College
(Nigeria by Theresa Odejayi, Yetunde Odeyemi, Helen Oloyede)

Cities in Nigeria are no different from other cities around the world in terms of their huge consumption patterns for natural resources. And water consumption is no exception to this rule. This is because the population in cities is usually large and there are very few restrictions on growth or expansion, especially in developing countries. The government of Nigeria is responsible for providing the people with water, and this usually takes the form of building a large dam or a central water-supply system, which generally does not seem to accommodate population expansion. So, after a few years of operation, such supply systems are not able to meet the demands and needs of the growing population.

The results are dry water taps and long queues at places with public water faucets, privately owned wells or boreholes the owners of which charge a certain amount (which is arbitrarily fixed) for all categories of users. Nor is it uncommon to find water for household use being taken from broken exposed pipes in unhygienic environments, streams, springs and any open cistern of water, sometimes several kilometres from people's homes, thereby making great demands on the time of women and girl children.

The acuteness of the situation is quite obvious at the household level and in places where people have to assemble for hours, such as school areas. This situation could be alleviated by decentralised water-supply systems enabling public places like schools, churches, markets and mosques to have their own water supply provided by the age-old system of rainwater harvesting.

Rainwater can be collected from various sources such as:

- Roofs
- Street channels
- Open ground surfaces, i.e. lakes or ponds

Despite the fact that most parts of the world make exclusive use of this technology to meet their water demands, it becomes cost-competitive when there are no other alternatives, no municipal water system nearby and no potable well water. The largest cost factor of any rainwater-harvesting system is the storage tank/cistern, which must be constructed to prevent flooding and water pollution and to ensure the water supply during drought periods. While water self-sufficiency is the primary aim, there is a need to ensure that the cistern is not oversized, thereby increasing the cost of water provision.

Rainwater-catchment systems provide a source of clean water without taxing local groundwater supplies. Rainwater is of high quality owing to its softness and almost total lack of dissolved minerals and salts, except for contamination from the environment either from the immediate atmosphere or the rooftops. Rainwater harvesting is an appropriate technology for the southern parts of Nigeria as this region has two rainy seasons (with maximum rainfall in June and October) and in most places no month devoid of rainfall.

A typical rainwater system has six components:

- The catchment area
- Gutters and downspouts
- Storage tank or cistern
- A conveyance system of pipes
- The roof washer (in some cases)
- Water treatment

## *Plan for Rainwater Harvesting for a Secondary School in Ibadan*

Loyola College in Ibadan is a school for college-age boys, founded in 1954 by Roman Catholics on 5,000m² of land. The school has a current student body of 2,500, a teaching staff of 120 and a non-teaching staff of 12. The school's development from its founding to the present shows population changes and indicates that whatever water supply it put in place is soon obsolete and inadequate, especially when there is no renovation or upgrading of the system. Nonavailability of water or low supply of water has forced the school authority to reduce student intake. This means that some children are deprived of the opportunity of schooling, which in turn means an increase in the number of street children and child-traffickers who roam the streets.

Also, the lack of drinking water and water for flushing toilets on the school premises has led to frequent outbreaks of diseases such as diarrhoea and typhoid among users of the school facilities. In view of the above, our aim is to provide drinking water from harvested rainwater for the school campus. This plan is based on the calculations below. Table 5.15 shows the annual average rainfall.

**Table 5.15.** Average Rainfall in Ibadan City (per year 1222.4mm)

| Jan | Feb | Mar | Apr | Mar | Jun | Jul | Aug | Sep | Oct | Nov | Dec |
|---|---|---|---|---|---|---|---|---|---|---|---|
| 10.1 | 22.3 | 85.9 | 134.1 | 149.4 | 182.1 | 158.3 | 92.8 | 176.7 | 162.6 | 40.2 | 9.5 |

Source: www.worldclimate.com

## *Calculations*

| | |
|---|---:|
| Total number of students | 2,000 |
| Drinking water demand (between 8.00 a.m. and 2.00 p.m.) | 1 litre/day |
| Total demand per day | 2,000 litres |
| In one month (20 days) | 40,000 litres |
| Roof catchment area | 400m² |
| 80% of the water is collected (owing to the high rate of evaporation) | |
| No. of persons · no. of litres per person · no. of days/month: | |
| Water demand for students per month = 2,000 · 1 · 20 = | 40m³ |

**Table 5.16.** Calculation for Rainwater Harvesting Plan (m³)

| Jan | Feb | Mar | Apr | May | Jun | Jul | Aug | Sep | Oct | Nov | Dec |
|---|---|---|---|---|---|---|---|---|---|---|---|
| 3.2 | 6.4 | 27.5 | 42.9 | 47.8 | 58.3 | 50.7 | 29.7 | 56.5 | 52.0 | 12.7 | 3.04 |

Monthly rainfall · per cent of collected water (0,8) · roof catchment area (400m²)

Calculating from the average rainfall per year (1222.4mm/12 months):
$1.224 \cdot 0.8 \cdot 400 = 391.68 m^3$ (the amount of available rainwater for the year)

- Q = quantity of water that can be collected yearly 391.68m³
- $Q_0$ = water requirement for duration of dry season (4 months)
- $3.2 m^3 \cdot$ 20 days (one month) · 4 months (duration of dry season) = 256m³

Since $Q_0$, which is the requirement for the dry season, is less than Q, which is quantity of water that can be collected, there is guaranteed to be enough water for use at the school compound. Therefore, with four big storage tanks situated around the school compound, there is guaranteed to be enough drinking water for the 2,000 students at the school, even during the dry season.

### Safety Measures

The gutters will be made of synthetic plastic and installed with the valve at the link point with the downspouts so as to allow removal of the flush of the rain. This is one good way of ensuring that the dirt on the roof does not get collected along with the water. Also roof washers will be installed to help get rid of the dust and leaf litter from the roof. The tanks will be of synthetic plastic. The rainwater collected need not be subjected to chemical treatment; rather it will be physically treated by decantation, filtration and sedimentation. Traditional coagulants in the form of seeds could be used.

## 5.6 Summary

The International Women's University *ifu* was a genuinely international, interdisciplinary and multicultural event in terms of faculty selection and the junior scientist body. 15 June 2000 (the official start of *ifu*) was a day eagerly awaited by many who were anxious to experience *ifu*, to try out its truly global concept. I believed in *ifu* from the moment I landed in Hannover on 25 June. All of us were sceptical before the junior scientists arrived, but slowly the enthusiasm began to visibly grow and we were all charged up.

For nearly three and a half months, Suderburg (Lower Saxony) was home to nearly 150 women junior scientists from all over the world. It was the host city for the project area Water, and everyone contributed to making it an enduring success. There was information exchange, entertainment and also an element of contemplation. Dances, multicultural parties, festivities, sports and shopping, something for everyone to enjoy. There was excellence, brilliance, enthusiasm, creativity from around the globe. What I discovered there was a great deal of respect and curiosity among junior scientists about each other's countries, culture and heritage. The topic 'Water' was the element binding together the contributions of all junior scientists.

The project was designed around seven main objectives, as defined in the beginning of this chapter. A glance at the junior scientists' educational and professional qualifications showed that they had interdisciplinary backgrounds, and this fostered intercultural understanding among them. Junior scientists provided insights into their activities (research/studies) in their home countries, and discussions were invariably lively and interactive. They included informal talks on traditional/cultural/social/political aspects of their countries. Junior scientists had the opportunity to work in small groups on assigned topics, problems and experiments, to find ways of linking their own research interests to the focus of the project.

The project emphasised the need for researchers to constantly share experiences both within their countries and across national boundaries. It was concerned with the development of practicable solutions to tangible, real-world water problems at the grassroot level. The contributions of all junior scientists focused on the key topic "Rainwater Harvesting" and a constructive workplan were drafted for their respective countries. There was an abundance of technology, humanity, culture and nature, forming an umbrella network among many of the junior scientists. Young and like-minded women shared their immensely rich experience and indigenous knowledge across all cultural, political and geographical boundaries. The project allowed first-hand experience of innovative ideas, offering a unique and intellectually enriching experience.

*ifu* was a fascinating opportunity to witness a truly world-wide workshop of ideas on the common and complex issues of the future of water. It was creative, stimulating and bursting with ideas, innovations, deliberations and thoughts, shaped by its highly motivated and talented junior scientists in harmony with nature and technology. It was a local event (held in Germany) with a global impact on the entire educational enterprise. *ifu* provided first-hand experience of exchanging ideas in the information age and carried a special value for the junior scientists. It served as a platform for the exchange of knowledge to help understand the global issues, challenges and problems that are opening up new research opportunities.

*ifu* served as a model for concentrated international educational partnerships, promoting and encouraging gender equality in higher education. I see it as a pioneer for the future of women's university education across the globe.

## References

1. Beers SK (1998) Sourcing Water from the Sky. www.edcmag.com
2. Thomas TH Current Technology for Storing Domestic Rainwater (Part I). www.eng.warwick.ac.uk
3. European Commission (1990-1994) Recommendations for Using Rainwater in Public and Private Buildings. Theme Programme Action-B 158., Directorate-General for Energy (DG XVII): 1,2,5.
4. Tata Energy Research Institute Delhi (1998) Looking Back to Think Ahead. India.
5. Reddish GG (ed) (1957) Antiseptics, Disinfectants, Fungicides and Chemical and Physical Sterilization. Iesa and Febiger 975, Philadelphia, PL
6. American Public Health Association (1995)Standard Methods for the Examination of Water and Wastewater. APHA,AWWA/WPCF: 860-897, 19$^{th}$ edition
7. www.provinceco-op.com/waterfaq/waterfaq.htm
8. www.worldclimate.com

# 6 Wastewater Treatment

Sabine Kunst, Artur Mennerich, Marc Wichern

This chapter presents the results of the wastewater treatment project. The main aim is to give a comprehensive though condensed overview of the state of wastewater treatment in Europe. As discussed in chapter 7, it is characterised by a high degree of centralisation. The systematic development of systems for treating domestic and industrial wastewater began in Europe about hundred years ago.

This process was based on the insight that rivers could no longer cope with the amount of discharged oxygen-consuming substances (dissolved pollutants and waste) and were on the point of biological collapse, i.e. being rendered void of oxygen. The discharge of such quantities of pollutants and waste into the rivers was a consequence of the development of urban sewerage systems and sanitation. (Those engaged in the construction of such systems were commonly referred to as health engineers).

About 50 years after industrialisation began, complete sewer networks were constructed in the European countries (London 1840, Hamburg 1842-1853, Berlin 1873) to remove water, excrement and waste from the cities. In Germany, the construction of sewer systems in the large cities was completed only a few decades ago during the huge construction boom following World War II. Once the centralised disposal of wastewater had become a fact, the centralised wastewater-treatment concept was developed. According to Dunbar (quoted in Bouwer, 1966), this comprises

- removal of coarse dissolved substances using rakes, rack screens, etc.
- removal of all undissolved components
- removal of the dissolved components of wastewater - using biological methods to prevent the wastewater from assimilating

These basic elements, defined in 1902, are today still crucial to wastewater treatment; only the biological methods have been further refined. This refinement is explained in detail below.

For decades, water management and sanitary engineering has been carried out without public participation. It is no wonder, then, that in Europe there is very little appreciation of wastewater treatment problems. Discussions usually centre on the cost of wastewater disposal, but there is no discussion of how and where the wastewater is to be treated.

However, there is a decade-long convention that the disposal of wastewater and the supply of drinking water cost an average of approx. DM7/m$^3$. Efficient instruments are in place to check that the so-called minimum requirements for purification are met.

The present project brought together junior scientists from 12 different parts of the world: Nigeria, Ethiopia, Sudan, South Africa, Ecuador, Russia, India, Indonesia, the UK, Turkey, Ukraine and Germany. Their professional specialisms were chemistry, biology, environmental sciences and engineering.

The goal was for them to learn about

- techniques of centralised wastewater treatment (A. Mennerich)
- computer models that can be used to estimate the dimensioning of biological wastewater-treatment plants (M. Wichern)
- the evaluation of application options of centralised wastewater treatment from the perspective of the junior scientists (S. Kunst)

## 6.1 Mechanical Wastewater Treatment

### 6.1.1 Overview

About 30-40% of the pollutants in municipal wastewater occur in undissolved form. For semi-industrial and industrial wastewater, this ratio can vary considerably. The mechanical treatment methods are most cost-effective for eliminating these substances from the wastewater. The aim of mechanical wastewater treatment is to initially achieve as extensive an elimination of the undissolved wastewater components as possible.

Apart from this elimination function, the mechanical purification plants are also designed to prevent inhibiting and clogging substances from reaching the plant components of subsequent wastewater-treatment stages. As the implements for mechanical wastewater treatment come into direct contact with the untreated raw wastewater, the demands on the material are very high, both in terms of their construction (concrete corrosion) and their machine technology. The machine technology used must be sturdy and reliable, i.e. not prone to breakdown. Where implements such as rake or strainer units are chosen, tried and tested methods with a good operation record should be preferred.

We now discuss the most frequently used methods for mechanical wastewater treatment:

- removal of coarse and fibrous substances using rakes and strainers
- removal of sand using sand catchers
- removal of settable substances by means of preliminary treatment tanks

### 6.1.2 Rakes and Strainers

*Basic Methods and Treatment Goals*

Rake and strainer implements are designed to remove coarse suspended substances from the wastewater entering the wastewater treatment plants. These mainly comprise: paper, food waste, cellulose material and pieces of plastic.

The elimination effect of rakes and strainers is based on the fact that particles above a certain size are intercepted as the wastewater flows through the rake grid or strainer. The rake must be scraped to prevent it from clogging. This scraping can be done at intervals (discontinuously) or all the time (continuously).

The increase in inhibiting substances in wastewater in recent years has necessitated the use of increasingly fine rake implements. Also, there has been an increasingly widespread use of strainer machines designed to retain very fine particles.

## *Technical Design*

### Rake Implements

Rake units are mostly constructed as prong rake bar screens, the functional element of which is referred to as rake grid. The rake grid consists of a row of vertical rake prongs; the grid is installed in a channel through which the wastewater flows, perpendicular to the flow direction.

In the influents to wastewater-treatment plants, the most frequent rakes are prong rakes, installed at an angle of 70° to 90° in the rake channel. At larger plants, it is generally essential to install a coarse rake before the fine rake in order to maintain the required capacity of the fine rake (two-stage rake unit).

### Strainer Implements

Owing to the increasing amount of solids in the wastewater, particularly plastic and fibrous substances, there is a tendency to reducing the gap between the rake prongs. Apart from the fine rake, then, strainer implements are being increasingly used to remove the rake material.

There are a variety of technical designs for the strainer implements. Particularly in smaller wastewater-treatment plants (WWTPs), strainer drums with integrated strainer-material transport and press are generally used. Sufficient technical experience with this type of strainer implement is now available to enable proper dimensioning. The operational reliability of the above-mentioned implements has also been proven, at least as far as the products of well-known manufacturers are concerned.

Strainer implements without mechanical treatment are only used in special cases as they tend to clog very quickly if the wastewater contains fibrous material. They have not performed sufficiently well when used for straining raw wastewater in municipal WWTPs. However, they are a simple and sturdy alternative if used to remove substances without any fine or fibre-forming components. In some WWTPs, they are used instead of settlement tanks to separate sludge from biofilm plants (e.g. trickling filters).

When designing strainer implements, consideration must be given to the fact that large quantities of water – and thus fast loading of the strainers – can lead to serious hydraulic problems.

**Rake-Material Presses**

The rake material caught during raking is very bulky and has a high water content. Compacting and dewatering of the rake material is therefore needed to reduce its storage and transport volume. The most simple solution is to discharge the collected rake material into a container equipped with a dewatering device (perforated two-layer bottom, discharge aperture). The gradually extracted press water can thus be discharged; however, the volume reduction is rather low. A much greater reduction of the rake material can be achieved using a piston or screw press.

### 6.1.3 Sand Catchers

*Basic Methods and Treatment Goals*

Sand catchers are channels or tank constructions through which the wastewater flows; they separate out the mineral particles contained in it. The higher specific weight of the mineral solids is utilised to remove them from the wastewater. Apart from the weight of the particles, another important parameter affecting the separation performance of a sand catcher is their size. Generally, one can assume that the separation performance declines with decreasing particle sizes. Here, the same laws apply as for other settlement processes (Stoke's Law).

The sand contained in the influent to the WWTP stems mainly from the dewatering of streets and other fastened surfaces (mixed sewage), but also from the domestic sphere. Because of external wastewater influences, the installation of a sand catcher is thus indispensable for separate sewage as well. In the case of mixed sewage, another fact that must be taken into account is that rainfall will greatly increase the amount of sand because, on top of the sand rinsed off surfaces, mineral particles that have settled in the channel will be whirled off and transported to the WWTP ("flushing peak").

In many cases, the sand catcher is added to the rake or strainer device as part of the mechanical wastewater treatment unit. If there is a two-stage rake implement – consisting, for instance, of a rake with an aperture width of 20mm and a subsequent fine strainer – the sand catcher can also be inserted to separate the sand from the strainer material.

If the sand removal is insufficient, long-term problems are likely to occur in the WWTP, e.g.:

- sediment in the subsequent treatment tanks (activation tank, fermentation tanks, etc.)
- increased strain on all mechanical parts, e.g. pumps, stirring devices, scrapers
- clogging of pipelines through sand settling, especially in the vicinity of sludge funnels and sludge pipelines

The two most commonly used technological solutions are unaerated and aerated sand catchers. These are discussed below. Other devices used for sand catching are hydrocyclones, hydraulically revolved sand catchers and deep sand catchers. Hydrocyclones are not used for sand catching in municipal WWTPs; their application is restricted to special cases of industrial wastewater treatment or as part of so-called sand-washing plants.

**Fig. 6.1.** Aerated Grit Chamber

## *Technical Design*

### Unaerated Long Sand Catcher
Unaerated long sand catchers consist of channels with a rectangular or trapezoidal flowing cross-section. In their most basic form, they consist of two channels running parallel to each other, which can be operated alternately using a plug lever slide gate. The channel that is not in operation is emptied so that the sand can be dewatered. It is then removed by hand. After that, the channels are changed, i.e. the emptied sand catcher is now operated, while the one that was previously in operation is cleaned. Today, manual sand-catcher cleaning is no longer permitted for hygienic reasons. Scraping machines are now standard equipment for this job.

For an unaerated long sand catcher to function properly, a constant horizontal flow speed of 0.2-0.3m/s must be maintained. At this speed, the sand is almost completely separated, while the organic components remain suspended and are subsequently conveyed to the following treatment stages.

### Aerated Sand Catcher
The aerated long sand catcher is the type most frequently used in Germany today. Here, the sedimentation of the mineral substances is not controlled by the horizontal flow speed. Instead, the aeration creates an evenly revolving stream that

restricts sedimentation of the particles. The advantage of the aerated sand catcher is that the flow speed in the water body is fixed, irrespective of the aerator's current throughput. Thus the sand catcher's performance does not depend on the amount of incoming water. Here, then, the task of keeping separation performance independent of throughput is solved much better than in the case of the unaerated sand catcher.

Long sand catchers consist of a channel (15m < L < 60m) with the aerator devices installed on the long wall of the tank. Ideally, the bottom has a semicircular profile with the sand-collecting channel in the middle. In practice, only the channel is profiled, for cost reasons.

A fat-collector bag is integrated into the construction. This is a narrower channel, separated from the sand catcher proper by a submerged wall. Here, floating substances such as fats contained in the influent are separated, this process being supported by the floating effect of the air. Today, fat catchers should be an integrated feature of aerated sand catchers, in order to reduce the amount of light substances in the subsequent stages.

## Design Details

Aeration is performed by coarse-bubble aerators (aerator pipes), laid 0.5-1.0m apart. Their discharge apertures should be installed at approx. 0.7 of the height of the sand catcher. The aerators should be designed such that the amount of air can be reduced to a minimum of $0.1 m^3/(m^2 \cdot h)$. This makes it possible to avoid an air supply during operation that would be undesirably high for the denitrification process.

The expected separation capacity of an aerated sand catcher is

- 95% for mineral particles with a grain size $\varnothing > 0.3mm$
- 50% for mineral particles with a grain size $\varnothing > 0.1mm$

For particles finer than < 0.07, hardly any separation effect will be achieved.

## Scraping of the Sand-Catcher Material

Today, manual scraping of the sand catchers should not be permitted for operational and hygienic reasons, even in small plants. For this reason, automatic scraping devices are now generally used. Long sand catchers use mobile scraping bridges with lengthwise scrapers or mammoth pumps, apron scrapers or screw scrapers installed lengthways at the bottom. In the case of round sand catchers, the sand is scraped by revolving scraper arms. By means of these scraping devices, the sand is removed either into collector funnels or bags, and then directly conveyed by other devices or by a combination of different mechanical scraping implements or pumps.

In the case of the above-mentioned methods, the sand-catcher material is normally conveyed as a sand-water mixture which needs dewatering. For this purpose, the sand-water mixture can be stored on sand-drying lots which have drainage layers at the bottom for discharge of the water. Alternatives to dewatering of the sand-catcher material are:

- Sand settling containers; here, the sand-water mixture is conveyed to containers in which the material settles and where the surplus water can flow out through an overflow, a lifting device or through the walls of the container if these are designed as filters. The settled sand is then disposed of in the containers or in special transport trucks.
- Mechanical devices; here, preferably grit washers or spiral conveyors are used to convey the sand to transport containers or trucks, which then dispose of the sand. Particularly in the case of small sand catchers, these devices are integrated in the tanks, so there is no need for separate scraping.

### Amount, Composition and Disposal of Sand-Catcher Material

During dry weather spells, the influent to the WWTPs consists mainly of fine sand with a grain size between 0.1 and 0.3mm, the largest amount having a grain size of approx. 0.2mm. In the case of stormwater, however, sand with a diameter of up to 3mm, which has settled in the sewage system, can flow into the WWTP. The amount of mineral particles contained in the WWTP influent is 10-60g/m$^3$ in dry weather.

Generally, the values range from 20 to 200 litres per 1,000m$^3$ of wastewater. In relation to the connected inhabitants, for Germany one can assume a sand-catcher material amount of approx. 4 l/PE per year. However, the very strong short-term variations must also be considered.

## 6.1.4 Preliminary Treatment/Settling Tank

Settling or sedimentation tanks are used to separate organic and/or inorganic sludge, which tend to flocculate, from the wastewater using the force of gravity.

### Basic Methods and Treatment Goals

The wastewater normally flows through the settling tank continuously and at a very low velocity. During its passage, the settling solids are separated from the clear water because of the differences in density. The solids sink to the bottom of the tank and can be removed.

As a preliminary treatment process, the settling tanks are designed to separate from the raw wastewater the suspended substances that are not of mineral nature and have therefore passed the sand catcher. As the preliminary treatment tanks are mostly followed by further treatment stages, a separation performance of $\eta=70\%$ of the filterable substances contained in the influent is sufficient.

**Horizontal-Flow Tanks**

In wastewater technology, the settling tanks used are normally horizontal-flow tanks. They may take the form of round tanks with an inside-to-outside flow direction or rectangular tanks through which the wastewater flows lengthways. All these tanks have the following construction elements in common:

- Wastewater distribution in the tank influent, which with round tanks is in the middle of the tank, and with rectangular tanks on one of the narrow sides. The distribution device must be designed in such a way that the inflowing wastewater can be distributed evenly over the entire cross-section of the tank. The kinetic energy must be converted as extensively as possible. Peak flow and short circuit flow must be avoided.
- Clear-water extraction via the tank diameter for round tanks, or at the side opposite the influent for rectangular tanks. Overflow weirs designed as clarifier weirs can be used for this purpose. Using long weirs allows consistently low overflow heights of only a few centimetres even for varying throughput. Other possible solutions, such as perforated pipes installed just below the water level, are being increasingly used. Crucial for guaranteeing effluent with a low solids content is the even extraction of the clear water over the tank perimeter or the tank width.
- Sludge-scraping system: as the sludge settles over the entire tank bottom, it must be collected mechanically and removed. The removed sludge should have as high a solids concentration as possible in order to minimise the sludge volume flow.
- Removal of floating substances: in some cases, it cannot be ruled out that some of the particles to be separated do not settle but float on the water surface. Settling tanks should therefore be equipped with devices to collect and remove floating substances.

**Vertical-Flow Tanks**

Tanks with a vertical flow are mostly round tanks. They are very deep in relation to their diameter, so that the wastewater flow is mainly in a vertical direction. For such tanks, only small tank diameters of up to 12m are used. The bottom is often designed as a funnel into which the settled sludge sinks without mechanical scraping.

## *Dimensioning of Settling Tanks*

The crucial parameter for the dimensioning of all settling tanks is the surface load:

$$q_A = Q/A \tag{6.1}$$

where: $q_A$     surface load in m/h
$Q$     flow volume in $m^3/h$
$A$     surface area of the settling tank in $m^2$

In simple terms, this is the amount of effluent wastewater in relation to the surface area of the settling tank. It must be lower than the sinking speed of the sludge particles that are to be separated.

A further dimensioning criterion is the contact time in the settling tank, which is related to the surface load via the tank depth according to the following equation:

$$t_R = Q/V = h/q_A \qquad (6.2)$$

where: $t_R$ average contact time in h
$V$ volume of the settling tank in m$^3$
$h$ water depth of the settling tank in m

Table 6.1 contains some rules of thumb for the dimensioning of settling tanks for particular application areas. It is apparent that the figures for preliminary treatment tanks are relatively high. This is only meaningful if they are followed by a further wastewater-treatment stage, for instance a biological stage. If, during the gradual construction of a WWTP, only the mechanical stage is initially built and operated for a certain period, so that the effluent from the preliminary treatment stage is discharged into the receiving water, the following values would apply:

$$t_R > 2h \text{ and } q_A < 2m/h \qquad (6.3)$$

**Table 6.1.: Rules of Thumb for the Dimensioning of Settling Tanks**

| Application as | Surface load $q_A$ (m/h) | Contact time $t_R$ (h) | Tank depth $h$ (m) |
|---|---|---|---|
| Preliminary treatment | 2.5 to 4.0 | 0.5 to 1.0 | 2.0 to 2.5 |
| Secondary treatment after activation plants | 0.5 to 1.0 | 4.0 to 7.0 | 3.0 to 4.5 |
| Secondary treatment after trickling-filter plants | 0.4 to 1.0 | 1.5 to 4.0 | 2.5 to 3,. |
| Secondary treatment after anaerobic plants | 0.2 to 0.5 | 8.0 to 15.0 | 3.0 to 4.0 |
| Sedimentation after precipitation/flocculation | 0.7 to 1.2 | 2.5 to 6.0 | 2.0 to 4.0 |

Common dimensions for tanks with horizontal flow are:

1. Rectangular tanks: L < 50m; W < 10m; ratio L/W = 5:1 to 10:1; bottom inclination 1:100
2. Round tanks: Ø < 60m; bottom inclination 1:15

In order to guarantee even effluent conditions, the feeding of the overflow rims q should not exceed the following values:

Preliminary treatment tank: $\quad q \leq 30 m^3/(m \cdot h)$

### Amount and Composition of the Residues

The kind and amount of the produced primary sludge are determined by the composition of the wastewater. The solids load can be estimated with the help of a mass balance if the solids concentrations in the influent and the separation capacities of the settling tanks are known. The volume of sludge that must be further treated is lower, the higher the solids concentration of the sludge. For the preliminary treatment tanks of municipal WWTPs, it ranges from 30 to 50kg/m$^3$.

## 6.2 Biological Wastewater Treatment

### 6.2.1 Overview

The aim of biological wastewater purification is to remove those substances contained in the wastewater that are hazardous to human beings or the environment, or to process them so that they lose this hazardous potential. Biological wastewater treatment utilises a principle that is prevalent in nature: naturally formed pollutants are converted into inorganic and organic end products by micro-organisms. In WWTPs, these conversion processes occur in the activated sludge tank. In this treatment method, wastewater and activated sludge are mixed in the activated-sludge tank and aerated. The oxygen required for biological degradation is injected through aeration devices.

The wastewater is purified by micro-organisms, the wastewater's organic substances being taken up and, in different ratios, respired or converted into settable biomass. From the activated-sludge tank, the wastewater/sludge mixture flows into the secondary clarifier, in which the activated sludge separates itself from the purified wastewater. The sludge settling in the secondary clarifier is returned to the activated-sludge tank as recirculation sludge, while the purified wastewater flows off. As the activated sludge increases during the biological processes, the surplus sludge is extracted.

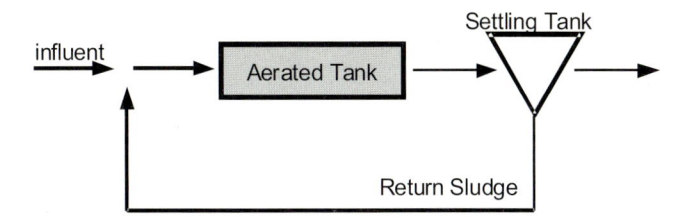

**Fig. 6.2.** Basic Activated Sludge System

## 6.2.2 Legal Requirements for Wastewater Treatment in Europe

The EC (European Commission) Urban Wastewater Treatment Directive (UWWTD) 91/271 sets minimum standards for sewage treatment and sewage collection systems. It specifies secondary treatment for all discharges serving populations or equivalent greater than 2,000PE to inland waters and estuaries, and greater than 10,000PE to coastal waters. The main goal of wastewater treatment, besides the reduction of organic pollution, is nutrient (i.e. nitrogen and phosphorus) removal. The negative effects of excess nutrient discharge to the receiving streams are:

- ammonia oxygen demand, toxicity
- nitrite toxicity
- nitrate water supply impact
- total Nitrogen eutrophication
- total phosphorus eutrophication

The UWWTD addresses the risk of surface waters being enriched by nitrogen and phosphates from sewage. It requires that phosphate-removal facilities be installed at all sewage works serving more than 10,000 population equivalents and discharging into potentially nutrient-sensitive waters. Nitrogen removal down to 15mg/l outflow concentration is required at works in conurbation of less than 100,000 population equivalent, and 10mg/l in larger conurbation. The phosphate-removal standards are down to 2mg/l below and 1mg/l above 100,000 population equivalent, respectively. This has led to substantial efforts to construct centralised sewage systems, i.e. sewer networks and centralised sewage treatment plants. By way of an example, the following tables show the developments in Germany in this area. It should be noted that in eastern Germany this process has been under way only since 1990.

**Table 6.2.** Development of Wastewater Treatment in Germany

|   |   | Year | | 1975 | 1979 | 1983 | 1987 | 1991 | 1995 |
|---|---|---|---|---|---|---|---|---|---|
| 1 | Connection | Germany | | | | | | 90% | 92% |
|   | to Sewer | | - west | 86% | 89% | 91% | 93% | 94% | 95% |
|   | System | | - east | - | - | - | - | 75% | 77% |
| 2 | Biological | Germany | | | | | | 79% | 84% |
|   | Treatment | | - west | 56% | 69% | 79% | 85% | 89% | 93% |
|   | | | - east | - | - | - | - | 37% | 49% |
|   | Mechanical | Germany | | | | | | 6% | 4% |
|   | Treatment | | - west | 19% | 10% | 5% | 2% | 2% | 1% |
|   | | | - east | - | - | - | - | 24% | 17% |

1 = Connection to public sewer system
2 = Connection to centralised sewage-treatment plant
published in: "Daten zur Umwelt" 1997, Federal Statistical Office

|   |   | 1963 | 1969 | 1975 | 1979 | 1983 | 1987 | 1991 |
|---|---|---|---|---|---|---|---|---|
| 1 | Germany | | | | | | | 9,935 |
|   | - west | 3,274 | 6,048 | ,647 | 8,167 | 8,805 | 8,841 | 8,667 |
|   | - east | | | | | | | 1,268 |
| 2 | Germany | | | | | | | 8,141 |
|   | - west | 1,291 | 3,478 | 5,252 | 5,823 | 6,658 | 7,196 | 7,301 |
|   | - east | | | | | | | 840 |
| 3 | Germany | | | | | | | 8,512 |
|   | - west | 3,481 | 4,707 | 6,007 | 7,235 | 7,672 | 8,883 | 7,518 |
|   | - east | | | | | | | 994 |

1 = sewage plants ( x 1,000 )
2 = biological plants
3 = amount of sewage, treated (million m$^3$)
published in: "Daten zur Umwelt" 1997, Federal Statistical Office

Particularly sensitive waters, such as the Alpine Lakes, have been the object of sewage-phosphate-removal programmes working to even more stringent standards than those laid down by the EC Directive. In these cases, tertiary-treatment steps like filtration combined with phosphorus precipitation are used.

Additionally, in recent years, considerable research has been conducted on the effects of nutrients on surface waters and the restoration of lakes affected by eutrophication problems. These studies underline the importance of other factors (zooplankton populations, which graze on algae; fish populations; changes in flow rate, often related to man-made dams or weirs).

Different developments in sewage nutrient removal, and a growing understanding of the different sources of nutrients in surface water and the non-linear and ecosystem-dependent response to nutrients, have led to new approaches to surface-water protection and restoration. Recent programmes adopt an overall catchment approach, including land use, agricultural and other diffuse sources and sewage treatment. The use of phosphates in detergents is compatible with such programmes because phosphates from all sources are removed in sewage treatment. Environmental objectives for the protection of surface waters (e.g. the EC Water Framework Directive put into effect in December 2000) require improved treatment of sewage and animal wastes across Europe.

Consequently, it is now increasingly recognised that diffuse agricultural sources are the main phosphate source in most surface waters, in Europe and elsewhere.

## *Nutrient-Removal Mechanisms*

### Nitrogen Removal

A certain amount of nitrogen removal occurs in any biological wastewater treatment system owing to the uptake of nitrogen into the waste sludge produced in the process. Nitrogen is a component of waste biomass produced as a result of the biological treatment of carbonaceous organic matter. Organic nitrogen is also a component of the non-biodegradable particulate organic matter present in much wastewater. This material is generally flocculated and incorporated into the biological-treatment system, mixed liquor and subsequently removed from the process with the waste sludge.

However, the amount of nitrogen removed by this method is not sufficient to meet the European standards mentioned above. As a rule of thumb, only about 25% of the nitrogen load present in the raw sewage will be eliminated. Nutrient-Removal Systems (NRS) are therefore used in wastewater treatment. The difference between a typical biological wastewater-treatment system and a Biological Nutrient-Removal system (BNR) is that in a BNR system additional nitrogen removal is achieved by the combined action of the two biological reactions:

- nitrification
- denitrification.

### Nitrification

Nitrification is the biological conversion of ammonia nitrogen to nitrate nitrogen. It is accomplished by members of a group of bacteria called autotrophs. Autotrophic micro-organisms oxidise inorganic constituents to obtain energy for growth and maintenance, while they obtain carbon for the production of new biomass by reducing carbon dioxide. Organic matter is not required for the growth of autotrophic bacteria. Nitrification is actually a two-step reaction. The first step is oxidation of ammonia nitrogen to nitrite nitrogen by bacteria of the genus *Nitrosomonas spec.*. The equation for this reaction is presented here in simplified form:

$$NH_4 + 1.5\ O_2 \rightarrow NO_3 + 2H + H_2O \tag{6.4}$$

The second step is the oxidation of nitrite nitrogen to nitrate nitrogen by bacteria of the genus *Nitrobacter spec.*. The simplified equation for this reaction is:

$$NO_2 + 0.5\ O_2 \rightarrow NO_3 \tag{6.5}$$

Under constant conditions, these two reactions will be in balance and the overall reaction will basically run its course. The overall reaction, including the synthesis of new biomass (expressed as the typical composition of biomass), is:

$$NH_4 + 1.83\ O_2 + 1.98\ HCO_3 \rightarrow 0.98\ NO_3 + 0.021\ C_5H_7NO_2 + \\ 1.88\ H_2CO_3 + 1.04\ H_2O \tag{6.6}$$

This equation illustrates the stoichiometry of the nitrification reaction. Oxygen is required to oxidise ammonia nitrogen, 4.6mg of $O_2$ being required for each mg of $NO_3$-N generated. Bicarbonate alkalinity is also consumed in the reaction both to neutralise the acid produced (i.e. ammonia nitrogen is a base while nitrate nitrogen is an acid) and as required for the synthesis of new biomass (from carbon dioxide, which is present as bicarbonate alkalinity). The alkalinity requirement calculated from the equation is 7.2mg of alkalinity as $CaCO_3$ for each mg of $NO_3$-N produced. Biomass-yield values are typically low for autotrophic bacteria, and the nitrification reaction is no exception. The yield value for the nitrifiers (both Nitrosomonas and Nitrobacter) is 0.15mg of bacteria as total suspended solids (TSS) per mg of nitrate nitrogen generated.

The growth of nitrifying bacteria is affected by a number of factors. First of all, the specific growth rate of the nitrifying bacteria is generally lower than that of the heterotrophic bacteria which oxidise carbonaceous organic matter in biological wastewater-treatment systems. The specific growth rate of micro-organisms in biological wastewater-treatment systems is often expressed as the minimum mean solids-retention time (minimum SRT), which represents the SRT corresponding to the maximum specific growth rate for the subject micro-organisms.

When the process is operated at the minimum SRT, the micro-organisms grow at their maximum rate and are simply washed out of the system. The SRT must be longer than the minimum SRT if the organisms are to grow and survive in the system. The SRT in the part of the system where nitrification occurs is typically 1.5 to 2.0 times the minimum SRT. Operation at an SRT less than the minimum SRT will result in "washout" of the subject micro-organisms. In practice, most plants are operated with an SRT between 10 and 15 days, depending on boundary conditions like temperature. Because the nitrifiers' growth rate decreases as temperature drops, in Germany, for example, nitrogen removal is only required if the operating temperature at the plant is above 12°C.

The growth of nitrifying bacteria is affected by a number of environmental factors. Some of the most significant of these are dissolved oxygen (DO), pH and the presence of inhibitors. As illustrated in the equation, nitrification is an aerobic reaction requiring the presence of dissolved oxygen. The activity of the nitrifying bacteria is reduced if the dissolved concentration is decrease below about 2-3mg/L, and it is totally inhibited if dissolved oxygen is not supplied. Nitrification can occur when dissolved oxygen is supplied. DO-concentrations are low, the rate is reduced.

The activity of the nitrifying bacteria is also affected by pH. The optimum pH for growth of the nitrifying bacteria is generally about 7.5. The activity of the nitrifying bacteria is reduced somewhat as the pH is reduced below 7.0, and it is inhibited significantly when the pH drops below 6.5. Recent research distinguishes between the impact of pH on acclimated and unacclimated cultures. A wide variety of organic and inorganic compounds can also inhibit the growth of the nitrifying bacteria. This may be an important issue if certain industrial wastewater is being treated.

**Denitrification**
Denitrification is the utilisation of carbonaceous organic matter by heterotrophic bacteria, nitrate nitrogen being used as the terminal electron acceptor (i.e. the "oxygen source"). Many of the heterotrophic bacteria in biological wastewater-treatment systems are capable of using either dissolved oxygen or nitrate nitrogen as a terminal electron acceptor. Dissolved oxygen is preferred if both terminal electron acceptors are present. This is the case because slightly more energy can be obtained by the oxidation of carbonaceous organic matter using oxygen as the terminal electron acceptor rather than nitrate nitrogen.

However, dissolved oxygen and nitrate nitrogen provide essentially the same biochemical function. When nitrate nitrogen serves as the terminal electron acceptor (i.e. when denitrification occurs), the nitrate nitrogen is converted to nitrogen gas, which can then be released into the atmosphere. This reaction causes the removal of nitrogen from the wastewater stream.

Denitrification significantly impacts the stoichiometry of a biological wastewater-treatment system. For example, the fact that a portion of the carbonaceous oxygen demand is satisfied by the reduction of nitrate nitrogen means that the process oxygen demand is reduced. Theoretically, 2.86mg of carbonaceous oxygen demand is satisfied for each mg of $NO_3$-N that is reduced to nitrogen gas.

Denitrification also results in a reduction in process alkalinity consumption due to the removal of the acid nitrate. Theoretically, 3.6mg of alkalinity as $CaCO_3$ is produced per mg of $NO_3$-N reduced to nitrogen gas.

In municipal wastewater-treatment plants (WWTPs), the denitrification rate varies considerably as different fractions of the organic matter contained in the wastewater are oxidised. Initially, a relatively high denitrification rate occurs as the readily biodegradable organic matter is oxidised. When this matter is exhausted, the denitrification rate is reduced to that produced when the more slowly biodegradable organic matter is oxidised. Finally, when all the biodegradable organic matter has been oxidised, the denitrification rate is relatively low and is driven only by endogenous respiration. This means that denitrification rates will initially be relatively high as the readily biodegradable organic matter is oxidised. The rate will decrease with time, however, as first the readily biodegradable organic matter and then the slowly biodegradable organic matter is depleted. If the organic matter present in the influent wastewater is to be used as the primary carbon source, it must be used effectively if a relatively high degree of denitrification is to be achieved. This is accomplished by process configurations that use the influent wastewater for denitrification first, as discussed below.

## Process Design for Nitrogen-Removal Plants

A large number of biological nutrient-removal process options have been developed and are used in full-scale wastewater treatment plants across Europe. Which of this wide variety of process options potential users will choose depends on a number of factors like size of plant, sewage composition (industrial influence), energy-supply aspects, sludge-disposal options, etc. Differences between nutrient-removal process options can sometimes result in significant differences in process performance and/or operational characteristics. In other cases, they have little or no impact. Different process options may also simply represent different approaches to accomplishing the same objectives.

At most European WWTPs, a single activated sludge biological nitrification/denitrification process is used. There are several variants of process configuration, but basically process design follows two different approaches:

- Pre-denitrification
- Simultaneous or intermittent denitrification.

The pre-denitrification process takes place in two sections of the activated-sludge tank: the anoxic (denitrification) zone and the aerobic (nitrification) zone. An anoxic zone is a region of the biological reactor where dissolved oxygen is excluded and nitrate nitrogen is provided to serve as the terminal electron acceptor. Denitrification occurs in anoxic zones. An aerobic zone is one which is aerated to provide dissolved oxygen as the terminal electron acceptor. When dissolved oxygen is provided in the aerobic zone, both the oxidation of carbonaceous organic matter and nitrification can occur.

Denitrification (the second step in a biological nitrogen-removal process) occurs in the anoxic zone. Nitrate nitrogen is provided to the anoxic zone by the mixed-liquor recycle stream from the aerobic to the anoxic zone.

## Phosphorus Removal in Activated Sludge Systems

### Chemical Phosphorus Removal

The additional phosphorus and nitrogen removal is motivated by the problems associated with eutrophication of receiving waters. The simple and powerful idea is to eliminate the nutrients that are necessary for the growth of eutrophication-causing micro-organisms.

As mentioned above, most wastewater treatment facilities employ the activated sludge process. The total phosphorus-removal rate obtained in a conventional activated-sludge plant generally is less than 20%, and is even less in wastewater treatment plants where anaerobic digester supernatant is recycled to the head of the plant. Since it is not possible to achieve the 1mg P/L effluent limit with conventional biological wastewater treatment processes, additional or alternative treatment methods must be employed.

Traditionally, phosphate removal in activated sludge plants is accomplished by chemical precipitation. Depending on the point of chemical addition at which chemicals are added, the precipitation may be realised as

- primary precipitation (dosing point: upstream of primary clarifiers)
- simultaneous precipitation (dosing point: within the activated-sludge plant)
- tertiary precipitation (dosing point: in a separate part of the plant downstream of secondary clarifiers)

Simultaneous precipitation is the most common alternative because both process control and process stability are high and the additional construction costs are minimal.

Chemical precipitation increases the volume of sludge produced and often results in a sludge with poor settling and dewatering characteristics. Also, precipitation with metals salts can depress the pH. If nitrification is required, additional alkalinity will be consumed and the pH will drop further.

Biological phosphorus-removal systems can offer the following benefits:

- reducing or eliminating the need for chemical addition
- reduced sludge production
- improved sludge settlability and dewatering characteristics
- reduced oxygen requirements
- reduced process-alkalinity requirements

**Biological Phosphorus Removal**

For biological phosphorus removal by luxury uptake of phosphorus (BPR), a broad group of micro-organisms contribute to the uptake of soluble orthophosphate ($PO_4^{3-}$). These organisms are called phosphate-accumulating organisms (PAOs). Under cyclic anaerobic/aerobic ($O_2$ present) or anaerobic/anoxic ($NO_3$ present) conditions, they are able to sequester in the anaerobic reactor readily degradable organic substrates, which are stored as polyhydroxyalkanoates (PHA). The formation of PHA coincides with the degradation of intracellular polyphosphate and the intracellular consumption of carbohydrates such as glycogen. After conversion to orthophosphate, the degraded polyphosphate is released to the bulk phase.

In the subsequent aerobic (anoxic) reactor, the stored PHA is used for catabolism and metabolism. The dynamics for the internal storage products polyphosphate and glycogen are diametrical to the anaerobic phase. Recent research results show that at least some of the PAOs are able to use nitrate as an electron acceptor, which results in an anoxic uptake of phosphate and denitrification by the PAOs. Owing to the growth of PAOs in the process, an overall uptake of phosphate occurs, and the phosphate-rich micro-organisms are eventually separated by sedimentation.

The biological nitrogen and phosphorus removal makes the complexity of the process rather high. Anaerobic, anoxic and aerobic environments are needed and the operating conditions must ensure that nitrifiers, denitrifiers and PAOs are present in sufficient number while undesirable species (e.g. filamentous bacteria) are excluded. In the various subprocesses, the micro-organisms compete for the same resources such as organic substrate, which is needed both for denitrification and uptake of phosphate. Furthermore, the holding time of the process must be shared between the sequential processes of nitrification and denitrification.

Despite increasing knowledge about the different subprocesses, we still lack an overall understanding of the activated sludge process, especially with varying composition of the incoming wastewater. Better operational and control strategies for nutrient-removal activated-sludge processes are needed to improve design and enhance performance in normal and disturbed situations.

Typical disturbances of the process include – apart from daily variations in wastewater composition (depending on the respective municipality) – changes due to short-term weather events, such as rain and storms, and changes caused by long-term events such as seasonal weather periods. These can lead to a temporary increase in the effluent phosphate concentration; a lengthy period of normal wastewater composition is then required for recovery from these disturbances.

Several process configurations are currently being employed for biological phosphorus removal. In European wastewater treatment plants, process configurations normally combine nitrogen removal by nitrification and denitrification with biological phosphorus removal. However, all are based on the sequential exposure of micro-organisms to anaerobic and aerobic conditions in the biological reactor.

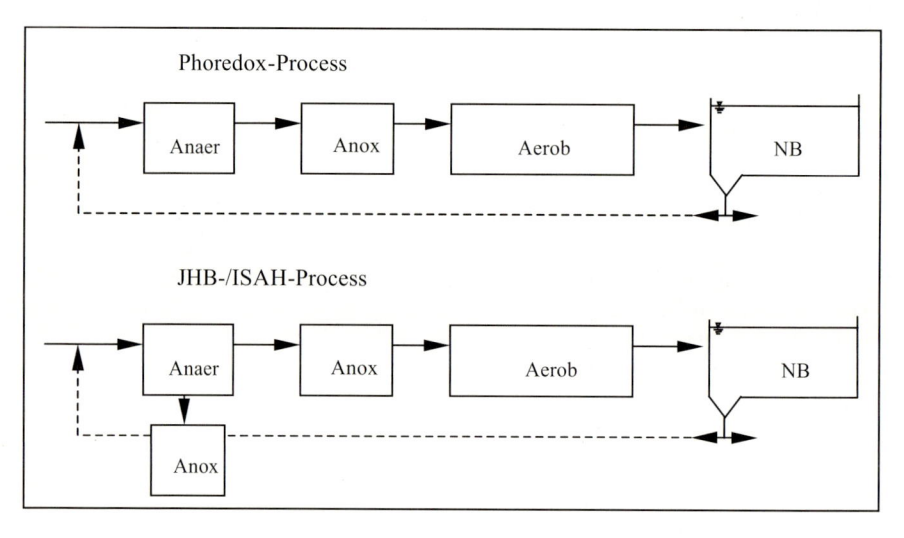

**Fig. 6.3.** Typical Treatment Processes for Enhanced Biological Phosphoros Removal (Phoredox and JHB)

## Wastewater Sludge Treatment and Disposal

The continuous increase in the total quantity of sewage sludge solids produced in German and European WWTPs has been accompanied by a charging regulatory and economic environment that is forcing many municipalities to develop new approaches to sludge disposal.

The expansion of both industrial and municipal wastewater treatment in recent years has resulted in a major increase in the quantities of by-product sludge produced and has exacerbated problems of waste-sludge treatment and/or disposal. The traditional method of waste-sewage-sludge treatment is anaerobic mesophilic digestion for sludge that is to be disposed of on agricultural land. Other disposal methods for untreated sludge includes incineration, an option receiving increased attention, and ocean dumping, an option that is environmentally unacceptable and will be no longer used within the European Union.

Sewage sludge can be considered a resource as far as its mineral nutrient composition is concerned. In many cases, however, this resource cannot be exploited in agriculture because sludge is inevitably contaminated with noxious chemicals that accumulated into the sludge during either primary or secondary wastewater treatment. The development of sewage-sludge treatment and disposal is currently under discussion throughout Europe.

## Centralised Wastewater Treatment

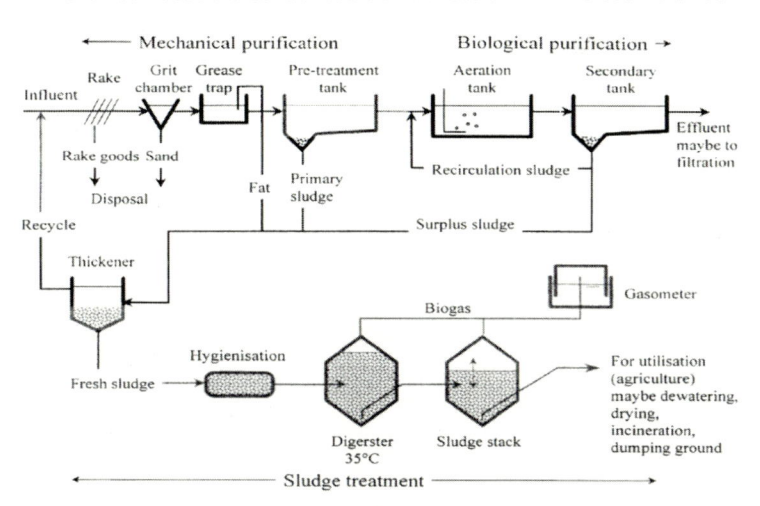

**Fig. 6.4.** General Outline of a Centralised Wastewater Treatment Plant

This diagram of centralised wastewater treatment shows the wastewater flow and the different treatment steps, which will be discussed in more detail.

## Biological Purification

In some WWTPs, the wastewater passes through an anaerobic tank where some of phosphorus compounds are removed. The efficiency of this stage depends on the load of easily degradable organic compounds in the influent. Another way of removing phosphorus is by precipitation. In the aerated tank (nitrification tank), organic compounds are degraded and ammonia is oxidised to nitrate. Nitrate is reduced to nitrogen by a denitrification process. There are different ways of realising the denitrification process:

- Pre-denitrification: The denitrification tank is situated before the nitrification tank. The activated sludge from the nitrification tank, containing nitrate, and the return sludge is pumped back into the denitrification tank.
- Cascade denitrification: A number of denitrification and nitrification tanks are arranged one behind the other (for example: denitrification – nitrification – denitrification – nitrification). The influent is distributed to the denitrification tanks. With this system, it is not necessary to pump the activated sludge from nitrification tank to denitrification tank because there is a denitrification tank behind the nitrification tank.
- Post-denitrification: The denitrification tank comes after the nitrification tank. This method is used if an additional source of carbon is needed.
- Simultaneous denitrification: The denitrification and nitrification processes take place in different zones in the same tank.
- Alternating denitrification: Two tanks are aerated alternately. The influent always passes first through the tank that is not aerated.

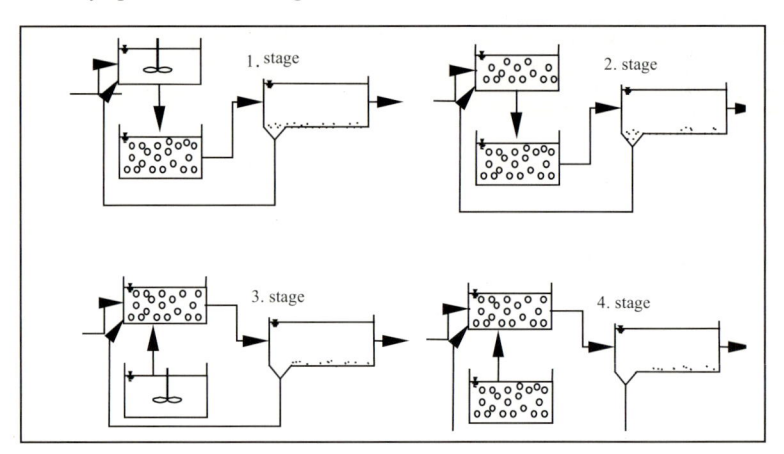

**Fig. 6.5.** Alternating Denitrification

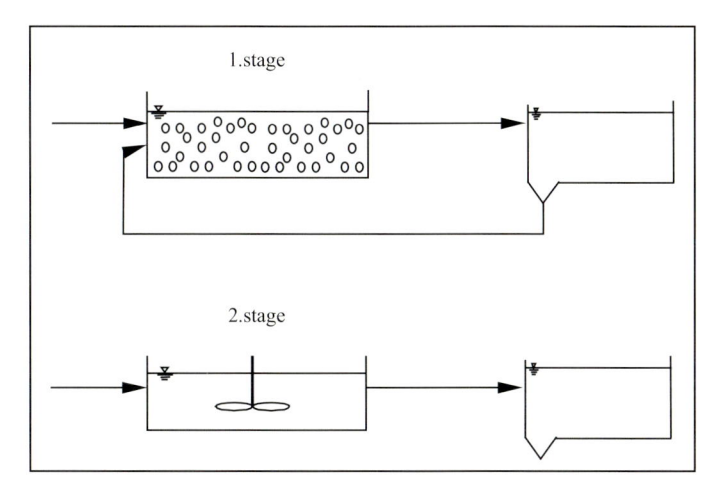

**Fig. 6.6.** Intermittent Denitrification

- Intermittent denitrification: Nitrification and denitrification take place in one tank at different times. The tank is aerated during the nitrification phase, aeration being stopped for denitrification.

In the secondary clarifier tank, activated sludge is separated from the purified wastewater. The sludge sediments on the bottom of the tank and is removed by a scraper. Part of the sludge (recirculation sludge) is pumped to activated-sludge tanks; the rest is surplus sludge and can be treated using different methods. The outlet of the secondary clarifier is situated near the water surface.

**Fig. 6.7.** Nitrification and Pre-Denitrification

## 6.3 Models for the Design and Simulation of Wastewater Treatment Plants (WWTPs)

### 6.3.1 Overview

Ecological and economic aspects are becoming increasingly important in the dimensioning and simulation of wastewater treatment plants (henceforth abbreviated to WWTPs). In view of more stringent environmental requirements and refinement of the methods used for wastewater purification, complex mathematical and biological models have been developed that allow the optimal dimensioning of WWTPs while keeping construction costs to a minimum.

Apart from the dimensioning of new plants, another increasingly popular approach is to enhance the capacity of existing plants by increasing tank volumes and adding new technology and to optimise their efficiency using suitable simulation models to minimise operation costs. Mathematical models can only be meaningfully used in conjunction with computer programs. They help to deepen our understanding of the biological processes involved, e.g. by allowing us to modify the operation of a virtual plant on a computer or by enabling us to test different treatment systems and clarifiers to find an optimal operation point. A detailed description of the modelling fundamentals based on German experience reported here could also help to adapt existing equations to the wastewater situation in other countries. (Wichern a. Rosenwinkel 2000)

The fundamentals of these models are summarised below, starting with the steady-state models used for the dimensioning and extension of WWTPs, and then moving on to the dynamic models for the simulation and operational optimisation of existing plants.

#### *Basics of the Wastewater Treatment*

All treatment methods are characterised by a combination of aerobic (dissolved oxygen present) and thus nitrifying tank zones, and anoxic (no dissolved oxygen present, but bound oxygen in the form of nitrate) and thus denitrifying zones. Moreover, there are anaerobic tank zones (neither dissolved nor bound oxygen present) designed to facilitate an increased biological P-elimination.

**Nitrification**
In the nitrification process, ammonium ($NH_4$-N) is biochemically converted into nitrate. This process occurs through the bacteria – the nitrifiers – which oxidise the inorganic nitrogen into nitrate ($NO_3$). The obligatorily aerobic micro-organisms depend on $CO_2$ and thus do not need any organic carbon for the growth of their biomass. The first stage in the nitrification process is nitritation, the second stage nitratation, with different micro-organisms taking part in the different stages.

## Denitrification
Denitrification means the conversion (reduction) of nitrate into volatile nitrogen (N2) using micro-organisms. Many heterotrophic bacteria (denitrifiers) are able to respire the nitrate oxygen instead of the bound oxygen. If oxygen is available (aerobic milieu), micro-organisms always prefer $O_2$-respiration; only in the case of a lack of oxygen and the presence of nitrate and/or nitrite (anoxic milieu) will they switch to denitrification. Thus, nitrate nitrogen is only removed from the wastewater when there is a lack of dissolved $O_2$.

## Biological and Chemical P-Removal
Biological P-Removal: Apart from the elimination of nitrogen, the removal of phosphorus from the wastewater is of particular importance. The phosphorus elimination is performed biologically by bacteria which under certain conditions take up an increased amount of phosphate. By extracting these bacteria through the surplus sludge, it is also possible to increase phosphorus removal.

The increased phosphate uptake of the bacteria occurs where the organisms are first subjected to oxygen-free (anaerobic) zones, in which the bacteria dispense phosphate into the wastewater (P-release), and then enter the aerated or anoxic area of the WWTP, where they increasingly take up phosphate.

Thus, the increased biological P-elimination is coupled to the change between anaerobic and aerobic or anoxic states. In the anaerobic zone, organic acids are stored, the polyphosphate stores being used for gaining energy and the dissolved phosphate being released. In the following aerobic or anoxic zones, the BioP organisms (PAOs) grow on the storage substances and in turn fill their polyphosphate stores (cf. Kunst 1991).

Chemical P-Precipitation: Apart from biological P-elimination, chemical phosphorus precipitation is widely used. By adding multivalent metal ions (mainly $Fe^{3+}$, $Al^{3+}$, $Fe^{2+}$ and $Ca^{2+}$), undissolvable compounds are created. This leads to the destabilisation of colloids contained in the wastewater and to the accumulation of microflakes, which can later be removed from the wastewater as macroflakes, e.g. by sedimentation. Often, during macroflake creation, floating substances and organic material are incorporated into the flake. Today, chemical precipitation is frequently used as simultaneous precipitation – i.e. directly in the activated-sludge tank – but it can also be employed as a pre-treatment process in conjunction with the separation of floating substances in the pre-treatment unit, or as secondary precipitation, possibly in conjunction with filtration.

## *Model Basics*

### Nitrification
If the nitrifiers are kept together with heterotrophic organisms in the activated-sludge tank – as is usually the case in municipal wastewater-purification plants – a minimum dwelling time of the nitrifiers in the activated-sludge tank (aerobic sludge age) must be guaranteed in order that the nitrifier amount extracted from the surplus sludge does not exceed the amount that is subsequently grown.

When dimensioning, nitrification is designed according to the aerobic sludge age, i.e. the nitrification zone is designed taking into account the growth and decay of the micro-organisms so that the autotrophic biomass can establish itself almost constantly in the activated-sludge tank.

$$t_{TS,aer} = \frac{SF}{\mu_N \cdot f_T \dfrac{NH_{4-e,m}}{K_N + NH_{4-e,m}} - b_A \cdot f_T} \quad (6.7)$$

The equation results from the balancing of the autotrophic biomass and takes into account the fact that the amount of nitrifiers extracted with the surplus sludge must not exceed the net growth amount. In the equation, the denominator is comprised of the difference between the current growth rate as a function of the ammonium nitrogen and the decay rate of the nitrifiers. Additionally, a safety factor is used to take into account inhibition and peak loads.

### Carbon Degradation
In the case of municipal WWTPs, the crucial factor for the dimensioning of the aerators and the resulting surplus sludge is the elimination of carbons. The oxidisable carbon can be approximately described as BOD or COD. Using COD has the advantage that this term comprises all substances that can be oxidised into $CO_2$ and $H_2O$ and which can then also be accounted for.

### Surplus Sludge Production
Generally, the equations for ascertaining the surplus sludge production are very similar in the different approaches. The sludge production is mainly comprised of three parts:

- Active net surplus sludge amount

$$SS_H = \frac{Y_H \cdot BOD_{5,o} \cdot Q_o \cdot 10^{-3}}{1 + b_H \cdot t_{DS}}, \; [\text{kg}_{DS}/d] \quad (6.8)$$

- Inert production from the decay of active biomass

$$SS_I = f_I \cdot \ddot{U}S_H \cdot t_{DS} \cdot b_H, \; [\text{kg}_{DS}/d] \quad (6.9)$$

- Inert sludge amount of the influent

$$SS_I = 0{,}6 \cdot TS_o \cdot Q_o \cdot 10^{-3}, \; [\text{kg}_{DS}/d] \quad (6.10)$$

### Oxygen Demand
The oxygen demand in the plant consists of three components:

- Oxygen demand from the carbon degradation $OD_C$ which develops during the growth and decay of the heterotrophic biomass
- oxygen demand for nitrification

- oxygen gain from the denitrification of nitrate nitrogen and the connected conversion of carbon substrate in the denitrification tank

The oxygen demand of the heterotrophic biomass again consists of two parts: the substrate respiration occurring through the growth of the heterotrophic biomass, where about 1/3 of the COD are oxidised directly, and the endogenous respiration. The latter develops through the $O_2$-demand occurring during oxidation of the decayed organic ratio of the heterotrophic biomass (endogenous respiration). When dimensioning the model, the heterotrophic oxygen demand of the entire activated-sludge tank is, in a first approximation, distributed to the nitrification and denitrification zones according to their volumetric ratio.

Entire oxygen demand from carbon degradation

$$OD_C = a_{substrate} \cdot BOD_{5,o} \cdot Q_o \cdot 10^{-3} + a_{endogenous} \cdot SS_H \cdot t_{DS} \quad (6.11)$$
$$, [kgO_2/d]$$

The maximum oxygen demand for the carbon degradation in German plants is 1.6kg $O_2$/kg $BOD_5$.

**Denitrification**

Taking into account the reduction of the heterotrophic biomass in the anoxic zone, the following equation is obtained for the oxygen demand in the denitrification tank for simultaneous or intermittent denitrification:

Oxygen demand in the denitrification zone of the dimensioning

$$O_{2-Demand} = OD_C \cdot \frac{V_D}{V_{AT}} \cdot f_D, [kgO_2/d] \quad (6.12)$$

If – as in the process – the supply of nitrate nitrogen and the demand in the anoxic zone are equal, the entire returned and fed nitrate nitrogen is removed. In the effluent of the denitrification tank, the nitrate load is nil. The oxygen demand for the carbon degradation is equivalent to the returned and fed nitrate nitrogen.

Oxygen Balance in the Denitrification Zone in the Dimensioning: Oxygen demand in the denitrification tank from BOD degradation = Returned nitrate nitrogen amount

$$OD_C \cdot \frac{V_D}{V_{AT}} \cdot f_D = 2,9 \cdot \left( \frac{RF}{1+RF} \cdot NH_4N_n \cdot Q_o \cdot 10^{-3} \right), \quad (6.13)$$
$$[kgO_2/d]$$

For the steady-state simulation of existing plants, it is crucial to be able to predict the nitrate reduction in the denitrification tank. Now, one can no longer suppose that the relation between oxygen demand (carbon degradation) and nitrate nitrogen supply is balanced. There are two possibilities:

- Oxygen demand shortage (BOD restriction). The consequence is incomplete nitrate reduction, but complete oxygen demand degradation.

- Oxygen supply shortage (NO3-N restriction) in the denitrification tank. The consequence is complete nitrate reduction, but incomplete oxygen demand degradation in the denitrification tank.

**Biological P-Removal**

The growth of biomass requires phosphorus, which is incorporated into the biomass and then removed from the WWTP during surplus sludge extraction. Generally, the models work with highly simplified calculations for the phosphorus incorporation to prevent calculations from becoming too complex.

Phosphorus Incorporation: Generally, there are two ways to describe the phosphorus incorporation. According to the ATV 131 (2000), the incorporated amount is defined as a function of the incoming BOD; in the HSG approaches, for instance, it is related to the active heterotrophic biomass.

P-incorporation as a function of the incoming BOD (ATV 2000)

$$P_{SS} = BOD_5 \cdot i_P, [g_P/m^3] \qquad (6.14)$$

With normal biological P-removal, the factor $i_P$ assumes the value 1%; with increased biological P-elimination (EBPR), it assumes the value 2.5% (if a preliminary anaerobic tank is used). P-incorporation is dependent on the active heterotrophic biomass (HSG, Böhnke et al 1989).

Depending on the active heterotrophic biomass, approximately $i_P$=0.03gP/gaVSS is incorporated into the biomass in the case of normal P-elimination, and 0.06 with increased biological P-elimination. The resulting phosphorus amount incorporated in the surplus sludge is then:

$$P_{SS} = i_P \cdot SS_H, [kg_P/d] \qquad (6.15)$$

Anaerobic Volume: The anaerobic volume used with the increased biological P-elimination is, in Germany, normally ascertained on the basis of the ATV; the anaerobic volume ratio in relation to the nitrification and denitrification tanks must be defined depending on whether favourable or unfavourable wastewater conditions dominate. Depending on the overall conditions (easily degradable substrates, etc.), the contact time (anaerobic volume per inflow and return sludge) of the anaerobic tank should be between 0.75 and 2h.

Points of Criticism: The above-described phosphorus elimination is given consideration on the basis of empirical values. Important parameters for the increased biological P-elimination – such as the volume of the anaerobic tank, the amount of easily degradable COD in the influent or the nitrate nitrogen input into the anaerobic tank – are only considered in more complex calculations (e.g. Wichern 2000)

**Chemical P-Precipitation according to ATV-A131**

For the chemical P-precipitation, there are several tried approaches available. Here, we will discuss exemplarily only the model based on the ATV standard 131. The sludge production caused by precipitation is calculated as follows:

Iron as precipitaiton agent:
$$SS_P = 6.8 \cdot P_{eli}, [\text{kg}_{DS}/d] \tag{6.16}$$

Aluminium as precipitation agent:
$$SS_P = 5.3 \cdot P_{eli}, [\text{kg}_{DS}/d] \tag{6.17}$$

### Treatment Methods

The preliminary denitrification provides the basis for calculating different treatment processes in Germany. With this method, the wastewater to be purified first flows through the denitrification tank. The TKN nitrogen content is not changed during passage through this tank, except for the nitrogen incorporation into the biomass. After the denitrification tank, the wastewater is conveyed to the nitrification tank in which the ammonium is nitrified.

The created nitrate is conveyed back to the denitrification tank via the recirculation sludge and, additionally, via an intensive circulation (internal recycle) from the effluent of the nitrification tank. The nitrate fed into the denitrification tank is converted there into nitrogen ($N_2$) by oxidation of the BOD. The efficiency of the denitrification depends on the entire recirculation ratio (sum of the recirculation sludge ratio and the internal recirculation ratio) and on a sufficient oxygen demand from the carbon degradation.

Besides preliminary denitrification, there are other well-known systems like simultaneous denitrification in a recirculation ditch, intermittent and alternating aerated systems and cascade denitrification. All these systems could be used for simultaneous aerobic stabilisation, operated with high sludge ages of 20-25d to reduce the organic part of the sludge.

**Secondary Settling Tank**

Activated sludge is separated in the secondary settling tank by sedimentation. For this, the hydraulic loading of the surface of the secondary clarifier is the crucial factor in the design of the SSTs, to ensure the sedimentation of the sludge and guarantee effluent values of dry solids lower than 20mg/l.

$$q_A = \frac{Q_m}{A_{SST}}, [\text{m/h}] \tag{6.18}$$

Typical values of $q_A$ should be lower than 1.2m/h with horizontal water flow and lower than 1.6 m/h with vertical flow. In addition, the sludge volume index is the most important factor for the concentration of suspended solids in mixed liquor (MLSS) in the activated-sludge tank. The concentration that could be expected in the return sludge is calculated as follows:

$$DS_{RS} = (0.5 \text{ up to } 0.7) \cdot \frac{1000}{SVI} \cdot \sqrt[3]{t_E} \tag{6.19}$$

## 6.3.2 Dynamic Models

While stationary models are generally used for the dimensioning of new plants or the extension of existing ones, dynamic models are mostly used for the optimisation of plants already in operation. Such models can take into account the new load situations caused by additional input or, for instance, mixed water loads. Dynamic simulation differs from the stationary models in that, instead of working with daily average influent values, it is based on 2h average values or 15-minutes online measurements, for instance, which makes it possible to examine short-term peak loads and their impact on the effluent values or the required aerator performance.

Apart from the mathematical model structure, the fractionating of the influent COD-load into detailed ratios is of particular importance (This could also be successfully done with BOD, Kayser 1997). A distinction is made between dissolved (S) and undissolved (X) substances, biomass ($X_H$; $X_A$), non-degradable ($S_I$, $X_P$) and degradable substances ($S_S$, $X_S$). For the denitrification and the P-elimination (EBPR), the easily degradable COD is of particular importance. The following figure shows a typical fractionating.

**Typical COD-Fractioning of the influent**

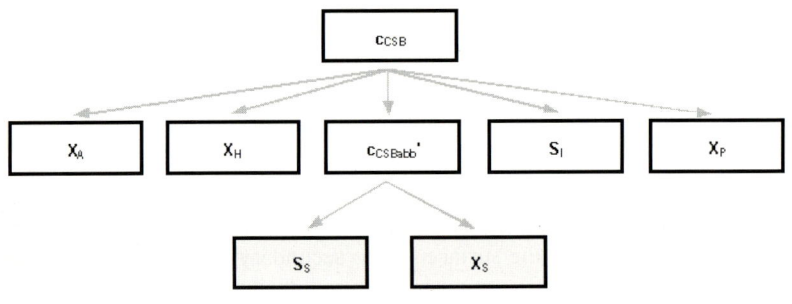

[Diagram: $C_{CSB}$ branches to $X_A$, $X_H$, $C_{CSBabb}'$, $S_I$, $X_P$; $C_{CSBabb}'$ branches to $S_S$ and $X_S$]

**Fig. 6.8.** COD Fractioning of the Influent

### *Matrix Notation of the Dynamic Models*

Normally, the so-called matrix notation is used for the dynamic models. This establishes a connection between the process speeds and the degradation amount of single substances. The following table shows a small excerpt from the Activated Sludge Model No.1 (ASM 1, Henze et al 1987)

# 6 Wastewater Treatment

| substances | | | | | rates (velocities) |
|---|---|---|---|---|---|
| $S_{NO}$ | $S_{NH}$ | $S_{ND}$ | $X_{ND}$ | $S_{ALK}$ | |
| | $-i_{XB}$ | | | $-\frac{i_{XB}}{14}$ | $\hat{\mu}_H \left( \frac{S_S}{K_S+S_S} \right) \left( \frac{S_O}{K_{O,H}+S_O} \right) X_{BH}$ |

degradation of substances with certain rate

**Fig. 6.9.** Extract from a Matrix of a Dynamic Model for COD Degradation

The right-hand side of the table shows the velocity at which the conversion can be expected to take place during aerobic heterotrophic growth. The columns headed 'Substances' shows the conversion of the different substances that are affected by the growth of the heterotrophic biomass. Here, for instance, there is a decrease in COD caused by biomass growth and a decrease in ammonium nitrogen due to nitrogen incorporation. The left-hand side is the so-called stoichiometric matrix, comprising the stoichiometric equations known from biology and chemistry. This means that one line of the matrix always contains the entire conversion process, from the change of substance concentrations (left side) to the process speed (right side). The different lines of the matrix cover the different conversion processes.

**Principle of Growth and Inhibition**
So-called monokinetics is used to describe the impact of the concentration of a dissolved nutrient on the process speed (e.g. bacterial growth). Generally, two different terms appear in the matrix to describe the process speed: the growth term, i.e. the increase in process speed with increased substance supply; and the inhibition term, which with increasing substance supply reduces the process speed to zero.

Growth term:

$$\mu = \mu_{max} \cdot \frac{S}{K_S + S} \tag{6.20}$$

It becomes apparent that the growth rate $\mu$ increases with increasing substance concentration S. The saturation coefficient $K_S$ marks the substance concentration at which the growth speed reaches half the maximum growth speed – thus, it is often also referred to as half-saturation constant. At low $K_S$, the organisms work at a high conversion speed, even in the case of low substance concentrations. One example of a growth term can be found in nitrification, which with a high oxygen concentration (S) reacts at a higher degradation speed and with an improved ammonium nitrogen degradation.

Inhibition term: Some of the processes occurring in the activated sludge plant are slowed down by the presence of certain substances.

$$\mu = \mu_{max} \cdot \frac{K_S}{K_S + S} \tag{6.21}$$

This term gets smaller, and the inhibition hence stronger, the higher the concentration of the substance S. At very high S, the term becomes zero, so that the process breaks down. One typical example of such an inhibition in the model can be found in denitrification, which is inhibited by a high oxygen concentration (S).

### Activated Sludge Model No. 1

In 1987, an internationally composed IWA (International Water Association) working group on mathematical modelling published the Activated Sludge Model No. 1 (Henze et al 1987). It is a mathematical model for the dynamic simulation of activated-sludge plants. The ASM1 considers both the degradation of organic substance (COD) and the conversion of nitrogen compounds, i.e. it models nitrification, denitrification and carbon degradation.

The following figure shows the characteristic degradation paths described by the model for the degradation of COD and nitrogen.

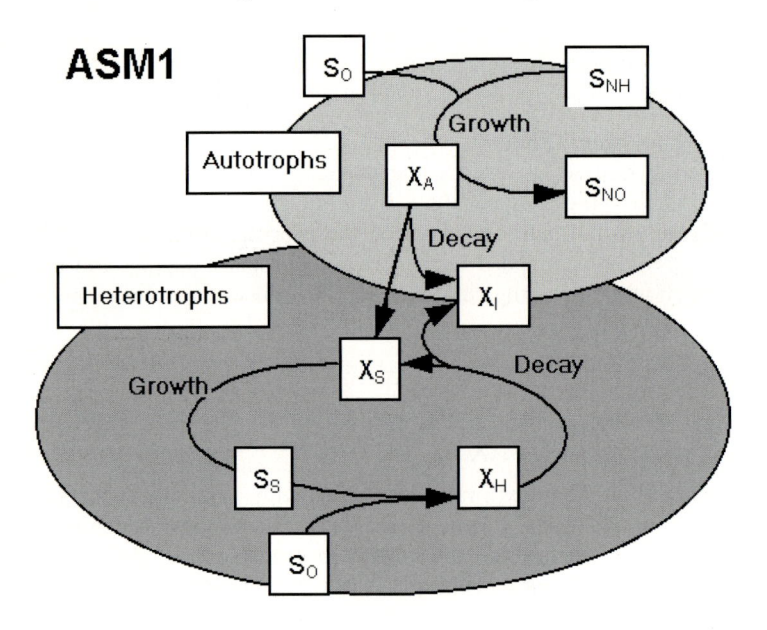

**Fig. 6.10.** Principle of Growth and Decay in the Activated Sludge Model No. 1 (Henze et al 1987)

**COD Degradation**
During the degradation of COD in the model, heterotrophic biomass $X_H$ grows, the energy stemming from the oxidation of easily degradable substances through oxygen. In the model, the growth takes place exclusively with easily degradable substances $S_S$, one part of which is contained in the influent to the WWTP, while the other must be provided by the conversion of barely degradable substances $X_S$ (hydrolysis). The decay of heterotrophic biomass is marked by the phenomenon that non-degradable (inert) COD is produce, and that, on the other hand, new substrate develops that is barely degradable. After this new substrate has again been converted into easily degradable COD (hydrolysis), heterotrophic biomass grows again.

**Nitrification**
Autotrophic biomass $X_A$ grows through the provision of energy from the nitrification. The basis for the cell substance is the dissolved $CO_2$ in the wastewater, which, after growth, is considered in the model as COD ($X_A$). Apart from the development of nitrate nitrogen during nitrification, which is degraded again during denitrification, the autotrophic biomass decays again as well. As with the heterotrophic biomass, inert substances ($X_I$) and barely degradable COD ($X_S$) develop during this process.

### *Activated Sludge Model No. 2d*

The Activated Sludge Model No. 2d (Henze et al 1998) is an extension of the Activated Sludge Model No. 1 (ASM1). ASM2d is a considerably more complex model, as it contains a number of additional components that are needed to more precisely characterise the wastewater and the activated sludge in order to model the enhanced biological phosphorus removal (EBPR). For the biological phosphorus elimination, the BioP organisms (PAOs), their polyphosphate ($X_{PP}$) and substrate stores ($X_{PHA}$) as well as the dissolved phosphorus ($S_{PO4}$) are modelled.

### 6.3.3 Use of Computer Programs

When using complex steady-state and dynamic models, computer programs are highly recommended. The main advantage they offer is the possibility of comparing different load situations, temperatures and treatment systems at a virtual WWTP. This enables plant investment and operation costs to be reduced and helps improve understanding of the complex biological behaviour of different competing organisms.

The well-known software packages DENIKAplus (Rosenwinkel et al 2001) and DENISIM (Neumann et al 1999) were specially developed for designing and optimising WWTPs using steady-state and dynamic simulation. To this end, the software has integrated the parameters relating to activated-sludge tanks and secondary clarifiers, chemical and enhanced biological phosphorus removal, dimensioning of aerators, pre-treatment, dosing of external COD and modules for industrial wastewater.

An example of how denitrification could be optimised is given in the following figure; degradation of NO$_3$-N is shown taking into account BOD and the nitrate recycle.

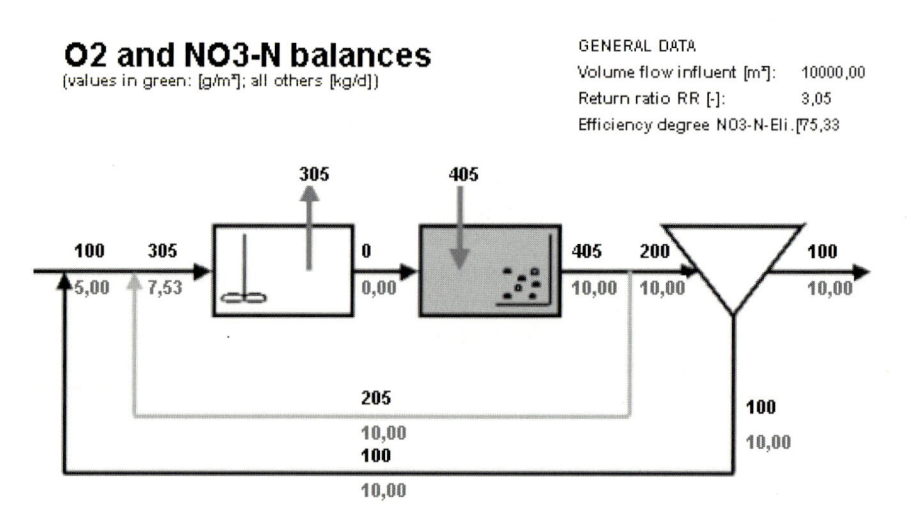

**O2 and NO3-N balances**
(values in green: [g/m³]; all others [kg/d])

GENERAL DATA
Volume flow influent [m³]: 10000,00
Return ratio RR [-]: 3,05
Efficiency degree NO3-N-Eli.[75,33

**Fig. 6.11.** Example of Nitrate Substance Balances in the Computer Software DENIKAplus (Rosenwinkel et al 2001)

## Abbreviations

| | |
|---|---|
| $\mu$ | Specific growth rate [d] |
| $\mu_{max}$ | Maximum specific growth rate [d] |
| $\mu_{N,T}$ | Maximum growth rate of the nitrifiers = 0.52 1/d (T=15°C) |
| $a_{endogenous}$ | Factor of the endogenous respiration for the oxidation of decayed biomass residues (energy gain for endogenous respiration/survival of the organisms) = 0.2-0.24 |
| $A_{SST}$ | Surface area of secondary clarifier [m²] |
| $a_{Substrate}$ | Factor for the substrate respiration (energy gain for growing) = 0.5-0.56 [gO$_2$/g$_{BOD}$] |
| $b_{A,T}$ | Decay rate of the nitrifiers = 0.05 1/d (T=15°C)) |
| $b_H$ | Decay rate of the heterotrophic biomass = 0.08-0.17 1/d (T=15°C) |
| $BOD_{5,o}$ | BOD influent concentration [g$_{BOD}$/m³] |
| $DS_o$ | Total suspended solids in the influent [g$_{DS}$/m³] |
| $DS_{RS}$ | Total suspended solids in the return sludge [kg$_{DS}$/m³] |
| $f_D$ | Reduction of respiration in the denitrification tank compared to aerobic respiration =0.75 [-] |
| $F_I$ | Inert (non-degradable) ratio of decayed biomass = 0.1-0.2 [-] |
| $f_T$ | Temperature factor for nitrifiers $1.103^{T-15}$ |

| | |
|---|---|
| $I_P$ | Quotient of BOD in the influent and incorporated phosphorus amount in the surplus sludge = 0.01-0.015 $g_P/g_{BOD}$) |
| $I_P$ | Quotient of heterotrophic active surplus sludge production and incorporated phosphorus amount in the surplus sludge = 0.03-0.06 [$g_P/g_{aDS}$] |
| $K_N$ | Saturation constant for ammonium = 0.7-1mg/l |
| $K_S$ | Saturation coefficient [g/m³] |
| $NH_{4e,m}$ | Ammonium nitrogen effluent concentration [$g_N/m^3$] |
| $NH_4N_n$ | Nitrified nitrogen concentration in relation to the influent volume flow [$g_N/m^3$] |
| $O_{2\text{-Demand}}$ | Required oxygen amount in the denitrification tank for the carbon degradation [$kg_{O2}/d$] |
| $OD_C$ | Oxygen demand from carbon degradation [$kg_{O2}/d$] |
| $P_{SS}$ | Phosphorus amount incorporated in the surplus sludge [$g_P/m^3$] |
| $q_A$ | Hydraulic loading of SST surface [m/h] |
| $Q_o$ | Influent volume flow [m³/d] |
| $Q_m$ | Influent rainy-weather volume flow [m³/h] |
| RF | Returned wastewater amount zone in relation to the influent volume flow [-] |
| S | Soluble substrate [g/m³] |
| SF | Safety factor [-] |
| $S_I$ | Inert dissolved COD [$g_{COD}/m^3$] |
| $SS_H$ | Heterotrophic surplus sludge production [$Kg_{DS}/d$] |
| $SS_I$ | Inert (non-degradable) biomass concentration [$Kg_{DS}/d$] |
| $SS_P$ | Surplus sludge amount from the P-precipitation [$Kg_{DS}/d$] |
| SVI | Sludge volume index [ml/g] |
| $t_{DS,aer}$ | Aerobic sludge age [d] |
| $t_{DS}$ | Entire sludge age [d] |
| $t_E$ | Thickening time [0.5-2.5h] |
| $V_D/V_{AT}$ | Anoxic volume ratio in relation to the entire activated-sludge tank volume [-] |
| $X_A$ | Autotrophic biomass [$g_{COD}/m^3$] |
| $X_{CODdegr.}$ | Degradable COD [$g_{COD}/m^3$] |
| $X_H$ | Heterotrophic biomass [$g_{COD}/m^3$] |
| $X_P$ | Undissolved inert COD [$g_{COD}/m^3$] |
| $Y_H$ | Yield of the heterotrophic biomass = 0.6-0.75$g_{DS}/g_{BOD}$ |

## 6.4 Evaluation of Centralised Wastewater Treatment

Knowledge about wastewater treatment eventually led to the realisation that centralisation has both advantages and disadvantages. Both are discussed in detail below.

**Hypothesis**
Centralised wastewater treatment plants have the advantage that hazardous substances, even toxic ones, are so diluted by being mixed with municipal wastewater that they do not cause any repression to the wastewater-treatment system.

In larger settlements, sewage consists not only of human excrement and water, it can contain chemicals (such as organic pollutants and metals) that enter the system from households, businesses and industrial enterprises. These toxic substances stem from solvents, detergents, cleansers, inks, pesticides, paints and a multitude of other materials used by modern households and businesses. Sewage also includes debris, such as gravel, grit, tampons, condoms, rags and hair. Tons of food waste from sink grinders add to the load. Small on-site treatment plants tend to become instable if varying amounts of problematic substances are fed to the plant. Decentralised treatment is more sensitive to 'shock loading' or sudden inflows of toxic substances that upset the bacterial balance in the treatment process, resulting in the release of virtually untreated sewage while the system is realigned. Thus, state-of-the-art centralised wastewater treatment plants can achieve better results compared with small-scale on-site solutions.

It is important to bear in mind that metals and other non-biodegradable toxic substances removed from the effluent usually end up in the sludge, which has to be disposed of. The only real solution to this problem is to remove them at their source, which is being practised to an increasing extent in Europe. Thus, centralised treatment also means controlling industrial discharges; this has proved to be much more difficult with decentralised treatment systems.

Contradicting the above opinion, we might argue that barely degradable or even toxic substances do not belong in the wastewater in the first place, but should be disposed of where they occur. Then, using different technologies, it would be possible to achieve targeted and individually tailored disposal solutions for these substances. This approach is already being practised as part of so-called process-integrated environmental protection. However, the joint treatment of municipal and industrial wastewater in the mainstream is still regarded as the best option for the extensive purification of wastewater, even that containing problematic substances. The above argument thus reflects the opinion of many engineers working in the field of sanitary engineering, but it also means the perpetuation of the existing centralised disposal infrastructure.

**Hypothesis**

Centralised wastewater treatment plants are the most efficient treatment systems for urban situations. The treatment is the more economical, the more wastewater is treated.

## *Conventional Treatment*

In centralised treatment plants, there are four levels of conventional treatment: preliminary, primary, secondary and tertiary.

**Preliminary Treatment**

Preliminary treatment is not really treatment at all; it consists of screens that remove large solid materials, such as condoms. Preliminary treatment has virtually no effect on the levels of TSS, BOD, bacteria or toxins.

**Primary Treatment**

Primary treatment is a physical process in which the sewage flow is slowed down in settling tanks. The part of the solids present in raw wastewater, known as sludge, is then removed from the bottom and disposed of in a variety of ways.

Floatable solids, oil and grease are usually skimmed off the surface before the remaining effluent is fed to the biological stage of the treatment plant.

**Secondary Biological Treatment**
Secondary biological treatment reduces the organic pollution of the wastewater (dissolved or in particle form) through biochemical oxidation by bacteria and other micro-organisms. Also, it further reduces the amount of solids.

**Tertiary Treatment**
Tertiary treatment aims to remove 85-95% of nutrients (nitrogen, phosphorus). In order to achieve nitrogen removal, biochemical processes like nitrification and denitrification are also used. In single-stage biological systems, secondary and tertiary treatment take place in the same tanks. Phosphorus may be controlled by either chemical or biological methods. In practice, a combination of both is often used.

Tertiary or equivalent-level treatment is the only practical option to produce "clean" water that meets European effluent standards. There are many environmentally sound alternative technologies that can provide treatment of tertiary level or better. Biological Nutrient Removal (BNR) is a general term for a number of different systems. There are some situations in which secondary-level treatment has been combined with natural systems to achieve satisfactory results. However, these systems require excellent source control, large areas of land and appropriate surrounding vegetation; and they fail to address the issue of sludge disposal.

Another common option is using constructed wetlands, which utilise the natural physical, chemical and biological processes of soil to treat sewage. These come in the form of constructed open water marshes or constructed rock marshes, which use submerged flow. None of these options are appropriate for densely populated settlements because of their extensive land requirements.

Even if there are different factors influencing the health, growth and economic development of society, the timely availability of water and sewer services can enhance growth and development in and around municipalities where other public services exist.

In any case, centralised water and wastewater systems for urban areas include piping networks that use transporting water and wastewater to/from individual water users/wastewater generators and remote centralised treatment plants. In discussions, it is often said that conventional treatment is expensive, energy- and resource-intensive. On the other hand, the cost of small on-site systems is underestimated. In fact, wastewater treatment itself becomes more economical, the larger the plant.

The only real problem is the enormous cost of the sewer network in rural areas that are sparsely populated. Actually, in terms of investment, the sewer networks account for a larger portion of the money spent than the treatment plants themselves. Hence, there are many situations, especially in developing countries, where such conventional practice is neither cost-effective nor sustainable for a variety of reasons (e.g. low-density development, rugged topography, limited water and energy supplies, lack of skilled labour). In such situations, decentralised strategies can be applied using on-site water and wastewater systems.

This argument reflects facts that have been verified for the situation in developed European countries: in Germany, for instance, running and extending a centralised wastewater treatment plant is the most efficient and economical solution for large cities. One fact contributing to this situation should be considered: investments for constructing complete sewer networks have been gradually made over the course of several decades, and energy-supply costs are relatively low. The following table shows a comparative evaluation of Europe and developing countries.

**Table 6.3.** Comparative Evaluation of the Investment Costs for Waterway-Protection Measures in Europe and Developing Countries

|  | Europe |  | Developing Countries |  |
| --- | --- | --- | --- | --- |
|  | Evaluation | Remarks | Evaluation | Remarks |
| Material costs | Relatively low | High-grade materials are used | Relatively expensive | Not available everywhere. Replaced by locally available materials |
| Building construction | Relatively low | Costs depend strongly on respective economic situation | Simple concrete constructions relatively cheap | Often, planning/supervi-sion by foreign companies necessary |
| MSR | Relatively cheap in case of mass production, expensive for individual solutions | Increasing ratio | Expensive compared to other costs | Still low ratio of overall investments |
| Machine technology | Relatively expensive | Increasing ratio | Very expensive compared to other costs | Still low ratio |

Considering wastewater treatment technologies, Dohmann (2001), among others, has pointed out that energy-saving solutions, such as the UASB technology (Upflow Anaerobic Sludge Blanket Process) with or without topped pond plants, have land requirements 20 times as high as the activated sludge plants favoured in Europe. But he also notes that such solutions have correspondingly lower requirements in terms of overall energy and cost than the "modern" European solutions.

The argument about the huge land requirements of so-called alternative solutions must be examined very closely. Even if activated sludge-treatment plants with a land requirement of 0.4-0.45m$^2$/PE offer benefits over pond plants (1.2-1.8m$^2$/PE ), combinations of UASB and pond plants with a land requirement of 0.5m$^2$/PE are extremely competitive. Even considering only the traditional arguments for sanitary engineering, it is apparent that many developments are possible along the lines of the wastewater-treatment systems presented here. These offer considerable advantages if we consider that in developing countries the costs for both monitoring/control technology and machine technology are very high compared to other costs.

## 6.4.1 Comparison of Wastewater Treatment Plants

We present below the results of the junior scientists' evaluation of the wastewater treatment plants they visited. These are followed by a final summary.

The sewage plants visited during the excursion week and evaluated in this study demonstrate well the available alternatives for centralised wastewater treatment. All of them use a combination of preliminary, primary and secondary treatment. Some also have a tertiary-treatment step. The effluent concentrations of all the plants all complied with the German standards. This means that

- every plant is equipped with nitrification/denitrification
- every plant provides for phosphorus removal by either chemical or biological means (or a combination of both)

**Table 6.4.** Basic Data of Sewage Treatment Plants in Lower Saxony, 2000

| City/ Community | Suderburg | Wrestedt | Uelzen | Lüneburg | Salzhausen | Celle |
|---|---|---|---|---|---|---|
| Size (P.E.) | 9,000 | 15,000 | 38,000 | 380,000 | 7,500 | 70,000 |
| Industrial Dischargers | Vegetable Processing | - | Dairy Fruit Juice Ice Cream | - | - | - |
| Daily Flow (m³/d) | 1,125 | 2,250 | 4,650 | 57,000 | 1,125 | 84,000 |
| Primary Clarifier | n | n | y | y | n | y |
| Biological Reactor | Activated Sludge | Activated Sludge | Activated Sludge | Activated Sludge | S B R | Activated Sludge |
| Aeration System | Surface, Fine Bubble | Fine Bubble | Surface | Fine Bubble | Fine Bubble | Surface |
| Denitrification | Simultan. | Intermittent | Pre-denitr. | Pre-denitr. | - | Pre-denitr. Post-denitr. |
| Phosphorus Removal | Biological | Chemical | Chemical | Biological Chemical | Biological | Chemical |
| Sludge Stabilisation | Extended Aeration | Extended Aeration | Anaerobic Digestion | Anaerobic Digestion | Extended Aeration | Anaerobic Digestion |
| Sludge Disposal | Storage, Agriculture | Storage, Agriculture | Dewatering, Agriculture | Dewatering, Agriculture | Storage, Agriculture | Dewatering, Agriculture |
| Tertiary Treatment | - | - | - | - | - | Denitrifying Biofilter |

The aim of preliminary treatment is to remove components that would affect the following stages of the plant:

- Paper, plastic material and fibrous matter that might clog pumps and other equipment
- Sand that would otherwise lead to a building up of sediments in tanks and pipes and that would also reduce the life of pumps, pipes, etc.

All the plants considered in this report are equipped with preliminary-treatment steps, i.e. some kind of screens or sieves followed by a grit chamber.

The removal of screenings and sand from grit chambers and their automatic preparation for disposal are standard procedures in Europe. Only the Suderburg plant still has a manually cleaned grit chamber. It should be mentioned that the technical equipment in preliminary-treatment units is quite often subject to operational trouble, particularly if the suppliers do not have sufficient experience in manufacturing this kind of machinery. Plant owners should be aware that buying cheap equipment, especially for this part of the wastewater-treatment plant, will in the long run probably turn out to be a poor decision.

**Primary Treatment**
Primary treatment is a physical process in which the sewage flow is slowed down in settling tanks, the so-called primary clarifiers. Primary sludge is not easy to handle because it is unstable and subject to fast degradation, which causes strong odour emissions. Primary sludge must therefore be stabilised before further treatment and disposal, e.g. in an anaerobic digester. Without stabilisation, proper handling of the primary sludge is impossible. This is why all plants without separate anaerobic digesters (Suderburg, Wrestedt, Salzhausen) are not equipped with primary sedimentation tanks.

**Primary Clarifiers: the Pros and Cons**
- Primary clarification can reduce organic pollution by 20-40% (but does not markedly reduce nutrient concentrations) without using complicated equipment or energy.
- Removal of organic substances reduces the load fed to the following biological step, thus saving reactor volume and energy consumption.
- Primary sludge fed to anaerobic digesters is readily biodegradable, leading to good gas production, which can be used as an energy source on site. However, the operation of anaerobic digesters requires skilled staff.
- Handling of primary sludge is problematic (clogging of pipes, pumps, etc. is possible).
- Anaerobic digestion is expensive in terms of construction and difficult to operate.

The use of primary sedimentation is therefore normally confined to large plants where the energy recovery from sludge digestion is big enough to operate gas-utilisation equipment and where skilled personnel is available to operate the primary sludge treatment and sludge digestion (in this report: Uelzen, Lüneburg, Celle).

**Secondary Biological and Tertiary Treatment**
Secondary biological and tertiary treatment reduce the organic pollutant load of the wastewater (dissolved or in particle form) through biochemical oxidation and nutrient removal by bacteria.

In all the plants considered in this study, a conventional activated sludge (AS) process is used for biological treatment. There is only one sequencing batch reactor (SBR) plant (Salzhausen), which was commissioned in the spring of 2000. Both of these technical solutions need little space compared to "natural" systems like lagoons, constructed wetlands, etc. All of the plants are designed and operated for nitrogen removal, i.e. nitrification and denitrification. The data evaluated shows that all plants meet German standards for nitrogen levels in the effluent. This is not surprising because plants in Germany have been using nitrogen-removal systems since the 1980s. However, good results depend on certain overall conditions:

- Both AS and SBR plants can only be operated properly if the electricity supply is reliable. Cutting off electricity for even a few hours a day will lead to severe operational problems and poor effluent quality.
- To control degradation processes properly and run the system efficiently, minimum online control of certain parameters is desirable: wastewater flow, dissolved oxygen, pH value.
- If effluent standards are high (and actually enforced, as they are in Europe), nitrate, ammonia and phosphorus levels must also be measured, either online or several times a week. However, such modern measuring and control equipment is normally only affordable for large plants and requires well-trained operation and maintenance personnel.
- Process conditions such as activated-sludge concentration or settling properties must be monitored daily in both AS and SBR plants. Without trained personnel, it is impossible to achieve good effluent standards.

## *Technical Details*

### Type of Aeration
All AS and SBR systems depend on a proper technical oxygen supply because natural aeration is not sufficient for these types of biological processes with a high degradation rate. There are two systems available: fine-bubble aeration and surface aeration. Throughout Europe, there is a preference for fine-bubble aeration systems because their better energy efficiency. It should be mentioned that construction costs are lower for surface aerators. Also, (high-quality) surface aerators have a longer life than fine-bubble diffusers, which normally have to be replaced after about 5 years of operation. Thus, depending on the local situation, surface aerators may actually be the better alternative.

### Denitrification Process
Some of the plants are designed for simultaneous/intermittent denitrification. The benefits are: only one tank and no nitrate recirculation are necessary, which increases the flexibility of the process. On the other hand, the benefits of pre-denitrification are: optimal utilisation of the influent substances as a carbon source for denitrification, the required aeration equipment can be smaller than for plants with simultaneous/intermittent denitrification. The best choice depends on several conditions and varies from case to case.

## Conventional Activated Sludge System (AS) or Sequencing Batch Reactor (SBR)

In recent years, SBR systems have become very popular throughout Europe, especially for small or medium-sized plants. This is mainly due to the simplicity of SBR design (a simple tank requiring no final clarifier), which makes economical construction possible. In addition, the process design allows flexible adaptation to changing influent conditions. At present, SBR is a well-known single-sludge BNR (Biological Nitrification Removal) system with high operational flexibility. Most SBR plants are dimensioned at below 10,000 PE.

Over the last decade, there have been various technical innovations in this area. Aeration systems, sludge recycling and removal devices have been improved and methods have been developed for monitoring the removal of C, N and P in a single-tank SBR. However, operation of SBR is only possible if reliable process-control equipment is used. A minimum requirement here is that the control and monitoring system must provide a fixed time schedule for the subsequent filling, reaction and extraction phases.

This involves extensive use of automatic valves and drives. However, most of the SBR processes are controlled conventionally by the sequential control method in which the duration of the various operation phases is fixed. This control method is not the best solution as regards resource conservation and optimal performance because it cannot respond to variations in the influent flow rate and concentrations, especially in plants that are connected to combined sewer networks. More recently, then, automatic control of SBRs has also been used for cost reduction and process optimisation. In the automatic control of the SBR process, online monitoring of the respiration rate and pH value has proved a useful technique for process optimisation, particularly when using an SBR for nutrient removal.

For instance, the time intervals of the aerobic and anoxic phases can be regulated dynamically by a real-time control strategy under different influent conditions: the reduction of the time intervals, the system capacity as well as the energy requirements for aeration. Depending on energy prices, this may be an important factor. (In Germany, electricity costs range from $50 to $80 per MWh, and in the future energy prices are expected to rise world-wide.) Moreover, higher total nitrogen-removal rates are achieved by real-time control with greater process stability than in the case of sequential control.

The plant at Celle is the only one featuring a separate tertiary step. The reason is that this plant must meet even stricter standards than usual. A biological filtration unit was therefore added in 1999, with an external carbon source for denitrification, phosphorus being removed by chemical precipitation. It should be mentioned that this type of treatment is very expensive, so that its application is restricted, even in Europe, to cases where the effluent is discharged into certain surface waters with a high risk of eutrophication.

## 6.4.2 Conclusions

As far as adapting technological developments from Europe to conditions indeveloping countries is concerned, it is the economic aspects that must first be considered. While in industrialised countries there has long been a tendency to automate operation processes to save personnel costs, such costs are not a problem in developing countries. Human resources are available in plenty and at low cost, which means that manual labour should not be replaced by automated processes. The following table gives a comparative evaluation of the investment cost ratios of wastewater-treatment plants, taking into account operation costs. It is evident that the technologies presented in this project cannot, or at least only in specific cases, be considered suitable for countries where

- personnel costs are low
- energy is relatively expensive and not regularly available
- machine technology is expensive and in some cases not yet available
- measuring, monitoring and control technology have yet to be developed or introduced

**Table 6.5.** Comparative Evaluation of the Operation Costs for Waterway-Protection Measures in Europe and in Developing Countries (Dohmann 2001)

|  | Europe |  | Developing Countries |  |
|---|---|---|---|---|
|  | Evaluation | Remarks | Evaluation | Remarks |
| Personnel | High | Tendency to reduce personnel demand | Low | Because of low wages, automation does not offer economic benefits |
| Residue disposal | High (20-35%) | Depends on the disposal method | Low | As a rule, sewage-sludge utilization in agriculture or at landfills |
| Energy | Average (15-25% ratio) | Tendency towards falling energy costs through deregulation of energy sector | Expensive compared to other costs | Low supply reliability |

If we compare, for instance, dimensioning data and costs for biological WWTPs (Kloss 1995/2000) in Germany and Bolivia, which were planned roughly at the same time, we find that, after economic analysis of several technological solutions, Bolivia opted for a multistage, unaerated lagoon plant with preliminary anaerobic treatment, while Germany (Dessau) chose a conventional mechanical-biological plant with topped (filtration stage) and anaerobic sludge stabilisation. For the latter plant, the investment costs for a construction size of 270,000 PE were DM 500/PE, whereas the costs for the lagoon plant were DM 25/PE.

Apart from the economic arguments, however, the discussion is also about new development options for wastewater disposal – options that are ecological and take into account and meet users' expectations. During the project, it became obvious that

- conventional wastewater treatment systems are well suited for biological treatment of industrial wastewater mixed with municipal wastewater, but that
- conventional systems are, for fundamental reasons, not suitable for or transferable to the home countries of the project junior scientists. There, such systems would mean wasting scarce water to transport faeces.

Only systems that use as little water as possible for transporting faeces can therefore be regarded as suitable. Key requirements, then, are separating wastewater into faeces and urine and minimising water requirements for other household uses such as washing, rinsing and general cleaning. For these purposes, an average consumption of 30 l/day is realistic. For concentrated industrial wastewater, a technical solution is sometimes the only option; for such cases, then, technical solutions should be integrated into the overall disposal concept.

In summary, it can be said that we need to look into the form that urban wastewater-disposal concepts with semi-decentralised structures might take. This involves treating industrial wastewater conventionally in technical treatment plants. The disposal of domestic wastewater, however, should be organised on a decentralised basis in catchment areas with only a few thousand inhabitants. With this approach, faeces, urine and "grey water" are collected separately, e.g. in vacuum or compost toilets, which helps to save a great deal of water.

The faeces are then subjected to an anaerobic digestion process. Then biological wastewater treatment is confined to the grey water generated by washing and rinsing. Urine could be co-treated anaerobically or used as fertiliser. Especially in countries with a relatively warm climate, this separation could enable biogas to be obtained from the faecal matter. For the treatment of grey water, e.g. in constructed wetlands, a comparatively small area of $1 m^2/I$ is sufficient; such wetlands can be well integrated into the urban landscape.

Parallel to the ideas developed and discussed during this *ifu* project, another project called "Wastewater in the Cityscape" (Seggern a. Kunst 2001) was conducted. This project was concerned with a fictitious alternative planning concept for a German city with approx. 100,000 PE. Here, exemplary in Salzgitter-Lebenstedt, it was assumed that industrial wastewater is treated conventionally, but that for the disposal of municipal wastewater vacuum toilets are used to collect faeces and urine. The grey water is then treated in constructed wetlands, serving semi-decentralised units of 3,000 PE, and subsequently discharged.

In the following Figure the different semi-decentralised units were illustrated by white points, the "new" semi-decentralised sewer system by black lines and the "old" centralised sewer system by light-grey lines (except the main sewer canal).

**Fig. 6.12.** Decentralised Wastewater Conception – Salzgitter-Lebenstedt

**The Main Conclusions Drawn from this Planning Concept Are:**
- in the urban landscape of this German town (100,000 inhabitants) there is enough space available to accommodate the semi-decentralised and natural treatment plants
- it is possible, under the mentioned conditions (treatment of grey water), to discharge the treated wastewater visibly into receiving waterways which in turn discharge into open water bodies that have been previously straightened and lined with piping

Based on the separation of wastewater into three bit-streams – grey water, yellow matter, and black water – (Otterpohl et al 1999; Niederste-Hollenberg a. Otterpohl 2000), different adapted technologies, such as that realised in Lübeck Flintenbreite, were presented.

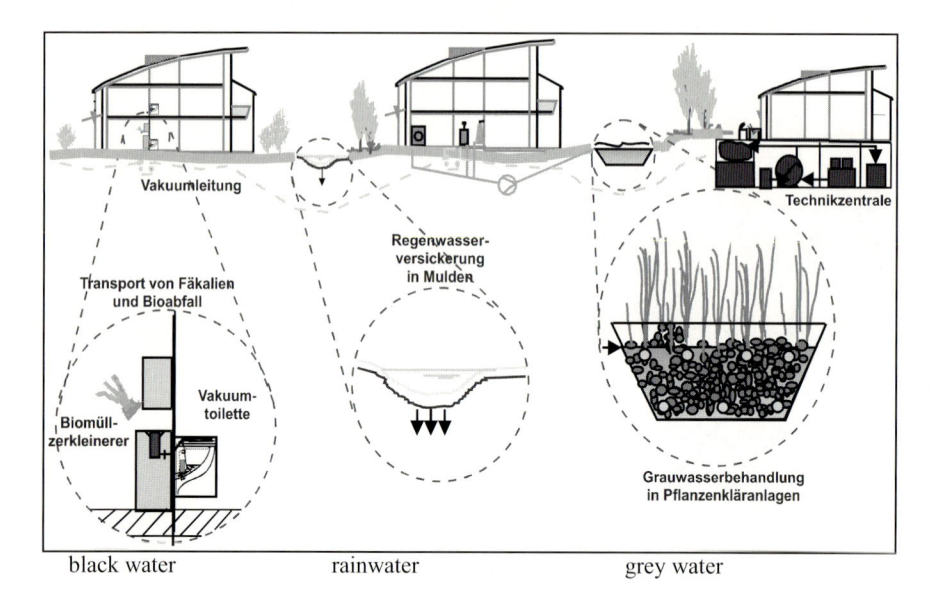

**Fig. 6.13.** Diagram of the Sanitation Concept at Lübeck Flintenbreite

Here, an integrated sanitation concept has been realised using vacuum toilets and a biogas plant for black water. In this pilot project, 300 inhabitants and an area of 3.5ha are not connected to the central sewer system. A block-type thermal power station is adapted to utilise the biogas from the anaerobic reactor and to produce the required heat for the settlement. Passive solar power is additionally used for heating the houses, and there are active solar systems for hot-water production. Grey water is treated in constructed wetlands and discharged into a nearby trench; stormwater is percolated.

The view of the project junior scientists is that the approach adopted in Europe, with its centralised wastewater disposal, cannot be copied. The existing political and infrastructural conditions make decentralised solutions the obvious choice for domestic wastewater treatment in the individual municipal districts, e.g. with semi-decentralised grey-water-treatment plants which could, for example, be built as constructed wetlands.

In the project, concepts were developed for individual areas, each serving approx. 3,000 inhabitants, who separately collect and partly reuse faeces and urine. The grey water is processed biologically on site. With this concept, determination of the wastewater would be much more effective than with the conventional technologies developed in Europe. A gradual development of wastewater disposal would be possible. The overall structure consists of a number of small semi-decentralised units interacting with one another. These continue to grow in number and eventually result in a comprehensive disposal network. Another advantage pointed to by the junior scientists is that a semi-decentralised and transparent structure of this sort would also facilitate public participation. The technical plants are of limited complexity, enabling non-experts to participate in their construction, maintenance and operation. This aspect in particular is crucially important for the junior scientists (see also the results of chapter 8) and will be the subject of further research and NGO campaigning.

## References

ATV-Arbeitsblatt A 131 (2000) Bemessung von einstufigen Belebungsanlagen. Abwassertechnische Vereinigung e.V., Hennef
ATV (1997) ATV-Handbuch Mechanische Abwasserreinigung. Ernst & Sohn, Berlin
Böhnke B (1989) Bemessung der Stickstoffelimination in der Abwasserreinigung - Ergebnisse eines Erfahrungsaustausches der Hochschulen. Korrespondenz Abwasser 9: 1046–1061
Deutsches Institut für Normung e.V.: Versch. DIN-Normen
- DIN 4040 (EN 1825) Abscheideranlagen für Fett
- DIN 1999 (EN 858) Abscheider für Leichtflüssigkeiten
- DIN 19551 Kläranlagen, Rechteckbecken
- DIN 19552 Kläranlagen, Rundbecken
- DIN 19554 Kläranlagen, Rechenbauwerke
- DIN 19558 Kläranlagen, Überfallwehr und Tauchwand
- DIN 19569 Teil 2, Kläranlagen, Baugrundsätze

Dohmann M (2001) Ziele und Tendenzen in der Abwasserreinigung. In: Deutsche Bundesstiftung Umwelt DBU (ed) Wasser im 21. Jahrhundert - Perspektiven, Handlungsfelder, Strategien. Bramsche
Franta J et al (1992) Mechanische Abwasserreinigung durch Siebung. Korrespondenz Abwasser 6/92: 907-909
Henze M, Grady CPL, Gujer W, Matsuo T, Marais GVR (1987) Activated Sludge Model No. 1, Scientific and technical report No. 1. IAWPRC (now IWA)
Henze M, Gujer W, Mino T, Matsuo T, Wentzel MC, Marais GVR, Van Loosdrecht MCM (1998) Activated Sludge Model No. 2d. Wat. Sci. Tech. 39(1): 165–182
Hosang/Tochof (1993) Abwassertechnik.Teubner, Stuttgart
Imhoff K (1993) Taschenbuch der Stadtentwässerung. R. Oldenbourg, München-Wien
Kayser R (1997) Ein neuer Ansatz zur Berechnung der Denitrifikationskapazität in Belebungsanlagen. gwf Wasser/Abwasser 138/5: 251
Klinger H, Barth H (1994) Entwicklung einer Sandrecyclinganlage auf Kläranlagen. Korrespondenz Abwasser 1/94: 48-53
Kloss R (1995) Oruro-Dessau – Zwei Städte, zwei Lösungen. ATV-Fortbildungskurs EL99, Fulda 1995
Kloss R (2000) Teichanlagen in Bolivien. persönliche Mitteilung
Kunst S (1991) Untersuchungen zur biologischen Phosphorelimination im Hinblick auf ihre abwassertechnische Nutzung. Institute for Sanitary Engineering and Waste Management, University of Hannover, Number 77

Neumann P, Spering V, Alex J (1999) DENISIM-Program for Simulation of Wastewater Treatment Plants. Institut für Automation und Kommunikation e.V. Magdeburg, Germany

Niederste-Hollenberg J, Otterpohl R (2000) Integrated Wastewater Management in Urban and Rural Areas including Source Separation and Reuse of Resources. Reader, Project Area Water, *ifu*, Hannover 2000

Otterpohl R, Albold A, Oldenburg M (1999) Source Control in Urban Sanitation and Waste Management for Different Social and Geographical Conditions. Water Science & Technology, No. ¾, Part 2

Rosenwinkel K-H, Wichern M, Lippert C, Arnold B, Fengler T (2001) DENIKAplus-Program for Design and Optimization of Wastewater Treatment Plants. Institute for Sanitary Engineering and Waste Management, University of Hannover, Germany

Seggern v. H, Kunst S (2001) Abwasser in der Stadtlandschaft, Workshop Projektergebnisse 4/2001, Hannover

Seyfried CF (1994) Mechanische Vorreinigung - Rechen, Sandfang, Vorklärung. ATV-Kurs H/2 (Abwasserreinigung) Fulda 05.10.-07.10.1994

Stein A (1992) Ein Beitrag zur Gestaltung belüfteter Sandfänge. Korrespondenz Abwasser

Wichern M, Rosenwinkel K-H (2000) Upgrading and Optimisation of Wastewater Treatment Plants using a Large Scale Plant Verified Model for enhanced Phosphoros Removal. IWA Conference on the Implementation of the Use of the EU Nutrient Emission Guidelines-Holland

# 7 Decentralised Wastewater Treatment - Wastewater Treatment in Rural Areas

Katrin Kayser, Sabine Kunst

## 7.1 Situation

Since the United Nations Conference on Environment and Development (Brazil 1992), an international consensus has emerged on the critical role of education in achieving sustainable development. This consensus has led to a new vision of education as a fundamental issue here. Our interdisciplinary programme addresses the relationship between population and environment, with special emphasis on water and development (for example, in UNESCO'S Environment and Population Education and Information for Development (EPD ) programme).

Different projects have determined, for instance, that only 45% of Africa's total population have access to clean drinking water, and less than one third to adequate sanitation. Furthermore, especially in rural areas, it has always been left to women to fetch the water for domestic needs. To find water that is more or less clean, the population, in particular women and children, have to cover increasing distances. This means that several hours a day are lost which could otherwise be used for education, training or other productive activities. A reliable and close source of water is therefore the key prerequisite for social development. Social development is, then, complemented by ecological viability. In the field of water management, there are hardly any women in responsible positions. It is therefore essential that more women be involved as advisors, planners and engineers in all areas from teaching to public administration.

According to the United Nations' most recent long-term forecasts, the world's population will grow to almost eleven billion people by the year 2050. Today, nearly 80% of the world's population live in developing countries, where sustained population growth is greatly increasing the pressure on water resources. It is assumed that by the year 2050 one quarter of the world's population will live with a chronic or recurrent shortage of water. On the other hand, it must be seen that the demand for water will continue to increase. As living standards rise, average water consumption and wastewater production will quickly reach levels similar to those in Europe. This will be due, among other things, to improved facilities and conditions in homes and communities. However, there will be only few options to cover the rising demand generated by urban centres and improved living conditions. There will be a need, then, to develop less costly technologies. Growing water consumption and increasing water contamination are two sides of the same coin: both reduce available water resources. Surface and groundwater are polluted, for instance, by both industrial and municipal wastewater. Currently, only 5% of wastewater is treated world-wide, and around 2 billion people lack safe water and are threatened by waterborne infectious diseases. It is estimated that some 80% of all diseases are water-related. Worst affected are the poor.

Globally, about 500 million kg of human faeces are generated daily in urban areas, and about 600 million kg in rural areas, bringing the total to more than one billion tons a day. Most of this biodegradable organic material is disposed of without treatment or with little treatment. The faeces thus pollute the environment with substances that may be highly detrimental to human health. The developed countries have efficient sewage systems and treatment plants ensuring that most human and industrial wastewater is properly treated and disposed of.

The is not the case in developing countries. There, some 65% of the population lack proper sanitation systems. Sanitation coverage varies enormously from region to region. The best-served regions are West Asia and Latin America including the Caribbean with approximately 65% coverage (WHO 1997). It is paradoxical that in Europe, as well as in countries where water is scarce, most sewage systems and wastewater treatment plants are water-dependent. Quite astonishing even that this should be the case in developing countries. This means that the transportation, and hence also the treatment, systems are water-dependent in that they use water as the transportation medium. Water is employed to transport originally organic waste from the point of emission to the point of treatment or reuse.

Domestic use, though crucial, accounts for only a small part (8%) of total consumption world-wide (250km$^3$). Yet, the loss is extremely high, with 30 to 50% of total water production. In most developing countries, industrial water consumption is still relatively low compared with that of industrialised countries. It is anticipated, then, that as industrial development continues in the developing countries, industrial demand for water will increase, along with water contamination.

## 7.1.1 Principles and Spheres of Action

Water is a key factor in economic development, growth and employment. If it is to be a positive factor, the world's scarce water resources must be utilised efficiently. In many countries, agriculture and industry are already strongly competitors for water. It is therefore especially important to help create a favourable environment and improve institutional performance capability by providing carefully tailored packages of measures and adapted technologies. Operational water management should take place within river watersheds and groundwater reserves, not within administrative boundaries. Those operating water-supply and wastewater-treatment systems must be able to take independent commercial decisions and be accountable first and foremost to their customers and owners. In rural areas, the emphasis is on appropriate decentralised water supply and sanitation systems and on the development of suitable wastewater treatment facilities. Awareness raising measures also play a very important role in successful sustainable development. As early as 1992, several principles were adopted at the International Conference on Water and the Environment held in Dublin, the so-called Dublin Principles. In relation to wastewater-treatment processes, it was stressed that

- water development and management should be based on a participatory approach, involving users, planners and policy-makers at all levels
- women play a key role in the provision, management and safeguarding of water
- water has an economic value and should be recognised as an economic good

The notion of integrated water resource management, dealing also with special aspects of water cycles, is of crucial importance. The most obvious meaning here is integrating the different uses that are made of water. Drinking water needs influence the situation with regard to wastewater treatment. Use of water in one domain affects the quantity and/or quality of water available for use in other domains.

## 7.1.2 Decentralisation and User Participation

The subsidiary principle also applies to the administration and operation of water and wastewater facilities. In many developing countries, this means transferring responsibility for water from centralised to decentralised institutions. Decentralisation offers considerable opportunities. At the local level, users can participate. Decisions on problems and priorities are taken locally with the participation of all those involved, especially women. A further aim in this context is to promote local forms of self-help, traditional supply structures and people's own initiatives. Users can participate in decision-making, e.g. on fee collection, maintenance, or rehabilitation of water infrastructure. They can also play a part in developing appropriate decentralised wastewater technologies. Decentralisation and user participation is a major factor in achieving sustainability in the areas of water. This can be summarised by focusing on the basics of Agenda 21:

---

**Basics for sustainable wastewater treatment in Agenda 21:**

Perusal of *Agenda 21, Chapter 18* on the subject of wastewater treatment in rural areas yields the following: Based on **increased consideration of under-provided rural areas**, the major factors in the utilisation of technologies for drinking-water supply and sanitation measures are **feasibility, acceptance, and sustainability. Adapted technologies** are meant to contribute to the **protection of public health** and to realise the **hygienically unproblematic re-use of process water and wastewater.** For the **cost-effective and environment-friendly disposal of wastewater, traditional methods** are to be promoted in order to **guarantee long-term local participation.**

---

Discussions in Europe on the disposal and treatment of wastewater in rural areas have frequently been heated. One major point has often been the question whether wastewater should be collected "conventionally" by a sewage system and treated in a large, centralised wastewater treatment plant, or whether a (semi-)decentralised solution dispensing with a costly sewer system and using instead small wastewater treatment plants (WWTPs) would be better. While centralised solutions generally use technical WWTPs such as aeration plants, smaller

WWTPs, acting as "rivals" to the technical ones, often use environment-friendly systems such as wastewater lagoons or planted soil filters. From the scientific point of view, there is no one solution or method that is right in all cases. The approaches must be as varied and manifold as the situations they are designed to meet (settlement structures, hydro-geological conditions), and concepts must be adapted to the respective situation. In this, the basic conditions for decent scientific and engineering performance are the same the world over. The approaches differ to the extent that in Europe, in particular Germany, nature-based methods are still fringe solutions. For instance, 95% of all wastewater in Germany is collected and disposed in centralised WWTPs.

The situation is quite different in some of the other European countries, but here, too, decentralised solutions tend to be the exception rather than the rule. The prevailing technology is that of centralised disposal in large, technical WWTPs (cf. chapter 6). Such technology is preferred because it would appear to be easier to control. Surveys conducted as part of a research project (2000) revealed that, in Germany, representatives of water management offices, the responsible authorities and also many engineering offices regard decentralised – so-called nature-based – methods as less reliable and less stable in operation (winter vs. summer operation, peak loads) than the seemingly better controllable technical plants. It was also argued that the former are less economical in terms of land requirements, investment and operation costs. However, a good deal of other research (Fehr a. Schütte 1990; UAN 2000) offers objective evidence that, even for Germany, the operation costs for ponds as well as planted soil filters are considerably lower than those for conventional technical WWTPs, and that realisation of the wastewater-treatment plant per PE (population equivalent) is actually on the same level concerning the requirement of land as that of centralised plants.

On the other hand, it is, of course, true – given the appropriate infrastructure – that in densely populated regions like Germany it may be more cost-effective for larger villages with concentrated and dense building to construct a centralised technical WWTP. These take up less space that decentralised, nature-based solutions, and the wastewater treatment can be controlled at a central location. Large parts of rural areas are, however, marked by a low settlement density, scattered settlements and single houses. The inhabitant-specific costs for sewer construction alone make central connection prohibitive in such areas. The vacant areas available in these thinly populated regions make them a suitable site so-called nature-based solutions such as wastewater lagoons and planted soil filters.

Wastewater lagoons, and particularly planted soil filters, are technologies adapted to rural areas and thus an ideal option for treating wastewater in an environment-friendly, cost-effective and highly efficient way: they have low energy requirements and utilise natural materials such as sand and reed. There are no operation costs for aeration, and only low costs for pumping processes. The state of the art and possible technical implementation of such solutions are discussed in the next chapter.

## 7.2 Nature-Based Wastewater and Sludge Treatment Methods as Components of Sustainable Concepts in Rural Regions – State of the Art

### 7.2.1 Introduction

Nature-based wastewater treatment methods such as planted soil filters and wastewater lagoons is particularly suitable for rural regions. Such systems may need a relatively large space but, unlike other small WWTPs, they achieve superb results, are less trouble-prone, easy to maintain and insensitive to load variations and times without wastewater input. The construction of planted soil filters and wastewater lagoons is relatively simple, which means that they can often be built and maintained by the municipality and with support of small local companies.

Self-contained sustainable concepts with direct irrigation of the purified wastewater and faecal sludge composting on location make it possible to achieve small closed water and nutrient cycles. Such concepts are especially suitable for small villages or communities using one WWTP and one sludge treatment plant, sharing the costs and responsibility for running the plants among the inhabitants. Figure 7.1 shows some nature-based wastewater- and sludge treatment methods as components of sustainable semi-decentralised concepts in rural areas.

Of course, there are many other sensible wastewater concepts, not least the reduction or avoidance of wastewater. In areas with water scarcity, for instance, it is sensible to collect and treat faeces separately, e.g. in composting toilets. The remaining wastewater from kitchens and bathrooms (grey water) is thus greatly reduced in quantity and has a comparatively low pollution level, as it contains neither faeces nor urine. Here, we refer to the chapter 8, which dealt with the possibility of using dry toilets.

In this chapter, we discuss in more detail the methods presented in Figure 7.1. We give an overview of

- the technical fundamentals
- treatment mechanisms
- dimensioning and
- efficiency

of different pre-treatment methods, wastewater lagoons, planted soil filters as well as sludge composting beds. In addition, we report on experience with the respective methods and detail necessary maintenance work. Finally, we look at possible ways of implementing these methods, giving practical examples.

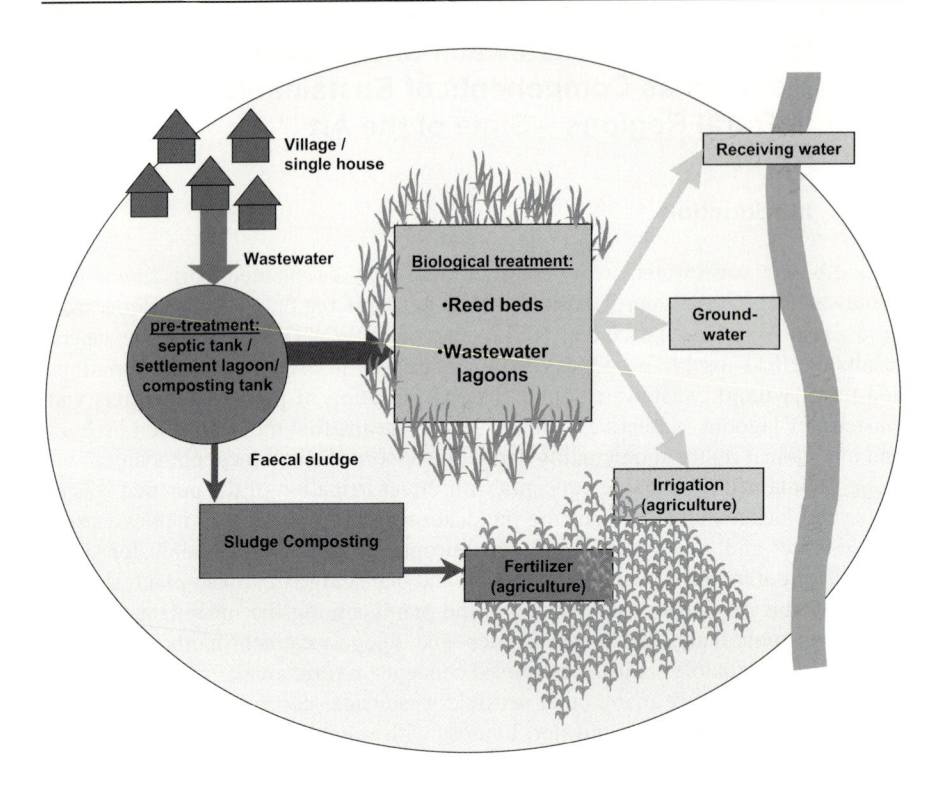

**Fig. 7.1.** Nature-Based Wastewater and Sludge Treatment Methods as Components of Sustainable (Semi-)Decentralised Concepts in Rural Areas

### 7.2.2 Overview - Wastewater Quantities and Wastewater Agents in Rural Areas

Wastewater pollution is normally described using the parameters COD (Chemical Oxygen Demand), $BOD_5$ (Biochemical Oxygen Demand), N (Nitrogen) and P (Phosphorus).

The parameters COD and $BOD_5$ are sum parameters which are used to measure the dissolved and non-deposable organic pollutants in the wastewater. They comprise the amount of oxygen necessary for the microbiological or chemical oxidation of the organic pollutants up to the inorganic final product. To determine the $BOD_5$, one measures the amount of oxygen consumed by the microbial conversion processes during the degradation of pollutants over a period of 5 days in the aerobic milieu. The COD is determined by chemical oxidation of the pollutants. Other important wastewater-relevant substances are the plant nutrients nitrogen and phosphorus. While the organic pollutants in the receiving water affect the oxygen balance, phosphorus and nitrogen promote the growth of algae and plants in the waterways, thus playing a crucial role in their eutrophication. Particularly problematic for the waterways is ammonium nitrogen, which also affects their oxygen balance. In the case of high pH-values, ammonium ($NH_4$-N) produces ammonia ($NH_3$-N), which has a toxic effect on fish.

Let us consider pollution loads and concentrations for domestic wastewater in rural areas of Germany, these are presented in Table 7.1.

**Table 7.1.** Pollution of Domestic Wastewater in Rural Areas (Germany)

|     | Load[1] [g/(PE·d)] | Concentration[2] [mg/l] |
| --- | --- | --- |
| COD | 120 | 1000 |
| $BOD_5$ | 60 | 500 |
| N | 12 | 100 |
| P | 2,5 | 21 |

[1] Taken from LUA NRW, 1994
[2] Water consumption in rural district of 120 l/(PE·d)

These values are to be understood as scales and can vary considerably in individual cases. The domestic wastewater, which would be treated in small wastewater treatment plants, corresponds to the normal domestic sewage in terms of agents. However, higher concentrations are the result of lower water consumption in rural areas (well below the normal consumption of 150 l/(PE·d) in Germany). Our estimate is a cautious one based on insufficient data (actual measurements do not exist). We take the generally accepted value of 120 l/(PE·d), although we know that – especially in small villages with only a few hundred inhabitants – real consumption is between 80 and 100 l/(PE·d).

The numbers presented above indicate that domestic water consumption varies considerably within a particular region, depending on the respective settlement structures. This variation is also evident in international comparisons: exclusive domestic water consumption in industrialised countries, for example, ranges from 120 l/(PE·d) in Belgium, to over 213 l/(PE·d) in Italy, to 295 l/(PE·d) in the USA (OECD: Household water pricing practices in OECD countries, Paris 1998). Especially in warmer regions, water demand is strongly dependent on water supplies. For instance, according to WHO estimates, average per capita water consumption in Central Africa, a region with scarce water resources, is only 39 l/d. By contrast, Farley (1999) reports water consumption levels between 400 and 500 l/(PE·d) in areas of the South Pacific with plenty of water, e.g. the island of Samoa.

The substantial differences in the water supply situation world-wide make it impossible to generalise about wastewater quantities and concentrations. The nature-based wastewater-treatment methods discussed below are based mainly on experience in Germany. However, the concepts and approaches presented are transferable to many other countries and regions, taking into account the different overall conditions and local water consumption rates as well as the local pollutant loads.

### 7.2.3 Pre-Treatment

The first stage in any WWTP is preliminary treatment, the main task of which is to allow settlement of residual solids. The effluent from this first stage is then treated in the biological stage. Both treatment stages must be considered as a unit.

For the mechanical pre-treatment of domestic wastewater, multi-compartment septic tanks are mostly used. The wastewater is conveyed to the first compartment, where the coarsest solids are separated out. Via a calmed overflow, the wastewater is then transported to compartments 2 and 3 for further treatment.

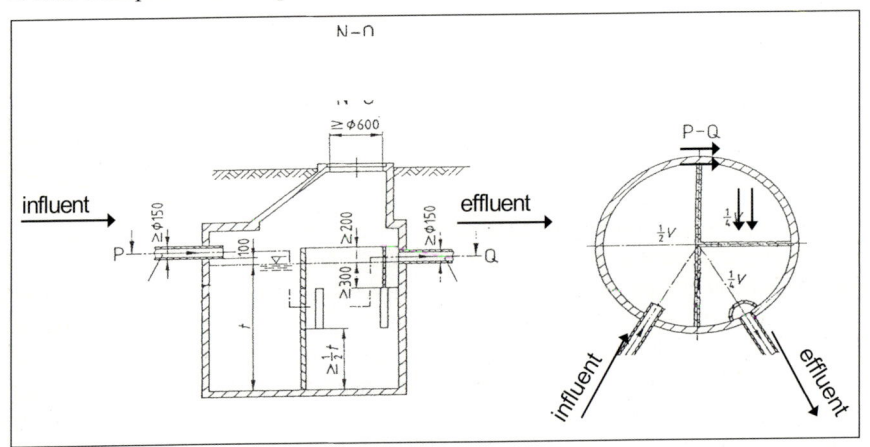

**Fig. 7.2.** 3-Compartment Septic Tank (DIN 4261)

For dimensioning purposes, a distinction is made between multi-compartment settlement-tanks and multi-compartment septic tanks (Table 7.2). The multi-compartment septic tanks provide for a considerably longer retention time of the wastewater. A wastewater quantity of 120 l/(PE·d) and a specific volume of 1,500 l/PE would result in a theoretical retention time of more than 12 days. In multi-compartment settlement tanks, however, the retention time is only 2-3 days. The long retention time in the multi-compartment septic tanks means that a part of the organic pollutant load is converted by anaerobic processes. The sludge produced is largely digested. Moreover, unlike multi-compartment settlement tanks, multi-compartment septic tanks have a higher sludge-storage capacity and hence a longer sludge-storage time. For these reasons, septic tanks should be preferred to settlement tanks.

# 7 Decentralised Wastewater Treatment - Wastewater Treatment in Rural Areas   145

**Table 7. 2.** Dimensioning of Septic Tanks and Settlement Tanks

|  | Volume |  | Compartments | Water depth |
|---|---|---|---|---|
|  | [l/P] | [l] |  | [m] |
| Settlement tank | 300 | ≥3000 | ≥2 | 1.2-3 |
| Septic tank | 1500 | ≥6000 | ≥3 | 1.2-3 |

Table 7.3. shows typical effluent concentrations from septic tanks in Germany. When this effluent data is compared with the concentrations found in untreated wastewater (Table 7.1), it becomes apparent that the preliminary treatment can already achieve a considerable reduction in the pollutant load (approx. 50% for COD and $BOD_5$). However, the effluent values are unacceptable for direct discharge into the receiving waters; secondary biological treatment is indispensable.

**Table 7. 3.** Average Effluent Concentrations of Septic Tanks (Kunst a. Kayser 2000)

|  |  | COD | $NH_4$-N | $NO_3$-N |
|---|---|---|---|---|
| Effluent concentrations | [mg/l] | 400 | 70 | 0 |

Another preliminary treatment method – though one which is not yet very common – are so-called composting tanks. Here, concrete tanks with 2 or 3 chambers are generally used (Figure 7.3.).

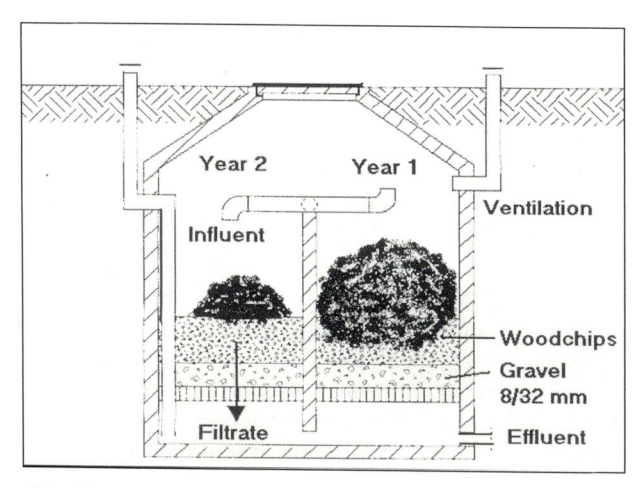

**Fig. 7.3.** Two-Chamber Composting Tank (Otterpohl 2000)

The wastewater/solids mixture is conveyed on to a filter layer, which is constructed on a bed of, say, aeration and drainage stones. Alternatively, retention of the solids can also be accomplished by a filter bag. The solids retained on this filter layer or in the filter bag and can, then, be (pre-)composted in the system. By adding organic structuring material (for instance, wood chips or straw), intensive ventilation and a speeded-up composting process are achieved. One chamber is fed about a year before it is filled, then the second chamber is fed. During this period, the matter in the first chamber can be further dewatered and composted. After two years, the first chamber is emptied. The material is then used for further composting, but it can also be used directly as fertiliser (Otterpohl 2000).

Another method for the preliminary treatment of domestic wastewater are settlement lagoons. As a rule, these are built prior to wastewater lagoons. The dimensioning of such settlement ponds is discussed in more detail in chap. 7.2.5.

### 7.2.4 Planted Soil Filters

*Fundamentals and Dimensioning of Planted Soil Filters*

To begin with, let us define and clarify the terminology. The generally used expression **constructed wetlands** is the generic term for a large number of different systems all of which are based on plants, soil and wastewater, but which otherwise have very different designs. For instance, there are surface and subsurface flow systems as well as systems with cohesive or sandy soils. In some methods, the purification performance of the plants plays a crucial role in treating the wastewater; in others the plants only have a minor part.

The latter also applies to the **planted soil filters**, which are discussed below. These are proven systems in wide use. The principal operation characteristics are:

- subsurface flow
- the utilisation of sandy soil substrate and
- the use of the soil filter as the actual biological-treatment stage

In some cases, the term **reed bed** is used to designate the same kind of treatment system. The term **planted soil filter** is preferable because the name already implies the crucial function of the soil filter in wastewater treatment.

Planted soil filters consist of beds that are sealed off against the ground and through which the mechanically pre-treated wastewater is conveyed. There are several sets of rules for dimensioning planted soil filters. Based on long-term experience with the building and operation of planted soil filters, today directives (e.g. ATV A262) are available that describe the generally accepted rules and the state of the art. The connection sizes range from four to several hundred PE, in exceptional cases to higher than 1000 PE. In Germany, there are some 7,000-10,000 planted soil filters with a size of <50 PE as well as 100-200 plants with a size of >50 PE (Geller 1997).

In planted soil filters, the wastewater, after its preliminary treatment, flows through a filter of sand grown with water and swamp plants. In Germany, the following plants are mostly used: common reed (*phragmites australis*), reedmace (*thypha latifolia*), and rushes (*juncus spec.*). Depending on the climatic conditions, a great number of other plants are suitable as well. An extensive list of these is given in Cooper et al (1996). When selecting the plants, one must make sure that they are indigenous and characterised by strong root growth. A large number of roots in the soil is particularly conducive to high permeability of the filter during long-term operation. Planted soil filters can be broadly classified according to flow direction and feeding methods.

In a **horizontal-flow planted soil filter**, the wastewater flows through the soil filter from one side to the other, i.e. horizontally. There is a continuous wastewater influent from the pre-treatment stage to the soil filter. In the filter, mainly anaerobic conditions prevail. Such systems can often be operated in a free slope, i.e. without pumps and thus without electrical energy.

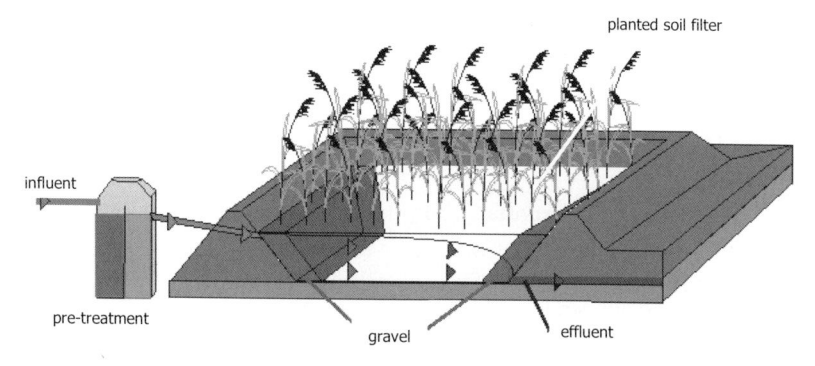

**Fig. 7.4.** Basic Sketch of a Planted Soil Filter, Horizontal Flow (Fehr et al 2000)

In a **vertical-flow planted soil filter**, the wastewater flows through the soil filter from top to bottom, i.e. vertically. The intermittent feeding of the soil filter with pre-treated wastewater is done several times a day via a pump. The intermittent feeding and the vertical flow ensure the filter is well supplied with oxygen.

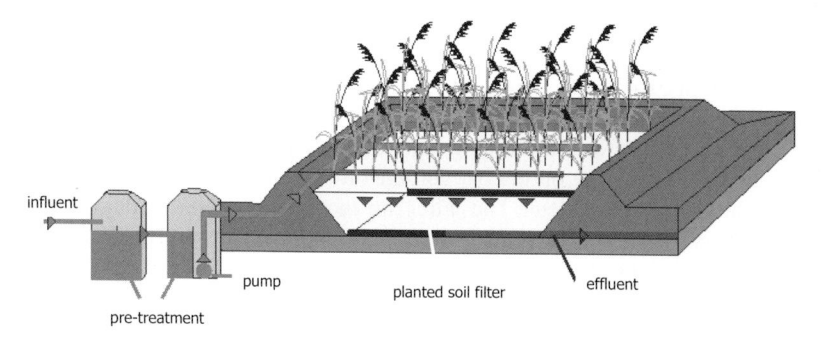

**Fig. 7.5.** Basic Sketch of a Planted Soil Filter, Vertical Flow (Fehr et al 2000)

The dimensioning of such plants is generally geared to the inhabitant-specific area, according to Table 7.4.:

**Table 7.4.** Dimensioning of Planted Soil Filters for Wastewater Treatment

|  | Vertical filter | Horizontal filter |
|---|---|---|
| ATV A262 (1998) | $\geq 2{,}5 m^2/PE$ | $\geq 5 m^2/PE$ |
| Various authors e.g. Kunst & Flasche (1995) | $\geq 3 m^2/PE$ | $\geq 5 m^2/PE$ |

## *Treatment Mechanisms and Purification Capacity*

The effect of planted soil filters is based on a combination of complex physical, biological and chemical processes, resulting from the interaction between soil, micro-organisms, plants and wastewater (Börner 1992). Accordingly, the primary degradation mechanisms are

- the mechanical retention of suspended and dispersed substances by the filtering effect of the soil
- the adsorption of colloidal substances in the mineral body of sand and to organic substances
- the ion exchange of clay minerals, humine substances and iron oxides
- the degradation and conversion of wastewater components by micro-organisms
- the utilisation of wastewater components for building up plant biomass

However, the effects of the different purification factors cannot be regarded as equal. The micro-organisms settling on the soil substrate, thus forming a biofilm on the single grains of soil or sand, play by far the most important role in the conversion and degradation processes.

Below, we give a brief overview of the respective degradation and conversion processes for the parameters nitrogen and phosphorus as well as for the organic compounds.

**Organic Compounds**
The degradation of carbon compounds (recorded in the sum parameters $BOD_5$ and COD) occurs mainly through the microbial growth on the soil particles, with filtration and adsorption effects increasing the availability of the substances for the micro-organisms. Depending on the oxygen conditions, anaerobic or aerobic degradation conditions are prevalent. The latter are more favourable for degradation performance because then a fast and extensive metabolism occurs up to the organic final products ($CO_2$, $H_2O$). By contrast, during anaerobic degradation, which is slower, mainly organic intermediate products or reduced final products are generated (Börner 1992).

## Phosphorus Compounds

The phosphorus compounds contained in the wastewater are mainly fixed in the soil as metal complexes by physicochemical reaction mechanisms. The binding of phosphorus depends, among other things, on the sorption ability of the soil, the amount and availability of reaction partners, the pH and the redox potential (Schütte a. Fehr 1992; Scheffer a. Schachtschabel 1998). Based on his own research and comparisons with other authors, Börner (1992) estimates the ratio of phosphorus compounds eliminated by conversion into plant biomass during harvesting to be less than 3% of the entire annual load.

## Nitrogen Compounds

In domestic wastewater, nitrogen is mainly found as urea and in the form of different types of proteins (Bahlo a. Wach 1995). These N-compounds are already hydrolysed by enzymes or ammonified by micro-organisms in the preliminary treatment, before flowing through the soil filter. Thus, nitrogen is conveyed to the soil filter mainly as ammonium nitrogen ($NH_4$-N).

There may occur an ammonium fixing to the clay components of the soil and to the living or decayed plant and bacteria mass. Moreover, nitrogen can be removed from the wastewater by possible plant harvesting or by $NH_3$ volatilisation under alkaline conditions (Börner 1992). None of these physicochemical and botanical elimination methods plays a major role in nitrogen elimination, as they remove in all only 18% of the nitrogen influent load.

It is the microbiological conversion and degradation processes that play the crucial role in the nitrogen conversions in the soil filter. The elimination of nitrogen is largely achieved by **nitrification** and **denitrification.**

### Nitrification

Nitrification is the microbial oxidation of the mineralised ammonium nitrogen. **Autotrophic nitrification**, which is also referred to as the "classical" form of nitrogen oxidation, is a two-stage, necessarily aerobic process. In the first step, the bacteria species Nitrosomonas oxidises the ammonium into the intermediate product nitrite (nitritation). In the second step, the produced nitrite is oxidised by bacteria of the Nitrobacter species into nitrate (nitratation).

$$NH_4^+ + 1{,}5\ O_2 \rightarrow NO_2^- + H_2O + 2\ H^+ \quad \text{(nitritation)}$$
$$NO_2^- + 0{,}5\ O_2 \rightarrow NO_3^- \quad \text{(nitratation)}$$
$$\overline{NH_4^+ + 2{,}0\ O_2 \rightarrow NO_3^- + H_2O + 2\ H^+} \quad \text{(nitrification)}$$

The reaction equation of nitrification shows the production of $H^+$-ions (acid), which lead to an increase in the soil acidity (Scheffer a. Schachtschabel 1998).

### Denitrification

Denitrification is the microbial reduction of the oxidised N-compounds (nitrate, nitrite) into elementary nitrogen ($N_2$), which is a gas and escapes from the water-soil system. During denitrification, mainly facultative anaerobic micro-organisms, so-called denitrifiers, use the nitrate ion instead of the oxygen as terminal hydrogen acceptor (nitrate respiration). The prerequisite for this is an anoxic milieu, i.e. oxygen must only be available in bound form (ATV-Handbuch 1997).

$$2\,NO_3^- + 2\,H_2 \rightarrow 2\,NO_2^- + 2\,H_2O$$
$$2\,NO_2^- + H_2 + 2\,H^+ \rightarrow 2\,NO + H_2O$$
$$2\,NO + H_2 \rightarrow N_2O + H_2O$$
$$N_2O + H_2 \rightarrow N_2 + H_2O$$
$$2\,NO_3^- + 5\,H_2 + 2\,H^+ \longrightarrow N_2 + 6\,H_2O$$

Regarding the conversion and degradation processes explained above, it becomes apparent that the availability of oxygen in the treatment process plays a crucial role. The characteristics of the two process variants, vertical- and horizontal-flow planted soil filters, with regard to their purification capacity are listed in Table 7.5. Owing to the good oxygen supply in the smaller vertical-flow filters, the processes of nitrification and carbon elimination are more efficient than in the horizontal-flow systems. By contrast, denitrification performance is better in the horizontal filters.

**Table 7.5.** Process Characteristics of Horizontal Flow and Vertical Flow Planted Soil Filters

|  | Vertical filter | Horizontal filter |
| --- | --- | --- |
|  | (good oxygen supply) | (poor oxygen supply) |
| C-degradation rate | high | satisfactory |
| nitrification performance | high | low |
| denitrification performance | low | high |
| P-fixing | Depends on the sorption ability of the soil (crucial factors would appear to be the iron, aluminium and calcium content of the substrate used) ||

Table 7.6. shows characteristic influent and effluent values of planted soil filters (domestic wastewater, preliminary treatment in a septic tank). The values are based essentially on feedback from a questionnaire circulated to several districts in Lower Saxony and on actual experiments and investigations conducted at the Institute of Sanitary Engineering and Waste Management, University of Hanover, Germany. The values should be understood as average concentrations, which arguably are feasible for normal operation of many such plants. The results should not be taken as effluent values achievable under optimal conditions.

The choice of a horizontal or vertical filter depends on different factors. One consideration is that vertical-flow systems require less space than horizontal systems. On the other hand, horizontal systems have the advantage that they are easier to build (especially the feeding device) and can be operated without pumps.

**Table 7. 6.** Characteristic Average Influent and Effluent Values of Planted Soil Filters (Kunst a. Kayser 2000)

|  | COD [mg/l] | $NH_4$-N [mg/l] | $NO_3$-N [mg/l] |
|---|---|---|---|
| Influent | 400 | 70 | 0 |
| Effluent |  |  |  |
| • Vertical filter | 70 | 10 | 40 |
| • Horizontal filter | 90 | 30 | 5 |

Furthermore, the decision depends largely on local conditions and the planned further use of the purified wastewater. If, for instance, the effluent of the WWTP is to trickle into the ground, horizontal filters may be more suitable in terms of groundwater protection because of their lower total nitrogen concentration. During passage through the soil, a further degradation of the carbon and nitrogen compounds occurs. If the wastewater is to be discharged into a receiving waterway, vertical filters are preferable because it is particularly the oxygen-consuming wastewater components such as carbon compounds and ammonium that pollute the waterways. If the WWTP effluent is to be used for irrigation, the vertical-flow method is again more suitable because the slightly higher total nitrogen content of the effluent can be used to supply nitrogen to the plants. In case of very high demands on the wastewater treatment, for instance - in ecologically sensitive area - the strengths of both plant types can be combined in multistage plants. Such plants allow extensive nutrient elimination (Bahlo 1996; Platzer 1997; Schütte et al 2000).

The discharge of biologically treated wastewater into surface waters raises the question of its hygienic condition. Wastewater has traditionally been tested primarily for the occurrence of indicator organisms (such as Escherichia coli, faecal streptococci, coliform germs), because these germs form part of the human intestinal flora. In a recent very extensive study, the 'Federal Environmental Agency' (UBA) addresses the issue of germ elimination in planted soil filters. Indicator organisms as well as pathogens (such as salmonella, Campylobacter/Arcobacter, Cryptosporides, Giardien) are determined and their behaviour analysed in multistage planted soil filters (Hagendorf et al 2001).

With regard to the indicator germs, reduction factors of $10^{1.5}$-$10^{2.5}$ per soil filter are obtained for colony-forming units, E. coli, coliform bacteria and faecal streptococci, which means that a reduction factor of up to $10^5$ can be achieved with multistage vertical and horizontal planted soil filters. For the pathogens, the picture is generally the same for the respective plant type. For instance, *Campylobacter/Arcobacter, Cryptosporides* and *Giardien* are reduced in planted soil filter systems by a factor of $10^4$-$10^5$. *Yersinia*, which like *salmonella* occur only sporadically and in low concentrations in the influent to planted soil filters, could not be detected in the effluent. These results confirm that it is possible to largely eliminate indicator germs by using certain types of planted soil filters. The effluent values of planted soil filters using sand or gravel are well below the limit and reference values of the EU Bathing Waters Directive and reach the permitted germ loads laid down in the Surface Water Directive Reference Values. Some effluent values of horizontal and vertical planted soil filters are even within/below the permitted maximum concentration laid down in the Drinking Water Directive. Thus, planted soil filters show much better elimination performance for indicator organisms and pathogens than classical biological wastewater-treatment plants (Hagendorf et al 2001).

## Operational Stability and Experience

Planted soil filters are low-technology solutions providing largely trouble-free operation. Different studies have confirmed their high operational stability. They also operate consistently in cases of pollution fluctuation and at times when there is no wastewater in process. This means that they are also well suited for use with small inhabitant equivalents (Kunst a. Kayser 2000).

Strongly fluctuating or extreme climatic conditions do not preclude their use either. A large number of planted soil filters in use in Germany have shown good operational results for both summer and winter. Of course, like all biological wastewater-treatment methods, they cannot be expected to provide optimal performance in wintertime. However, the normal wastewater temperature together with the insulating effect of the plants generally support stable operation and enable the required effluent values to be met. In summer, increased evaporation leads to a decrease in effluent or, in extreme cases, to a "no effluent" state. In such cases, shading the plants would prevent the filter drying out. This state does not result in a particular accumulation of pollutant loads, as the input pollutant load is not influenced by it. A longer period must be allowed for the start-up phase of a planted soil filter than for a technical plant because the root system has to develop first. During this period, there is also a greater risk of soil-clogging, which is one of the problems of planted soil filters.

Soil-clogging is a blockage of the filter's pores that occurs sometimes. Suspended solids and the biomass accumulation in the filter can cause a diminution of the hydraulic effective volume between the pores. This will, in the worst case, result in a rapid break of the hydraulic and purification capacity. The danger of soil-clogging grows with increasing amounts of argillaceous silt and clay in the filter. For these reasons, planted soil filters should be operated with middle-to-coarse sand or sand-gravel soil substrate (Kunst a. Flasche 1995). Another major parameter affecting the occurrence of soil-clogging is the input of solids. A sufficiently dimensioned pre-treatment system in which most of the solids are separated out is therefore of crucial importance.

The life-span of a sufficiently large planted soil filter using sand material has been put by Geller (1995) at from several years to several decades, assuming consistent purification performance. Through regular maintenance, the basically good operational stability of the plants can be further enhanced.

## Maintenance and Control of Constructed Wetlands

The simple technical design of planted soil filters makes them low-maintenance systems. Maintenance is necessary, of course, but it is relatively undemanding and requires little skill. Below, we give an overview of the main parameters influencing operation and treatment efficiency and of the work required to ensure the operational reliability of septic tanks and planted soil filters.

**Preliminary Treatment - Septic Tank**
The main parameters influencing operation and treatment efficiency of septic tanks are the thickness of the layer of scum and the bottom sludge level. The work required to ensure operational reliability consists mainly of controlling the sludge levels and the tank itself as well as the influent and effluent. Every 2 to 4 years, the faecal sludge in the first compartment has to be removed. It can then be transported to the nearest centralised treatment plant or treated in situ by sludge-composting (cf. chapter 7.2.6).

**Planted Soil Filter**
The main parameters influencing operation and treatment efficiency of soil filters are the permeability of the soil substrate and the efficiency of the feeding device. To ensure operational reliability, the feeding device must be controlled and the wastewater distribution and effluent must be inspected visually. Equally important is visual inspection of the filter surface for possible clogging. If there is evidence of incipient clogging, it may be helpful to stop feeding wastewater for some weeks (which is why it is useful to run two filters in parallel).

## 7.2.5 Wastewater Lagoons

### Fundamentals and Dimensioning of Wastewater Lagoons

Wastewater lagoons (also called wastewater ponds) are one of the oldest and most traditional forms of wastewater treatment. A widely employed method, they make use of biological self-cleaning processes. Here, too, the first stage is preliminary treatment in which the settable solids are separated out. In wastewater-lagoon plants, this first stage is generally a settlement lagoon. This anaerobic settlement lagoon is followed by the wastewater lagoons proper (which may be anaerobic) that serve as a biological treatment stage (cf. Figure 7.6.).

To achieve optimal purification performance, the required lagoon area must be divided into two or three units of equal size. According to ATV (1989), a rectangular design facilitates even flow, which makes for better contact between wastewater and bacteria. A topped gravel filter or a planted wetland prevent algae from drifting off.

Provided that sufficient space is available, wastewater lagoons can be well integrated into the surrounding landscape. They must be sealed off against the ground. In the case of soil conditions with a permeability coefficient of $k_f \leq 10^{-8}$, it is possible to dispense with artificial sealing (plastic sealing or artificially integrated layers of cohesive soil).

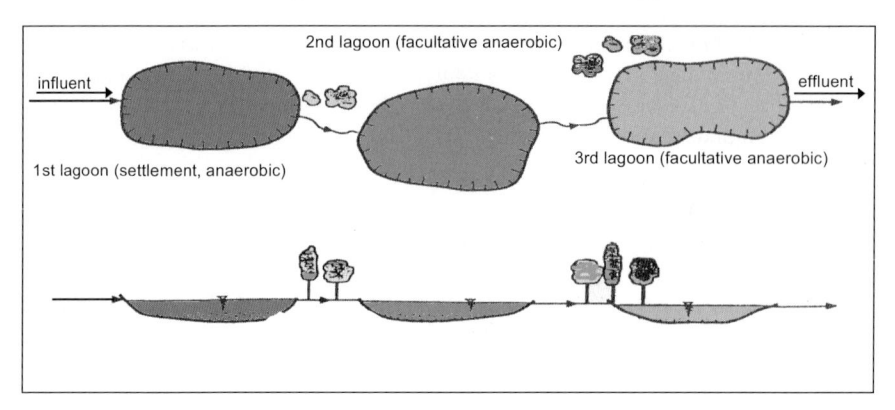

**Fig. 7.6.** Wastewater Lagoons, Basic Sketch

## Dimensioning of Wastewater Lagoons (ATV 1989, A201):

*settlement lagoon:*
volume:  $\geq 0.5 m^3/PE$, water depth: $\geq 1.5m$

*facultative lagoon*:
BOD$_5$ loading rate: 4-6g/(m²·d)
surface: $\geq 10m^2/PE$, for nitrification $\geq 15m^2/PE$
retention time (dry weather) $t_R$: $\geq 20d$
water depth: approx. 1m

## *Treatment Mechanisms and Purification Capacity*

The preliminary settlement lagoon serves to separate and digest the settable solids contained in the raw wastewater. Owing to the high BOD$_5$ load, oxygen consumption prevails in this settling tank; conditions in the water and the ground area are anaerobic.

Unaerated wastewater lagoons have a large surface and are shallow; aeration is achieved mainly via the water surface and by photosynthesis, circulation occurring through the impact of wind or temperature. If wastewater lagoons are sufficiently dimensioned, they have such low loads that aerobic conditions prevail, at least in the upper water layers. In the lower water layers and the ground area, however, anaerobic conditions prevail. In summer, the strong development of algae may lead for short spells to extreme oxygen peaks.

The purification processes in wastewater lagoons are rather complex, their biocoenoses covering the entire range of species found in natural lagoons. The following organisms, then, participate in the purification process:

- Heterotrophic micro-organisms, which convert the organic pollutants aerobically or anaerobically

- Autotrophic organisms (phytoplankton, algae, higher plants), which extract salt from the water and introduce oxygen, depending on the light conditions
- Animal species (such as zooplankton, fish), which are at the end of the food chain in the wastewater lagoon

Table 7.7. shows characteristic effluent values for wastewater lagoons. As in the case of the effluent concentrations of planted soil filters listed in Table 7.6, these values must be regarded as average concentrations, which arguably are feasible for normal operation of many such plants. The results should not be taken as effluent values achievable under optimal conditions.

Such lagoons have been shown to achieve relatively good purification results in terms of carbon compounds. Nitrogen elimination also occurs to some extent. It must be taken into account, though, that in many lagoons treat rainwater as well, which leads to dilution effects, as in the case of natural lagoons.

The purification capacity of wastewater lagoons must be seen as lower than that of vertical-flow planted soil filters, particularly with regard to nitrification.

**Table 7.7.** Characteristic Effluent Concentrations of Wastewater Lagoons (Kunst a. Kayser 2000)

|  |  | COD | $NH_4$-N | $NO_3$-N |
|---|---|---|---|---|
| Effluent concentration | [mg/l] | 80 | 15 | 5 |

### Operational Stability and Experience

Like planted soil filters, wastewater lagoons achieve good purification performance and consistent operating conditions. As this type of wastewater treatment requires little in the way of technical equipment (where elevation conditions are favourable, a wastewater lagoon can operate without any mechanical feeding devices), there is very little likelihood of malfunction. However, wastewater lagoons are prone to performance drops in the case of low temperatures (Neumann 1990).

The use of wastewater lagoons involves taking into account the sort of problems that may be caused by open wastewater surfaces (e.g. possible stenches, hygienic considerations in the case of direct contact with the wastewater as well as the increased occurrence of midges, particularly in regions with warmer climates).

Provided there is sufficient space available, wastewater lagoons can be well integrated into the surrounding landscape.

### Maintenance and Control of Wastewater Lagoons

Maintenance requirements for this nature-based purification method can be assumed to be very low. Here, again, maintenance *is* necessary, but it is quite simple, requiring skills than can be easily learned. The main parameter influencing the operation and treatment efficiency of wastewater lagoons is the sludge level. The work required to ensure operational reliability consists mainly of controlling the sludge level. The sludge has to be removed approximately every ten years. It can then be used as fertiliser in agriculture. Alternatively, it can first be treated and dewatered in situ by sludge composting. Equally important are visual checks for massive growth of *lemna spec.* or other plants that may cover the surface preventing contact between the water and the air.

## 7.2.6 Sludge Composting in Reed Beds

### *Fundamentals and Dimensioning of Sludge Composting Plants*

One advantage of planted soil filters and wastewater lagoons over technical methods is that they do not produce any surplus sludge. Thus, no cost or effort is involved for transporting and disposing of sewage sludge. Only the faecal sludge from the septic tank or the settlement lagoon has to be removed at regular intervals, as described above. This sludge can be treated either in a central WWTP or in situ in a sludge composting bed and then used in agriculture.

Sludge composting beds offer a sustainable and cost-effective option for the treatment of sewage and faecal sludge. The sludge put on such beds is largely dewatered, without conditioning agents or energy input, and rendered hygienic. Through the degradation of organic components in the beds, a considerable reduction in the overall quantity of sludge is achieved (Platzer a. Müller 1999). In the course of several years, then, a compost is produced that is suitable for use in agriculture.

**Fig. 7. 7.** Sludge Composting Plant at Wienhausen before Commencement of Operation

The structure of sludge composting beds resembles that of vertical-flow planted soil filters. They are constructed as sealed basins into which a sand-gravel filter is built. At the bottom of the beds, drainage pipes are installed which can take up the produced leachate. The sludge is evenly distributed over the surface of the beds by a feeding system. As a rule, the sand-gravel filter is planted with reed. The feeding intervals depend on the dry solids (DS) content of the spread sludge. The rest period between feeding phases should not be too short (at least one week) (Neemann 2000). After an operating period of 6 to 10 years and a subsequent idle time of six months to a year, the composted material can be removed.

Existing plants for sewage sludge composting are fed with annual loads of between 15 and 50kg DS/(m²·a) (Nudig a. Biehler 1998). Considering the slightly increased organic load of the faecal sludge compared to municipal sludge (Crössmann 1995), it would appear sensible to restrict the surface load to 20kg DS/(m²·a) for the composting of faecal sludge (Approach: Sludge composting Wienhausen – cf. chap. 7.2.6).

Data in the literature on the amounts of faecal sludge production and the corresponding surface-area requirements for composting beds varies. While for municipal sludge amounts in a range of 18-22kg DS/(PE·a) must be assumed (Platzer a. Müller 1999), the relevant faecal sludge amounts are considerably lower. Moreover, faecal sludge production is strongly influenced by the age of the sludge in the septic tanks. According to Schütte (2000), annual sludge production in a septic tank was 335 l/(PE·a) with annual emptying of the tank and 231 l/(PE·a) with a two-year removal cycle. The ATV-A123, however, assumes an annual sludge production of 1,000 l/(PE·a). Schütte (2000) and ATV-A123 (1985) agree in their estimate of the faecal sludge's DS content, putting this at 13 and 15g/l, respectively.

Based on the above data, the land requirements for faecal sludge composting can be calculated approximately as follows:

---

Faecal sludge production: approx. 200-1,000 l/(PE·a), DS content: approx. 15g/l

→**DS load: 3-15kg/(PE·a)**

Surface load of the faecal sludge composting plant: 20kg DS/(m²·a)

→ **Surface area of the faecal sludge composting plant: 0.15-0.75m²/PE**

---

Platzer a. Müller (1999) put space requirements at between 0.1 and 1.5m²/PE for the composting of municipal sewage sludge; due to the smaller amount of faecal sludge, the above mentioned values are in the lower section.

### *Treatment Mechanisms and Efficiency*

When the composting plant is fed with sludge, the sand-gravel filter first separates the sludge from the unbound water. The leachate is discharged via the drainage system and conveyed to a wastewater treatment plant. In the next phase, the natural drying process of the sludge begins. The reed plants with their transpiration capacity support this process by withdrawing moisture from the sludge. Moreover, the plants, with their marked root growth, create a pore system which allows air to enter the sludge layer, speeding up discharge of the produced leachate. The reed plants, which die in winter, also contribute a considerable amount of stable structuring material. A well-developed reed population with a dense growth of roots and rhizomes thus plays a very important role in the composting process (Platzer a. Müller 1999).

The achievable DS content of this material is in the region of 50%. Such effective dewatering cannot be achieved by conventional technical, energy consuming dewatering methods (Pabsch a. Pabsch 2000). Mineralisation causes a reduction in the organic content of up to 60%; thus, the DS content of the sludge may decrease by up to 50% (Platzer a. Müller 1999).

The composted sewage sludge should have a crumbly structure and smell of earth. So far, relatively few composting beds have been scraped, but initial experience with the product "composted sludge" show that this soil has been largely rendered hygienic (Pabsch a. Pabsch 2000) and that, despite the considerable reduction in quantity, there is no accumulation of heavy metals (EKO-PLANT 1999, quoted in Platzer a. Müller 1999). Thus, the utilisation of composted faecal sludge in agriculture would appear to be unproblematic.

## Operational Stability and Experience

The operation of a sludge composting plant is relatively simple, but it requires some finely-tuned skills, particularly during the initiation stage (Platzer a. Müller 1999). If its operation is controlled by the sludge distribution so as to produce an absolutely dense reed population, the crucial basic conditions for composting are met. If too much sludge is spread, the composting is insufficient – the plants become too wet and are damaged or decay.

The space requirements for sludge composting plants are not nearly as high as for nature-based wastewater treatment methods, which means that this method is not specifically confined to rural areas with decentralised wastewater treatment. In Germany, sludge composting plants starting from relatively small connection sizes of up to 90.000 PE are currently in use.

The composting of faecal sludge in reed beds does not, however, appear to be sensible solution for individual houses, as the reed plants need a relatively constant input of sludge and moisture, which is not ensured if the septic tanks are emptied every two years. The connection size should be large enough to enable the sludge bed to be fed in at least 3-4 times during the vegetation period, with each septic tank being emptied every two years at the most.

Another way of sludge composting using the so-called "sequential" drying method is currently being extensively tested (Pabsch a. Pabsch 2000). With this method, grass is used instead of reed. The sludge to be treated is distributed in a 20-40cm thick layer on the drainage layer of a bed. After it has dried for some weeks, grass is sown on the surface of the sludge layer. The grass serves to further dewater and loosen the sludge. After 6-9 months, a new sludge layer can be spread. This method does not depend on continuous sludge input and can therefore be used for very small connection sizes as well.

### 7.2.7 Practical Examples

In this section, we present several exemplary plants that were visited and examined as part of this practical project.

#### Wastewater Lagoons and Vertical-Flow Planted Soil Filters at Ettenbüttel

For decades, the wastewater of the village of Ettenbüttel was treated in unaerated wastewater lagoons. Analysis of the operational data of wastewater lagoons yields relatively good effluent data with regard to organic pollutants and shows that they achieve reasonable denitrification. However, it also indicates that they discharge relatively high ammonium loads (approx. 20mg/l $NH_4$-N) into the connected waterways.

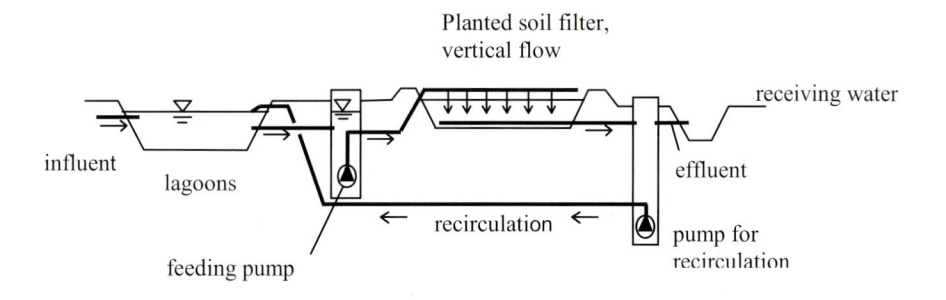

| | |
|---|---|
| **Design size:** | 1,000 PE |
| **Surface area of the lagoons**: | |
| Settlement lagoon 1: | 1,727m² |
| Lagoon 2: | 2,148m² |
| Lagoon 3: | 2,200m² |
| Surface area of the planted soil filter: | 2,250m² |

**Fig. 7.8.** Basic Sketch and Dimensioning of the Wastewater Treatment Plant at Ettenbüttel

Moreover, the wastewater lagoons at Ettenbüttel are extremely small, with a surface area of approx. 4,400m² for some 1,000 inhabitants (i.e. 4.4m²/PE), so that the effluent concentrations were accordingly high.

To improve the ecological situation, the wastewater-lagoon plant at Ettenbüttel was extended by a topped vertical-flow planted soil filter for nitrification.

Figure 7.8 shows – in principle - how the combined plant at Ettenbüttel basically works. First, the wastewater flows through lagoons 1 and 2. From lagoon 2, the pre-treated wastewater is conveyed by a pumping station to the vertical-flow planted soil filter. From the soil filter, the purified wastewater is discharged into the receiving water.

In the case of insufficient degradation performance or low receiving-water levels, the effluent from the reed bed can be pumped back into lagoon 1 in order to protect the discharge waterway. Apart from intermediate storage, the recirculation serves to achieve a more extensive purification, in particular denitrification. A third lagoon provides a sufficient buffer volume for the necessary water-amount management.

Table 7.8. shows the average influent and effluent values of the treatment plant at Ettenbüttel. It is evident that the topped planted soil filter achieves very good purification performance. While, after pond 2 the ammonium concentrations reached an average of 23mg $NH_4$-N /l, with a COD of 92mg/l, in the summer of 2000, the effluent values of the vertical filter showed constant ammonium concentrations below 2mg $NH_4$-N /l. In the winter of 1999/2000, when the planted soil filters began operation, the $NH_4$-N effluent values were still slightly higher, with an average of 9mg/l. The average COD effluent values of the planted soil filter were, from the start, at a consistently low level of approx. 35mg/l.

**Table 7.8.** Purification Performance of the Combined Wastewater Treatment System at Ettenbüttel – Average Values for the Investigation Period 10/99-8/00

|  | COD | $NH_4$-N | $NO_3$-N | $PO_4$-P |
|---|---|---|---|---|
|  | [mg/l] | [mg/l] | [mg/l] | [mg/l] |
| Influent Pond 1 | 816 | 60,5 | 0,6 | 10,02 |
| Effluent Pond 2 / Influent planted soil filter | 92 | 22.8 | <0.23 | 3.3 |
| Effluent planted soil filter |  |  |  |  |
| 10/99 until 3/00 (starting period – winter) | 35.65 | 9.9 | 13.4 | 1.26 |
| 3/00 until 8/00 | 33.4 | 1.9 | 9.96 | 0.87 |

## Multistage Planted Soil Filter at Konau

An improvement in the wastewater disposal situation was urgently needed in the municipality of Amt Neuhaus, situated in the Lower Saxony nature reserve Elbtalaue. Special solutions were considered for cost reasons (connection to the centralised treatment plant would be too expensive!) and because of the special requirements of this ecologically sensitive area. Multistage planted soil filters were chosen for the villages of Konau and Strachau because of the very specific requirements of this ecologically sensitive area. According to an existing analysis of special combination plants (Platzer 1997; Bahlo a. Ebeling 1995; Kunst a. Flasche 1995; Schütte 1991), high nutrient elimination could be expected.

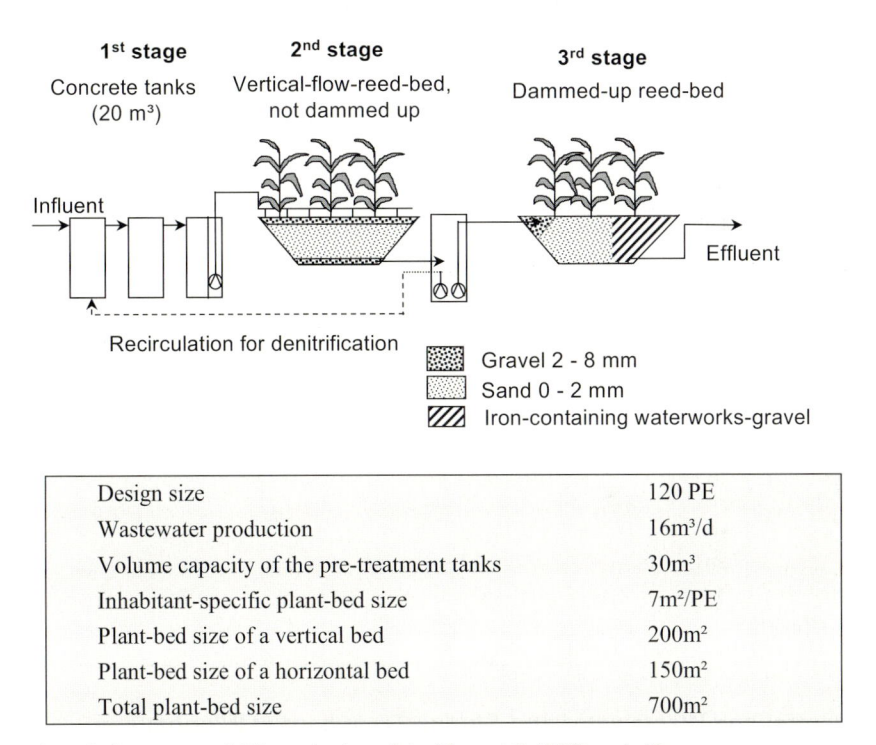

| Design size | 120 PE |
| --- | --- |
| Wastewater production | 16m³/d |
| Volume capacity of the pre-treatment tanks | 30m³ |
| Inhabitant-specific plant-bed size | 7m²/PE |
| Plant-bed size of a vertical bed | 200m² |
| Plant-bed size of a horizontal bed | 150m² |
| Total plant-bed size | 700m² |

**Fig. 7.9.** Structure and Dimensioning of the Planted Soil Filters in Konau

The planted soil filter at Konau/Popelau was designed for 120 PE and a wastewater production of 16m³/d. The existing old septic tanks of the individual houses are used as initial pre-treatment stages. From there, the wastewater is conveyed via the vacuum dewatering system to the central pre-treatment unit, before the planted soil filters. From there, the wastewater is distributed evenly with the help of a pump over the surface of the vertical filters. After trickling through the beds, the effluent is collected in a recirculation tank and part of it is fed back via the recirculation pipe into the pre-treatment stage for denitrification. The remainder of the effluent from the vertical filter is conveyed to the horizontal filter, from where it is discharged into the receiving water.

The horizontal filters contain a special ferriferous substrate (waterworks gravel) for the phosphorus sorption. Since, for system-specific reasons, the oxygen supply is nowhere near as good than in the vertical filter, the horizontal filters' function – apart from the phosphorus sorption – is the residual denitrification and the degradation of the remaining carbon compounds. In order to supply more carbon for further denitrification, raw wastewater from the pre-treatment unit can be pumped via a bypass directly on to the horizontal filter.

Table 7.9. shows average influent and effluent values for the plant at Konau. The results are subdivided into winter and summer operation. The plants were found to achieve effluent values for organic load, nitrogen and phosphorus far exceeding normal requirements. In summer, nitrification in the second treatment stage (vertical-flow filter) is very effective. However, it should be taken into account that in winter the nitrification performance is considerably lower because of the lower temperatures and that correspondingly higher ammonium concentrations occur in the effluent.

**Table 7. 9.** Average Influent and Effluent Concentrations for the Treatment Plant at Konau

|  | Winter | | Summer | |
| --- | --- | --- | --- | --- |
|  | 13/11/99–18/2/00 | | 19/4/00–11/6/00 | |
|  | influent | effluent | influent | effluent |
| COD [mg/l] | 446 | 37 | 682 | 37.5 |
| $NH_4$-N [mg/l] | 77.0 | 17.8 | 10.7 | 1.9 |
| $NO_3$-N [mg/l] | 0 | 33.3 | 0 | 30.3 |
| $PO_4$-P [mg/l] | 16.5 | 1.2 | 13.2 | 2.3 |

### Decentralised Wastewater and Sludge Treatment at Wienhausen

In the Lower Saxony municipality of Wienhausen, a closed and sustainable wastewater- and sludge treatment concept was realised as part of an EXPO project. The three small villages Offensen, Schwachhausen and Nordburg (municipality of Wienhausen), have a total population of approx. 560, living in some 120 households.

The treatment concept in these villages comprises the following steps (cf. Figure 7.10.):

- Operation of small WWTPs on individual properties to treat domestic wastewater. A total of 107 small WWTPs have been built, of which 88 are vertical-

flow planted soil filters. The other plants are compact technical systems, such as trickling filters.
- The purified wastewater is directly recharged into the local water cycle by trickling in situ.
- Building of a community plant for composting the faecal sludge.
- Local organisation of faecal-sludge removal and operation of the composting plant.
- Local utilisation of the composted sludge for soil enhancement.

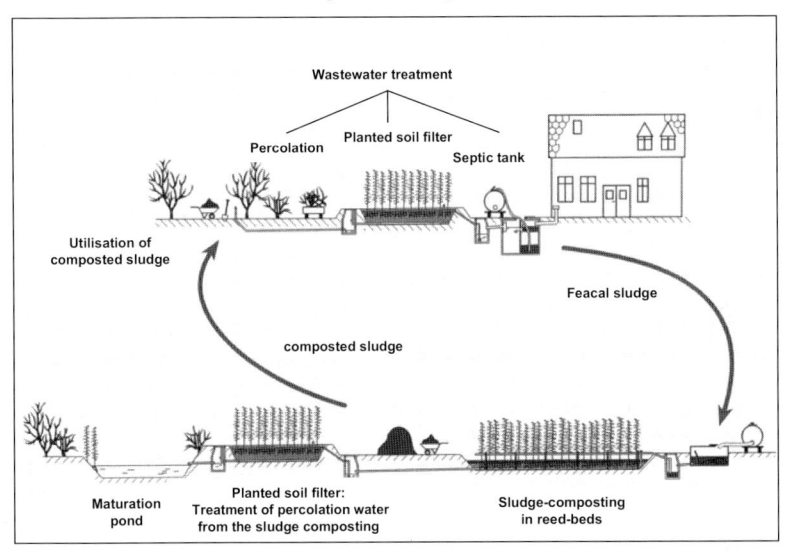

**Fig. 7.10.** Decentralised Wastewater and Sludge Treatment Concept at Wienhausen (Neemann 2000)

Experience with the project was very positive throughout. The planted soil filters used achieved consistently good purification performance. The villagers developed the high degree of commitment necessary for such a project and were happy to accept responsibility for "their" WWTPs, avoiding in their own interest water-endangering substances (Meyer 2000).

Experience with the composting plant has so far been fairly limited, as it only commenced operation in the spring of 2000. Below, we briefly explain the concept and dimensioning of the faecal-sludge-composting plant according to Neemann (2001).

The composting plant was designed to handle the faecal sludge from approx. 144 households with some 615 inhabitants. The sludge is collected in the first compartment of conventional 3-compartment septic tanks on the individual properties. Currently, about 80% of the compartment 1 volume is assumed to be removed every two years. This makes for an annual faecal sludge volume of approx. 275m$^3$.

During the summer months, the sludge is delivered in some four times from April to September (max. amount per campaign approx. 69m$^3$). The faecal sludge is pumped on to the composting beds (filter basins planted with reeds) using a coarse-material pump; the trickling water is then treated in a vertical-flow planted soil filter. The composting basins are designed in such a way that the surface load does not exceed a DS value of 20kg/(m$^2$·a), based on an average DS concentration of the retained sludge of 3%. The target surface load requires a basin size of approx. 410m$^2$, which is distributed over two basins of equal size. The two basins are fed alternately. Each basin is fed anew after a minimum break of six weeks.

The basins are sealed off against the underground and have a depth of 1.7m, of which 0.9m are available for storing the composting matter. The bottom 0.55m are filled with three sand/gravel layers of different grain sizes. The reeds are planted in these gravel layers. At the top, the basins have a free board of at least 0.25m. Overall, they are designed such that they can collect the sludge produced over a period of 14 years. After that, sludge removal is required. The removed material will be similar to humus; it is to be used in the three villages as fertiliser surrogate by farmers or gardeners.

## 7.3 Conclusions for Design Parameters

The following general planning parameters for wastewater purification plants in areas with rural structures contain two terms that need explaining before being used.

The term "rural regions" must be understood in terms of the different conditions prevailing in Europe and in the developing countries.

The term "region with rural structures" is used for European countries whose infrastructure in terms of wastewater disposal is characterised by large technical WWTPs operating in more or less the same manner. In its A 200 Brochure "Principles for Wastewater Disposal in Rural Areas", the German ATV/DVWK lists the following criteria as defining a region with rural structures (ATV/DVWK,2000):

- Small, sometimes highly remote villages and parts of towns
- Large properties because of loosely structured, open settlements, isolated farms, scattered settlements
- Low population density (up to 25 inhabitants/ha)
- Small consistent sewage networks, at times with gaps
- Small number of sewage plants; sewers often only fed from rainwater drainage
- Primarily agricultural structure; as a rule little trade and industry
- Small low-capacity receiving waterways, often pre-loaded with diverse input
- Leisure facilities frequently marked by seasonal variations

The general planning principles include a list of technological measures for wastewater treatment, sometimes also a catalogue of measures referred to as the wastewater-disposal concept, which is adopted in the respective community by those responsible for disposing of the wastewater. Dewatering and wastewater treatment plants together form a single unit and should be examined in relation to the receiving waterway into which the wastewater is eventually discharged.

The size of WWTPs in rural regions is generally below 5,000 PE (so-called small WWTPs). However, given the current development in rural areas, the limit for small WWTPs should rather be set at connection sizes of 10,000 PE. The new definition distinguishes more size categories than before; the current number is 5. In Category 3, connected BOD loads of 300-600kg/d and 5,000 to 10,000 l/d are treated, the listed BOD loads being realistic and normal for areas that meet the definition "rural region".

Defining a given WWTP according to the number of connected inhabitants is no longer a valid approach anyway – connected loads and water quantities are much more crucial criteria.

When considering developing countries, there are some important aspects that are not given much attention in Europe. In a critical discussion, however, it should be stated that the following points also apply to wastewater-purification concepts in Europe:

- Traditional wastewater management concepts need to be complemented by an equally powerful tool that serves areas of low population density in both industrialised and developing countries.
- Decentralisation of municipal sanitation can be considered an economically and ecologically interesting alternative to the traditional concept, provided that the individual treatment systems reliably produce high quality at a reasonable price.

As mentioned above, the current situation in many developing countries is that very different wastewater ratios must be dealt with. World-wide, 5% of wastewater is treated, the figure ranging from 0 to 50-65% of the wastewater collected. Around 50% is treated in many European countries (France 60%, Spain 50%, Poland 23%,) but also in some countries referred to as developing countries.

There are, however, some special criteria in the developing countries that must be considered when discussing which wastewater purification concepts would be sustainable.

Rural areas in developing countries are not defined in terms comparable to such areas in industrialised countries (as pointed out above). The population density there may be much higher than 25 PE/ha, but a common characteristic of all rural areas is that they have a mainly agricultural structure. There are differences in the amount of small trade and industry, which can be very high in certain regions, particularly in South-East Asia.

- Wastewater disposal concepts are often not required as planning basis. As a rule, only part of the produced wastewater is collected. How much, depends on random aspects, not on decisions based on options available in the context of an integrated water management system.
- As a rule, the wastewater in developing countries has higher pollutant concentrations than in Germany or Europe. This is due to the lower water consumption. However, the BOD amounts produced per inhabitant are less constant.

These are approx. 60g BOD/(PE·d) for Europe as a whole, whereas in the developing countries the amount varies from 15 to 60g BOD /(PE·d).
- Existing sewage systems often have more leaks. In Germany, some sources state that the sewage system built after World War II lost a maximum of 40% through leakage (19-40%, Härig, 1992). More recent studies by the Federal Environmental Agency (UBA) (Hagendorf a. Kraft 1996; Hagendorf 1997) show that there is in fact no evidence of high, continuously flowing losses from the sewage system, either in the municipal sector or in terms of industrial-wastewater discharge. Thus, figures of up to 40% discharge from the sewage system for specific substances must be regarded as an overestimate of the actual load. The latest research results on this subject are about to be published. In many countries around the world, losses through leakage from wastewater pipes, but also from drinking-water implements, are currently still about 30-50%.
- If the installation of sewage systems is dispensed with when developing new, alternative wastewater purification concepts, this can cut costs considerably (5/7 of the overall costs, Kreutzburg 1998).
- In the case of decentralised disposal, less untreated wastewater is discharged as pollutant into the groundwater.

Summarising experience with conventional wastewater treatment systems, we can point to five crucial advantages of a counter-model based on a decentralised approach:

- No or less sewer installations are necessary.
- There are numerous options for decentralised reuse of the wastewater.
- In smaller plants, malfunctions or possible breakdowns have less impact on the overall system, i.e. less damage is caused than in cases where a hazardous substance is fed into a large plant.
- The gradual updating of the disposal system can be organised in accordance with the financial resources of the respective municipality.
- The systems' low technology level allows participatory co-design and, in some cases too, the participation of interested users in the construction of the wastewater-treatment plants.

This leads us to the following conclusions:

- A new generation of highly efficient, compact, user-friendly and inexpensive treatment systems is urgently needed to meet the needs of most developing countries.
- Analysis of the requirements for low-cost, efficient wastewater treatment systems shows that the emphasis is generally on BOD/COD degradation, the elimination of so-called eutrophying substances having lower priority. Partial treatment is most common.
- Requirements for nutrient elimination are rarely available.
- Though legal controls are provided for in many countries, there is often no guarantee that they are actually carried out. Merely establishing limits is point-

less if no accompanying legal provisions are made to ensure that the limits are observed. Thus, in many developing countries, nitrogen and phosphorus elimination is, to all intents and purposes, omitted from the target list.
- The omission of nitrogen and/or phosphorus elimination is reasonable and understandable if the treated wastewater is to be used as fertiliser in agricultural production. This, however, is only the case in about 30% of all treatment plants and countries.
- If the treated wastewater is to be reused for agricultural irrigation, germ elimination is of overriding importance.
- Rainfall events must be evaluated and integrated into an adapted system. Rainwater that is not to be treated can either be utilised (rainwater harvesting) or must be conveyed to retention, trickling and/or discharge facilities.
- Special heed must be paid to the respective conditions in the different developing countries, e.g. mosquito occurrences, breeding ground for other pathogens. For the nature-based treatment processes, it may be said in summary that when using wastewater lagoons, particularly in warmer climates, consideration must definitely be given to problems that may be caused by open wastewater surfaces. Planted soil filters have the advantage that there is no open wastewater surface. Also, the germ-elimination rates they achieve are much better than those of conventional technical treatment processes .

Wastewater lagoons constitute the most common technology world-wide. They generally comprise a series of unaerated lagoons. After a mechanical preliminary treatment stage (not always present in such systems), the wastewater is subjected to preliminary treatment, followed by – in systems of the same construction –biological treatment. The first lagoons, in which biological digestion and pre-treatment takes place, are anaerobic and, optionally, anaerobic. The further the self-purification process progresses, the more aerobic the lagoons become. Bacteria and algae are the biological self-purification agents in the lagoons. Algae growth in the lagoons also changes as the purification progresses. In the case of strong pollution, greenish colours dominate (owing to green algae such as Chlorella spec.). The more the carbon is degraded, the more the lagoons take on a blue-green appearance, this being due to a shift in the algae population from purely green algae growth to a mixture of green and blue algae. In the case of carbon-compound shortages, i.e. after extensive purification, only blue algae are found in the lagoons.

Huge lagoon systems are in operation world-wide. Disposal systems serving populations of tens of thousands or even hundreds of thousands (e.g. in Lima/Peru) are not uncommon. The disadvantages of such lagoon systems are:

- stenches in the surrounding area, particularly around the first lagoon
- high space requirements
- breeding grounds for mosquitos, insufficient elimination of pathogens
- no nitrogen elimination
- no phosphorus elimination

Major advantages, on the other hand, are:

- simple, cost-effective construction
- low technology level and thus low maintenance costs
- long sedimentation times for worms, etc.; good elimination efficiency for typical tropical pathogens
- possible reuse for food production (fish lagoons) or for irrigating agriculturally used areas

Improvements are needed, then; what sort of improvements depends on the conditions prevailing the respective region.

In cases where lagoon systems discharge into other waterways such as running waters or oceans, the main requirement to be met is generally that less eutrophying substances be discharged, which will lead to a demand for nitrogen and possibly phosphorus elimination. In cases where the reuse of wastewater effluents as fertilisers in agriculture is planned, but also to enable the water to be reused for specific purposes, the optimisation preference will direct efforts towards improved germ elimination and improved COD degradation, enabling the wastewater to be further degraded (stabilised) and then discharged for further use.

These improvements should be accomplished by low-technology solutions with low or even zero energy consumption, operating without the need for a lot of pumps and control systems. In many areas, the acceptance of pollutants in water is a further burden, as the amount of the matrix to be treated increases. It must always be determined if and where water can be saved. In many cases, wastewater can then be turned into compostable solids (for instance, during the coffee bean harvest, using pulpa matter, etc.).

Planted soil filters are considered suitable for further improving the most widely used wastewater purification systems world-wide, i.e. wastewater lagoons. This is dealt with in greater detail below.

### 7.3.1 Characteristics of Decentralised Wastewater Treatment Systems which are Conducive to Sustainable Development

Referring again to the principles for sustainable wastewater treatment as outlined in Agenda 21 and mentioned at the beginning of this report, we focus on the following aspects that make the use of planted soil filters in sparsely populated rural areas particularly recommendable:

- Planted soil filters achieve good and stable purification results. The different process variants and their combination allow a broad range of requirements for effluent quality to be met.
- The building and maintenance of planted soil filters is relatively simple, which means it can be taken care of by the respective municipalities themselves with the support of small local companies.
- Natural materials such as sand or reed are generally used.
- Energy requirements and costs are relatively low; no sewage sludge is produced.

- The thorough desinfection of the wastewater means that it can be safely reused for irrigation purposes.
- Closed concepts using direct trickling or irrigation and faecal-sludge composting in situ facilitate the creation of small water and nutrient cycles.
- In particular the shared use of WWTPs by several municipalities or parts of towns allows closed sustainable concepts and promotes local participation.
- The plants can easily be integrated into the surrounding landscape (see Figure 7.11.).

**Fig. 7.11.** Multistage Planted Soil Filter at Konau

## 7.3.2 Impact of Gender Perspectives on Planning Criteria

Currently, decentralised approaches to wastewater treatment are still confined to rural regions. Despite the proven advantages of nature-based purification methods, such methods are regarded by water-management authorities as stopgap solutions for remote areas. No consideration is given to the design potential of these disposal technologies. This is due to the fact that the historically evolved infrastructure of city sewers and wastewater-treatment systems has led to a universalisation of wastewater technology. For more than a century, wastewater disposal in Europe has been marked by extensive centralisation. Historically, the reason for this is that the municipal authorities responsible for runoff and wastewater treatment had the task – as part of public health efforts – of guaranteeing local hygiene. The current state of affairs is shaped by the memory of the chaotic wastewater disposal situation in the 19$^{th}$ and early 20$^{th}$ centuries, which led to epidemics in conurbation, infestations of rats and great economic losses. This resulted in the insight that, in the interest of reasonable local hygiene standards and in order to effectively combat the spread of diseases, responsibility for this area could no longer be left to the individual. Part I of the original version of the DIN 4261 standard, dating from the war year 1942, documented this insight: "The joint servicing of several properties in terms of sewage disposal and wastewater purification by a central wastewater treatment plant is an economic necessity." Today in Europe, centrally organised and technology-based wastewater treatment is the rule. According to conventional wisdom, this is the only valid and optimal approach, given the interrelation with city planning. Normally, the creation of an infrastructure containing a decentralised discharge and sewage system is not even discussed, unless it is villages or scattered settlements that are being dealt with. Thus, the area "design of water supply and disposal" is a problem that city planning engineers fail to address.

Huge financial resources have been consumed by the existing technology. Their de facto existence has necessitated their continued maintenance. There has never been any objective discussion of this issue in Europe, let alone in Germany.

However, the opportunity to take a fresh look at this issue is still open to developing countries. Here, decisions can still be made, with the participation of socially active groups and taking into account the special interests of women, with regard to

- the design of water management
- regulation of the water supply
- general access to the natural resource "water"

By affording people greater insight into the functioning and problems of water disposal and wastewater treatment, decisions about the future are put into the public eye. This enables one principle of feminist planning to be realised: the idea of extending political participation to include technological and economic participation. In order to achieve such extended participation, everyday skills and professional expertise must be combined. This would then offer the chance for a critical discussion of the potential of environmental technologies. For instance, as we shall discuss below, the positive aspects of existing European technologies could be critically evaluated and utilised in new and creative ways. Such planning could integrate the positive aspects of decentralised solutions with the regional peculiarities of the different countries.

If we consider, for example, the situation in the megacities of Mexico, Latin America, India or Africa, we find that only very few areas in such cities actually enjoy regular and safe wastewater disposal and treatment. As a rule, only a small fraction of the wastewater is collected and treated. And yet there is an inability to see the necessity of designing the infrastructure of these growing cities in terms of decentralised units. This results in inadequately served areas on the outskirts of the cities – basically rows of small private dwellings with insufficient hygienic conditions.

This means that the insights resulting from a "different" perspective – different, that is, from the male-dominated mainstream of ideas – can offer solutions for problematic water disposal and wastewater treatment situations. This includes consideration of ecological, socio-economic and design aspects. The aim, then, is to create nature-based wastewater treatment plants - taking into account the specific structures of settlements in rural areas – which can also be applied in mainly densely populated areas as in urbanised areas. This helps to make the problems relating to water and wastewater visible and experienceable for the inhabitants. Participatory processes should govern the design, construction and operation of the plants. Consideration should be given to design options allowing the neat integration of such plants into the village or the surrounding landscape.

This brief outline of how to go about creating solutions also contains elements of a gender-specific view. One interpretation of the gender approach – among many others – would appear to be well suited to the discussion of innovative, sustainable water and wastewater concepts. This gender approach includes addressing both practical gender needs, such as improving women's position in society by the provision of water and sanitation closer to their homes, as well as strategic gender needs, such as improving women's position in society by increasing their awareness of their situation and their capacity to take decisions and influence change.

A gender approach also seeks to prevent further overburdening of women and stresses the importance of not automatically reinforcing and perpetuating traditional roles. This implies the need to address men as well as women, since to support this men must change their attitude and behaviour (WWV 1999). This gender concept has been criticised for attempting to reduce the complexity of reality and force it into a narrow conceptual framework. The complex discussion and analysis of gender as a social construction is missing here (Weller 1995).

This means that gender, as DOING gender, is seen as a society-structuring category. This structuring category refers to both everyday actions and to actions in the planning sphere. In the concept of DOING gender, gender is seen as personal characteristic, which can be used as precondition, but which can also be created/constituted anew in any situation. Thus, gender, in this DOING gender sense, is a complex construction of sensual, intellectual and behavioural dispositions for individual self-presentation and forms of interaction (Schultz 2001). So far, however, there has been no transfer of the spirit of DOING gender to management strategies for the resource "water".

Although the above-mentioned interpretation of the gender approach has been criticised, it would appear to be very useful for planning processes like decentralised wastewater treatment concepts. And its practical application is evidence of its acceptance.

When we talk about water and the role of the community or the general public, what we usually mean is:

- overall planning of the system
- construction of infrastructure
- management of the system
- maintenance of the system and the processing

Though women are key water managers, it has taken a long time for women to be given the opportunity to participate in the planning, construction, management and maintenance processes for water. As many examples show, only men – who were again approached by male administrators and technicians – have been invited to participate in these processes. Over the past decade, the participatory approach has gradually become more common. In villages, this meant that men and women formed village water committees. The members of such committees were expected to practically organise the entire system. This led to an unequal division of labour and responsibility between the sexes. While both men and women, particularly women, stand to benefit from paid and/or respected positions as technicians and managers, in reality women are often left to perform the less prestigious tasks, e.g. acting as members but not as managers of water committees. The trend towards more decentralised forms of resource management, especially user-based management, must properly incorporate gender concerns in the form of clear recommendations for planning and management.

The steps needed to achieve this objective vary considerably from one region to another depending on religious beliefs, the local water-balance situation and the respective cultural background.

Evaluation of the situation in the different world religions shows that the development options for a gender approach vary considerably.

Major differences are already evident as regards access to the resource "water". In south-east Asia, this is clearly demonstrated by women's rights with respect to land irrigation 90% of agricultural production depends on irrigation, and women are responsible for looking after the fields, but they have no right to take decisions on irrigation, such rights being directly tied to land ownership. Although several studies (Khan Tirmizi 2000; Zwarteveen a. Meinzen-Dick 2001) show that water management is more efficient if carried out by women, because their interest in multiple use of water is much more marked than that of men, existing gender relations preclude women's involvement in the development of sustainable water management. This situation might change if efforts were made to view resource management through the eyes of women.

## 7.4  Examples of Planning Ideas

The following project concepts are based on several guiding principles which were developed during discourse:

- simplicity of construction and operation
- possibility of achieving consistent water quality
- low operational costs and low maintenance requirements
- no need for long-distance transportation of sewage, cost reduction by treating wastewater close to its place of generation
- capability of withstanding both organic and hydraulic shock loading
- no need for continuous sludge handling and disposal facilities
- treated water can be reused, which actually reduces freshwater consumption
- emphasis on constructed wetlands combined with aerated lagoons or septic tanks

- tools and knowledge for planning, designing and building decentralised wastewater treatment systems applied to different conditions and situations

Evaluation of the situation in different developing countries showed that decentralised concepts are actually suitable for transfer to the home countries of the project participants. Decentralised concepts are those that

- use nature-based methods
- do not need any or only small wastewater-discharge systems
- can be built and operated with communal participation
- require no or only little electrical energy for their operation

These concepts enable different problem areas to be catered for.

## 7.4.1 Wastewater Purification for Remote Villages

### Vietnam:

**Actual Situation in Thi Tu Village on the Outskirts of Hanoi:**
- Topography: hilly, slope: 2-3%
- Annual average temperature: 23.4 degrees
- Annual average rainfall: 1,527mm
- Population: approx. 1,000
- Main product: noodles
- Generation of wastewater with a high concentration of organic compounds
- Ponds and lagoons: available
- Double-chamber toilets
- No sewage system

**Fig. 7.12.** Village on the Outskirts of Hanoi, Vietnam

**Proposed Constructed Wetland :**
- Grey and black water are conveyed to the ponds by gravity
- Vertical-flow planted soil filter
- Storage ponds for spray irrigation
- Participation of the local community in operation and maintenance

**Decentralised Industrial Wastewater Treatment**

Another crucial application is decentralised industrial wastewater treatment. In many countries, agricultural activities such as coffee growing and production, banana growing, tobacco or leather production are performed mainly by numerous small companies and industries. In Cuba, for example, there are 200 small coffee-producing industries which so far have no wastewater treatment. This poses a serious threat to the water quality of the waterways that serve as receiving waters because of the wastewater's high organic-pollutant load. The biodiversity in these regions and waterways is impaired. This is particularly tragic when it affects regions that should be under special protection. In Cuba, for instance, coffee is mainly grown and processed in a region that has potential for ecological tourism.

## *Cuba:*

**Escambry Plan for Wastewater Treatment :**
- Important biosphere reserve
- Great biodiversity
- Potential area for ecological tourism
- Surface-water contamination by the waste from family coffee industries
- More than 200 of such small industries in this area
- Cuban government greatly interested in finding a low-cost solution for wastewater treatment

**Fig. 7.13.** Area of Small Coffee Industries in Biosphere Reserve in Cuba

**Project:**
- Pilot plant for research on the treatment of wastewater from coffee industries
- This plant serving to demonstrate to other coffee communities the feasibility of such decentralised wastewater treatment

Similar problems with the wastewater and waste from coffee production also occur in San Salvador (Brazil). In the case of Brazil, a slightly different view is taken of the problem of realising decentralised wastewater treatment concepts.

**Module on Overall Concepts for Sustainable Resource Management**
In many industries, concepts are currently under consideration that seek to achieve the highest possible utilisation of residues and waste in combination with sustainable wastewater treatment. These concepts are concerned with wastewater and waste that cannot be prevented from re-entering the water cycle or water resources, or with those bit-streams that cannot be recycled in any other way. Crucial objectives of these wastewater purification concepts are:

- producing reusable wastewater for washing purposes, etc.
- zero- or low-energy operation of wastewater treatment

Here, the concepts for decentralised wastewater disposal can be integrated into the broader context of sustainable resource management at the industrial level.

**Fig. 7.14.** El Salvador; an Area in which Sustainable Resource Management is to be Investigated as an Overall Concept

### 7.4.2 Planning Ideas for Sensitive Regions in Rural Areas

*Brazil:*

**A Concept Was Developed for a Fishing Village with Seasonal Tourism:**
- Normal population: approx. 300 inhabitants
- Conflict between fishing and tourism as a source of income in the summer, resulting in river-water pollution through domestic wastewater
- High fluctuation in load and quantity of the wastewater
- Shock-loading
- Community participation and environmental education

Villages, such as fishing villages, with a population of a few hundred and considerable fluctuations in wastewater production and pollutant load during the tourist season has extensive problems with satisfying and constant treatment of wastewater. Maintaining the village's tourist "industry" is, in turn, directly dependent on assuring the quality of its water. To meet these requirements, a possible solution would be lagoons combined with planted soil filters (cf. chapter 7.2.7 – practical Example of Ettenbüttel)

**Fig. 7.15.** Fishing Village, Santa Catarina, Brazil

Although we do not know the exact frame conditions (fluctuation of wastewater amount and concentrations) in this special case we will give an rough estimation of the dimensioning of such plant for this situation. The dimensioning will be done for the normal situation with 300 PE considering a recirculation-rate of 100% from the planted soil filter back to $1^{st}$ lagoon and some security buffers. Regarding

# 7 Decentralised Wastewater Treatment - Wastewater Treatment in Rural Areas

the higher wastewater amount during the tourist season it has to be pointed out that lagoons and planted soil filters have a high tolerance regarding load fluctuations. In addition it will be possible to reduce or stop the recirculation to increase the absolute flow rate through the treatment plant.

**Dimensioning:**

*Frame conditions and assumptions*

normal situation: 300 PE
wastewater amount (assumption): 100 l/(PE·d)
total wastewater amount: 30m³/d
recirculation rate: 100%
→ flow rate 60m³/d

---

*settlement lagoon:*

dimensioning criteria: V ≥ 0.5m³/PE, water depth: d ≥ 1,5m

300 PE → 100m², d=1,5m
to achieve a good settlement the higher flow rate due to the recirculation should be considered:

$$\text{dimensions chosen: } 200m^2,\ d=1,5m$$

---

*facultative lagoon:*

dimensioning criteria:
surface:                              ≥ 10m²/PE
Retention time (dry weather) $t_R$:   ≥ 20d
water depth:                          about 1m
$BOD_5$ loading rate:                 4-6g/(m²·d)

300 PE → 3000m² (surface 10 m²/PE), depht d=1m
→ retention time with recirculation: 50d
→ there are no information concerning the $BOD_5$ concentrations – but with typical wastewater the inhabitant specific surface (10m²/PE) already fulfills the BOD-loading rate criteria. Even if the BOD-loading rate will be higher during the tourist season it has to be regarded that the effluent of the lagoons will pass the planted soil filter which is very effective in degradation of organic compounds.

$$\text{dimensions chosen: } 3000m^2,\ d=1,0m$$

*planted soil filter – vertical flow:*

dimensioning criteria: surface ≥ 2,5-3m²/PE

300 PE, 3m²/PE → 900 m²
→ hydraulic load without recirculation: 30000 l/d / (900m²) = 33 l/(m²·d)
→ hydraulic load with recirculation: 60000 l/d / (900m²) = 66 l/(m²·d)
Using sandy substrate a hydraulic load of 66 l/(m²·d) resp. mm/d is unproblematic. We made good experiences with short-term hydraulic loads up to 200mm/d during the warm seasons. With 200mm/d it would be possible to treat 180m³/d for some days/weeks

dimensions chosen: 900m², d=0,7m

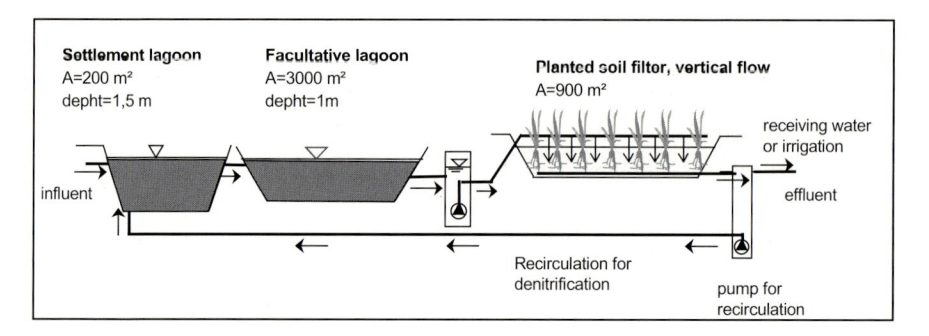

**Fig. 7.16.** Wastewater Lagoons and Planted Soil Filter – a Combined System as a Solution for the Fishing Village in Brazil

Fig. 7.16 shows a principle sketch of the dimensioned treatment plant. Out of the tourist season this plant can be used to treat the wastewater from the 300 inhabitants with a wastewater amount of about 30m³/d. In this case a recirculation-rate of 100% serves to achieve a more extensive purification and particularly the denitrification. With the same treatment plant it will be possible to treat 60m³/d (corresponding to 600 PE) without changing the flow rate only by stopping the recirculation. Wastewater amounts of up to 180m³/d (corresponding to 1800 PE) can be treated over a short period - for example during the weekends in the tourist season. Of course this will reduce the treatment efficiency compared with the "normal" situation (30m³/d), but also without the recirculation a sufficient treatment result can be expected.

## Infrastructural Measures to Combat Slums / Slum Development

Another important application is the management of the resource "water" in newly emerging urban areas. These areas are characterised by rapid growth and unregulated development. It may be regarded as a substantial achievement if such areas can be prevented from turning into slums. An absolutely essential infrastructural measure is proper organisation of the water supply, wastewater discharge, wastewater treatment and reuse of the purified wastewater. By managing the water resources, i.e. designing proper water-supply and wastewater-disposal systems, it should be possible to slow down or prevent the development of dwellings into slums, even where such a process is already in progress.

Numerous examples demonstrating this can be found not only in Brazil, but throughout the world, e.g. in several villages on the outskirts of Hanoi and Mexico City.

**Fig. 7.17.** New Urban Settlements to Replace Slums in Big Cities; Santa Catarina, Brazil

- Municipal governments have plans for new urban settlements
- Possibility of constructing separate sewerage systems for grey and black water
- Reduction of water consumption and wastewater discharge into receiving waters

## References

Abwassertechnische Vereinigung ATV (1985) Arbeitsblatt A 123, Behandlung und Beseitigung von Schlamm aus Kleinkläranlagen

Abwassertechnische Vereinigung ATV (1989) Arbeitsblatt A 201, Grundsätze für Bemessung, Bau und Betrieb von Abwasserteichen für kommunales Abwasser

Abwassertechnische Vereinigung ATV (1998) Arbeitsblatt A 262, Grundsätze für Bemessungen, Bau und Betrieb von Pflanzenbeeten für kommunales Abwasser bei Ausbaugrößen bis 1000 Einwohnerwerte. Gesellschaft zur Förderung der Abwassertechnik e.V.

ATV/DVWK (2000) ATV-DVWK-Kommentar, Abwasserentsorgung im ländlichen Raum. Hennef

ATV-Handbuch (1997) Biologische und weitergehende Abwasserreinigung. Ernst & Sohn Verlag für Architektur und technische Wissenschaften GmbH, Berlin

Bahlo K (1997) Reinigungsleistung und Bemessung von vertikal durchströmten Bodenfiltern mit Abwasserzirkulation. Disscrtation, Universität Hannover

Bahlo K, Ebeling B (1995) Betrieb von Pflanzenkläranlagen. In: Hauskläranlagen, Neueste Entwicklungen. Schriftenreihe der Kommunalen Umwelt-Aktion U.A.N., Heft 24, Hannover

Börner T (1992) Einflußfaktoren für die Leistungsfähigkeit von Pflanzenkläranlagen. Schriftenreihe WAR 58, TH Darmstadt (ed)

Cooper PF, Jobs GD, Green MB a. Shutes RBC (1996) Reed Bed Systems and Constructed Wetlands for Wastewater Treatment. Water Research Centre (WRC), Swindon

Crössmann G (1995) Klärschlämme aus Kleinkläranlagen landwirtschaftlicher Betriebe und deren Bewertung nach der Klärschlammverordnung (AbfklärV). 107 VDLUFA-Kongreß, VDLUFA-Schriftenreihe 39

Deutsches Institut für Normung e.V., DIN 4261, Kleinkläranlagen

Farley M (1999) Managing Water Loss in Developing Countries. In: Water 21, July/August

Fehr G, Schütte H (1990) Leistungsfähigkeit intermittierend beschickter, bepflanzter Bodenfilter. gwf 4/91

Fehr G, Tempel K, Welker B (2001) Projektmanagement und Öffentlichkeitsarbeit für das Verbundprojekt Bewachsene Bodenfilter, Homepage www.bodenfilter.de, AZ 14178-10, F&N Umweltconsult und uve Umweltmanagement und -planung, Hannover/Berlin, Teilprojekt im Rahmen des Verbundprojektes Bewachsene Bodenfilter als Verfahren der Biotechnologie, AZ 14178-01 gefördert durch die Deutsche Bundesstiftung Umwelt, Osnabrück,

Geller G (1995) Einsatz von Pflanzenkläranlagen zur Entsorgung im ländlichen Raum, langfristige Erfahrungen, Pflanzenkläranlagen -Stand der Technik, Zukunftsaspekte. Wiener Mitteilungen. Wasser Abwasser Gewässer 124

Geller G (1997) Jüngere Erfahrungen mit Pflanzenkläranlagen. Wasser-Abwasser-Praxis 5: 27-32

German Federal Ministry for Economic Co-operation and Development BMZ (2000) Water Resolving Conflicts, Shaping the Future. BMZ spezial 009, Berlin/Bonn

Hagendorf U, Bartocha W, Diehl K, Feuerpfeil I, Hummel A, Szewzyk R (2001) Mikrobiologische Untersuchungen zur seuchen hygienischen Bewertung naturnaher Abwasserbehandlungsanlagen, Teil 2 Untersuchungen durch klassische Kultivierungsmethoden, AZ 14178-07 Umweltbundesamt, Institut für Wasser-, Boden- und Lufthygiene, Dienstbereich Langen, Berlin/Langen, Teilprojekt im Rahmen des Verbundprojektes Bewachsene Bodenfilter als Verfahren der Biotechnologie, AZ 14178-01 gefördert durch die Deutsche Bundesstiftung Umwelt, Osnabrück

Hagendorf U (1997) Undichte Kanäle und ihre Auswirkungen auf Boden und Grundwasser Detektion, Quantifizierung und Bewertung. In: Hallesches Jahrb. Geowiss. 19: 149-158

Hagendorf U, Krafft H (1996) Erfassung und Bewertung undichter Abwasserkanäle. In: Umweltbundesamt (ed) Texte UBA 9/96

Kommunale Umweltaktion UAN (2000) Konzepte zur Abwasserbehandlung im ländlichen Raum. Vol III, Hannover

Kunst S, Kayser K (2000) Leistungsfähigkeit von Kleinkläranlagen. Schriftenreihe der kommunalen Umweltaktion U.A.N., Heft 36

Kunst S, Flasche K (1995) Untersuchungen zur Betriebssicherheit und Reinigungsleistung von Kleinkläranlagen mit besonderer Berücksichtigung der bewachsenen Bodenfilter. Abschlußbericht Forschungsvorhaben AZ 32-201 00091, Institut für Siedlungswasserwirtschaft und Abfalltechnik, Universität Hannover

Landesumweltamt Nordrhein-Westfalen (LUA NRW) (1994) Abwasserbeseitigung im Außenbereich (Kleinkläranlagen). Merkblätter Nr. 3, Essen

Platzer C (1998) Entwicklung eines Bemessungsansatzes zur Stickstoffelimination in Pflanzenkläranlagen. Schriftenreihe des Fachgebietes Siedlungswasserwirtschaft der Technischen Universität Berlin, Nr. 6

Platzer, Müller (1999) Klärschlammvererdung, eine nachhaltige und gleichzeitig kostengünstige Methode der Klärschlammbehandlung. Hamburger Berichte zur Siedlungswasserwirtschaft, Nr. 29

Scheffer F, Schachtschabel P (1998) Lehrbuch der Bodenkunde. Ferdinand Enke Verlag, Stuttgart

Schultz I (2001) Forschungen im Rahmen des Themas Gender & Environment, ein Blick auf die deutsche Debatte. Hintergrundtext für den Niedersächsischen Forschungsverbund für Frauenforschung in Naturwissenschaft/Technik und Medizin, http://www.isoe.de/ftp/Gund.E.pdf

Schütte H (1991) Untersuchung von Kleinkläranlagen im ländlichen Raum Niedersachsens mit der Zielsetzung der verfahrenstechnischen Optimierung bei Neubau und Sanierung. Im Auftrag des Niedersächsischen Landesamtes für Wasser und Abfall, Hannover

Schütte H (2000) Betriebserfahrungen mit Kleinkläranlagen. Korrespondenz Abwasser 10: 1499-1505

Schütte H, Brüdern U, Kayser K, Kunst S (2000) Extensive Nitrogen Elimination in Constructed Wetlands. Beitrag auf dem 1st World Water Congress of the IWA, 3.-7.7.2000, Paris

Meyer D (2000) Kreislaufsysteme in der dezentralen Abwasserbehandlung. Schriftenreihe der kommunalen Umweltaktion U.A.N., Heft 37

Neemann G (2000) Schlammvererdung, ein naturnahes Verfahren zur Behandlung von Abwasserschlämmen im schilfbewachsenen Filterbecken. Schriftenreihe der kommunalen Umweltaktion U.A.N., Heft 37

Neemann G (2001) Information about Planning, Processing and Design of the Faecal Sludge Composting Site at Offensen/Wienhausen. Information from Gerd Neemann, Büro für Landschaftsplanung und Umweltstudien (BLaU)

Neumann H (1990) Unbelüftete Abwasserteiche, Darstellung der naturwissenschaftlichen Verfahrensgrundlagen und Bericht über Erfahrungen in Niedersachsen. Kommunale Umwelt-Aktion U.A.N., Heft 4 Abwasserteiche unbelüftet - belüftet, Hannover

Nudig R, Biehler M (1998) Vererdung von kommunalem Klärschlamm in Schilfbeeten. Wasser Abwasser Praxis (WAP) 1: 49-55

OECD (1998) Household Water Pricing Practices in OECD Countries. Paris

Otterpohl R (2000) Design and First Experiences with Source Control and Reuse in Semi-centralised Urban Sanitation. EURO Summer School, DESAR Decentralised Sanitation and Reuse, June 18–23, Wageningen

Pabsch J, Pabsch H (2000) Klärschlammvererdung im sequentiellen Trocknungsverfahren. Schriftenreihe der kommunalen Umweltaktion U.A.N., Heft 37

Weller I (1995) Zur Diskussion der Stoffe und Stoffströme in der Chemie(-politik): Erster Versuch einer feministischen Kritik, Frankfurt

Wilderer PA, Schreff D (1999) Decentralised and Centralised Wastewater Management, a Challenge for Technology Developers. IAWQ 4th Specialised Conference on Small Wastewater Treatment Plants, Stratford upon Avon, UK

Zwarteveen M, Meinzen-Dick R (2001) Gender and Property Rights in the Commons, Examples of Water Rights in South Asia. Agriculture and Human Values 18: 11-25

# 8 Alternative Technologies for Sanitation, Recycling and Reuse

Sabine Kunst, Namrata Pathak

## 8.1 Overview

Globally, some 500 million kg of human faeces are generated daily in urban areas and some 600 million kg in rural areas – a total, then, of over one million tons per day. Most of this biodegradable organic material is disposed of untreated or after very little treatment, thus polluting the environment with substances that are highly detrimental to human health.

In the developed countries, sewage systems are efficient enough to ensure that most human waste is safely treated and disposed of. The situation is different in the developing countries. There, about 65% of the population are without adequate sanitation. Sanitation coverage in the developing countries varies greatly from region to region: the best-served regions are West Asia with 68%, and Latin America and the Caribbean with 63%; the least-served regions are Asia and the Pacific with 29%, and Africa with 35%.

Paradoxically, most conventional sewage systems in use in both the developed and developing countries are water-borne. This means that they usually rely on water to transport waste from the point of emission to the point of treatment. Since water itself is a very precious and increasingly scarce resource, the world cannot afford to waste it in this manner. Besides, what is regarded as human waste is itself a useful resource that is not being exploited.

Human faeces and urine contain huge amounts of nutrients and can easily be converted into biofertiliser. The use of conventional sewage systems means that the vast quantities of nutrients present in wastewater are constantly being lost because of the failure of these systems to recycle the waste. Given this situation, it is imperative that alternatives to water-borne sanitation be explored. Such alternative sanitation systems should be based on the principle of conserving and recycling earth's resources.

**Sustainable Sanitation**
The lack of adequate sanitation in large parts of the world – with the attendant problems of disease and pollution – can be remedied by tackling the issue at the grass-roots level. New integrated sanitation and waste-management systems will have to consider the different qualities of outputs from human settlements: black water, biowaste, grey water and storm water.

Pollution-prevention, sewage reduction and water conservation efforts should be maximised. A conservative technology is based on the principles: Preserve, Conserve and Protect. Health authorities are traditionally sceptical of people's ability to manage their own problems. The regulatory and sanitary-engineering authorities (generally one and the same body) also believe that matters are best left in their hands. Environmental groups, too, have tended to share the conviction that "end-of-pipe" treatment is the surest way of cleaning polluted water.

The interests of industry are another very important factor that must be taken into account: discharging its waste into public sewers is the cheapest solution for industry; and the system of central sewage collection is extremely expensive, thus ensuring the highest profits for engineering and construction firms.

For example, pipe-laying accounts for some 80% of the total cost of sewage collection and treatment, and engineering and construction firms get a flat 20% share of the total budget here. Fixing the 5-10% of septic systems that are in need of repair or renewal (i.e. overflowing or causing pollution) would never generate the sort of profits to be had from the blanket provision of central collection and treatment facilities.

A great deal of public resources are utilised to build large treatment plants to serve those people whose waste is conducted by a sewage system. This happens in spite of the fact that an average of 60 to 70% of the population of developing countries has no access to sewage piping systems nor to the amount of water required to conduct such human waste. The problem cannot be solved by investigating the situation in which there is access to a sewage system. Instead it is necessary to look at those who urgently need dry solutions or small wastewater treatment plants in order to treat their water as close as possible to their homes and in a way that the population can afford.

Combining decentralised technologies with recycling and reuse can lead to solutions such as the so-called SIRDO. SIRDO is an integral system for recycling organic waste (Mena-Abraham 2000). It is promoted by GTA (Grupo de Tecnologica Alternativa, S.C.), which is a non-profit organisation focused on developing alternative technologies for recycling liquid and solid domestic waste. GTA's president Josefina Mena-Abraham and her assistant of many years, Veronica Corella-Barud, provided material about the Mexican patent SIRDO. Its design is based on the biochemical process which characterises the indigenous Mexican "chinampa" (Mena-Abraham 2000). It comprises two main models for two different situations, the products generated are water for agricultural purposes and biofertilizer:

1. the wet treatment (with a collecting pipe system for grey and black waters, either mixed or separated) for situations in which water is sufficiently available;
2. the dry treatment (similar to a composting toilet) for water-scarce situations.

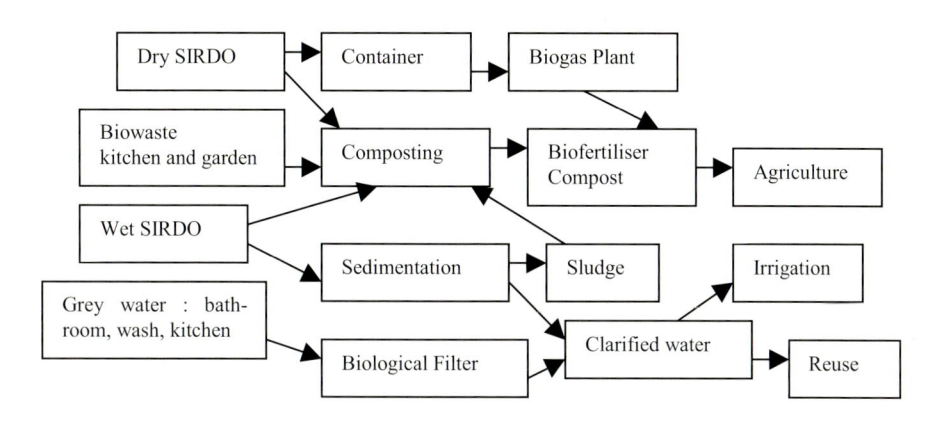

**Fig. 8.1.** Integral Domestic Wastewater and Waste Management

**The Potential for Sustainable Sanitation**

The potential of the concept of sustainable sanitation depends on the distribution and quality of organic load in the different flows of domestic wastewater. Domestic wastewater consists of three flows of totally different quality (1. and 2. can be defined as black water):

1. the urine (yellow matter), which is the most important component regarding nutrient load (87% nitrogen and 50% phosphorus and potassium),
2. the faeces (brown water), poor in nutrient load and high in organic carbon load.
3. grey water, which amounts to great volumes but contains little in terms of loads.

Often only black and grey waters are distinguished. Black water and kitchen waste contain the most nutrients: nitrogen, phosphorus and potassium. Grey water contains few nutrients and, provided phosphorus-free detergents are used, it can be easily treated as long as it does not come into contact with black water. Grey water often can have high COD concentration because the water is less diluted when used in small quantities.

**Table 8.1.** Components in Different Flows of Domestic Sewage (Otterpohl et al 2000)

|  | Black water |  | Grey water |
| --- | --- | --- | --- |
|  | Urine | Faeces | Kitchen, Bath, Cleaning |
| Amount [l/(PE · a)] | 500 | 50 | > 30,000 |
| Organ. Carbon [kg/(PE · a)] | 6 | 17 | 5.5 |
| Nitrogen (Kjedahl) [kg/(PE · a)] | 5 | 0.5 | 0.3 |
| Phosphorus [kg/(PE · a)] | 0.4 | 0.2 | 0.5 |
| Potassium ($K_2O$) [kg/(PE · a)] | 1 | 0.17 | 1.1 |

Given the various characteristics, many forms of sanitation and treatment concepts could be appropriate and applied in different situations (Otterpohl et al 1999). Meanwhile a number of various techniques are known to be implemented in source separation systems. They will be described later on.

## 8.2 The Composting Process

The composting process is an important basis upon which innovative sanitation concepts are developed. It will therefore be described extensively.

### 8.2.1 The Phases

In the process of composting, micro-organisms break down organic matter, producing carbon dioxide, water, heat and humus, the relatively stable organic end product. Under optimal conditions, composting comprises three phases: 1) the mesophilic, or moderate-temperature phase, which lasts for a couple of days, 2) the thermophilic, or high-temperature phase, which can last from a few days to several months and 3) a several-month cooling and maturation phase.

**Mesophilic Phase**
Different communities of micro-organisms predominate during the various composting phases. Initial decomposition is carried out by mesophilic micro-organisms, which rapidly break down the soluble, readily degradable compounds. The heat they produce causes the compost temperature to rise rapidly.

Once the temperature exceeds about 40°C, the mesophilic micro-organisms become less competitive and are replaced by others that are thermophilic (*Bacillus stearothermophilus, Rhizomucor pusillus*), or heat-loving. At temperatures of 55°C and above, many micro-organisms that are human or plant pathogens are destroyed. As temperatures over about 65°C kill many forms of microbes and limit the rate of decomposition, compost managers use aeration and mixing to keep the temperature below this point.

**Thermophilic Phase**
During the thermophilic phase (Chaetomium thermophile, Humicola insolens, Humicola (Thermomyces) lanuginosus, Thermoascus aurantiascus and Aspergillus fumigatus), high temperatures accelerate the breakdown of proteins, fats and complex carbohydrates like cellulose and hemicellulose, the major structural molecules in plants. As the supply of these high-energy compounds becomes exhausted, the compost temperature gradually decreases and mesophilic micro-organisms once again take over for the final phase of "curing" or maturation of the remaining organic matter.

**Maturation Phase**
During the maturation phase, once the temperature has dropped back into the mesophilic range, these microbes begin to dominate again. The drop in tempera-

ture is the best indication that a compost heap has entered the maturation or curing phase. During this phase, the remaining organic material becomes increasingly complex. These complex organic compounds resist further decomposition and are collectively referred to as humus.

The changes in temperature and population are shown for mesophilic fungi (broken line) and thermophilic fungi (dotted line) in a wheat-straw compost. (based on data from Chang a. Hudson 1967). The left axis shows fungal populations (logarithm of colony-forming units per gram of compost plated on to agar); the right axis shows the temperature at the centre of the compost.

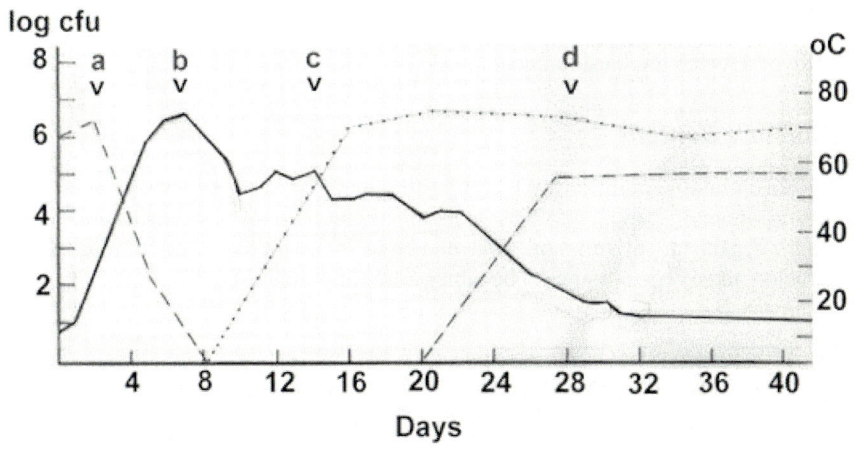

**Fig. 8.2.** Changes in Temperature (Solid Line) and Population in Compost

- **a**: Initial phase of rapid microbial growth on the most readily available sugars and amino acids.
- **b**: Peak heating phase that destroys or inactivates all the mesophilic organisms and leads to a prolonged high-temperature phase that favours other thermophilic species.
- **c**: At 40–60°C, a second group of thermophilic fungi start to grow.
- **d**: As temperatures drop, mesophilic organisms recolonise the compost and displace the thermophiles.

### 8.2.2 Environmental Factors in Composting

*Oxygen*

The aerobic organisms responsible for the composting process need free atmospheric or molecular oxygen to survive. Without oxygen, they die and are replaced by anaerobic micro-organisms that slow down the composting process and generate odours and potentially flammable methane gas. For composting toilets to work most effectively, the material being composted should be unsaturated with liquids and have a loose texture, allowing air to circulate freely within the heap.

Adding bulking agents, such as wood chips, shavings commonly available as pet bedding, coconut fibre, cottonwood, stale popped popcorn, etc., to increase pore spaces allows the influx of oxygen into, and the release of heat, water vapour and carbon dioxide from, the composting mass. Earthworms also assist in this process. Maintaining sufficient airflow through the material involves providing proper ventilation (e.g. using compressed air, convection or a fan) and/or frequently mixing the compost, either manually or automatically.

Note: Users are warned not to use bulking agents like leaves, which may either mat or introduce bugs into the system, or certain types of wood such as cedar or redwood which are hard to decompose.

## Moisture Content

Under optimum conditions, the composting material has the consistency of a well-wrung sponge – about 45-70% moisture. Less than 45% is insufficient for the micro-organisms to function, and at above 70% saturated conditions begin to develop and oxygen depletion becomes a limiting factor.

## Temperature

Four temperature ranges should be recognised when considering the composting process:

- Below 42°F – little or no active microbial processing takes place. Within this temperature range, the system merely serves as a storage vessel for excrement, toilet paper and additives.
- From 42°F to 67°F – psychrophilic micro-organisms dominate (e.g. actinomycetes and fungi), resulting in a mouldering process. Mouldering toilets are designed to operate within this temperature range. As the composting process is so much slower in this range, larger composter vessel sizes may be needed to compensate for the slow volume reduction of the composting mass.
- From 68°F to 112°F – mesophilic bacteria dominate. This is the typical temperature range of most composting toilets.
- From 113°F to 160°F – thermophilic bacteria dominate (atypical of most compost systems unless assisted by an external heating system).

## Carbon to Nitrogen (C:N) Ratio

Micro-organisms need digestible carbon as an energy source for growth, and nitrogen and other nutrients for protein synthesis. When measured on a dry-weight basis, an optimum C:N ratio for aerobic bacteria is about 25:1. Although important, the significance of the C:N ratio in composting toilets is often overrated. The main reason for adding carbon material such as wood chips to a composting toilet is to create air pockets in the composting material. However, adding a small handful of dry matter per person per day or a few cups a week is a good rule of thumb for maintaining a reasonable C:N ratio and ensuring that excess moisture is absorbed and that there are sufficient pores in the composting material.

## Pathogens

Composting toilets are designed to protect people from exposure to human excrement and to store such waste under aesthetically acceptable conditions until it can be safely removed for disposal or reuse. Two primary factors affecting the survival of human pathogens in composting toilets are temperature and time. As a rule, pathogens die off when subjected to temperatures above 122°F for a sufficient length of time. However, achieving such temperatures during the composting process depends on ideal conditions, including adequate air supply, moisture content and C:N ratio. They are rarely achieved during the composting of human excrement in a composting toilet. Pathogens die off anyway eventually, depending on the retention time in the unit. Most bacteria, viruses and protozoans die within several months. However, certain helminth (e.g., *Ascaris lumbricoides* – the common roundworm) and protozoan oocyst-producing species (e.g., *Cryptosporidium parvum* – a type of parasite that causes diarrhoea) are highly resistant to environmental stresses and survive for longer periods of time. Also, unless fresh excrement is completely separated from the finished product in the unit, contamination of the latter can occur. For these reasons, care should be taken if application of the residual product to land is considered. The product of a composting toilet is not normally suitable for use on lawns or gardens.

## Vector Management

Vectors (e.g. flies, beetles, mites and other arthropods) are attracted to human excrement to feed or reproduce, as conditions allow. Vectors that have been in contact with excrement can then carry pathogens from the composting chamber and transmit them to humans via various paths (e.g. food contamination by flies). It is therefore important to prevent vector populations from entering the composting chamber. Considerations for the design and management of composting toilets include:

- Screening ventilation openings (note: screens may become clogged and require periodic cleaning)
- Sealing cracks and holes (a smoke test can reveal these)
- Applying environmental-friendly insect repellents, such as pyrethrins and diatomaceous earth
- Capturing insects with insect strips
- Avoiding putting kitchen scraps into the composter

### 8.2.3 Composting Micro-Organisms

Composting micro-organisms produce enzymes that are extracellular – like a chemical aura outside the organism's body. They transform molecules of organic matter into less complex chemicals and energy. All organisms have intracellular enzymes to manage the diverse complexity of the life pores. Enzymes are unstable proteins or protein-containing compounds that, when present in small amounts, promote a chemical reaction. The enzymes – such as amylase, cellulose, lipase and protease – are some of the catalysts responsible for decomposition.

#### *Bacteria*

Bacteria are most numerous in compost. They account for 80-90% of the billions of micro-organisms typically found in a gram of compost. Bacteria are responsible for most of the decomposition and heat generation in compost. They are the most nutritionally diverse group of compost organisms, using a wide range of enzymes to chemically break down a variety of organic materials.

Bacteria are single-celled and structured as either rod-shaped bacilli, sphere-shaped cocci or spiral-shaped spirilla. Many are motile, meaning that they have the ability to move under their own power. At the beginning of the composting process (0-40°C), mesophilic bacteria predominate. Most of these are forms that can also be found in topsoil.

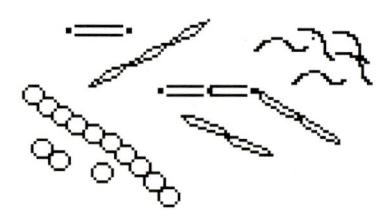

**Fig. 8.3.** Mesophilic Bacteria

As the compost heats up to above 40°C, thermophilic bacteria take over. The microbial populations during this phase are dominated by members of the genus *Bacillus spec.* diversity of bacilli species is fairly high at temperatures of 50-55°C but declines dramatically at 60°C or above. When conditions become unfavourable, bacilli survive by forming endospores, thick-walled spores that are highly resistant to heat, cold, dryness or lack of food. They are ubiquitous in nature and become active whenever environmental conditions are favourable.

At the highest compost temperatures, bacteria of the genus *Thermus spec.* have been isolated. Composters sometimes wonder how micro-organisms evolved by nature can withstand the high temperatures found in active compost. *Thermus spec.* bacteria were first found in hot springs at Yellowstone National Park and may have evolved there. Other places where thermophilic conditions exist in nature include deep-sea thermal vents, manure droppings and accumulations of decomposing vegetation that have the right conditions for generating heat as in a compost pile.

Once the compost cools down, mesophilic bacteria again predominate. The numbers and types of mesophilic microbes that recolonise compost as it matures depend on what spores and organisms are present in the compost and in the immediate environment. As a rule, the longer the curing or maturation phase, the more diverse the microbial community it supports.

### *Actinomycetes*

The characteristic earthy smell of soil is caused by actinomycetes, organisms that resemble fungi but are actually filamentous bacteria. Like other bacteria, they lack nuclei, but they grow multicellular filaments like fungi. In composting, they play an important role in degrading complex organic materials, such as cellulose, lignin, chitin, and proteins. Their enzymes enable them to chemically break down tough debris like woody stems, bark or newspaper. Some species appear during the thermophilic phase, others become important during the cooler curing phase, when only the most resistant compounds remain in the last stages of humus formation.

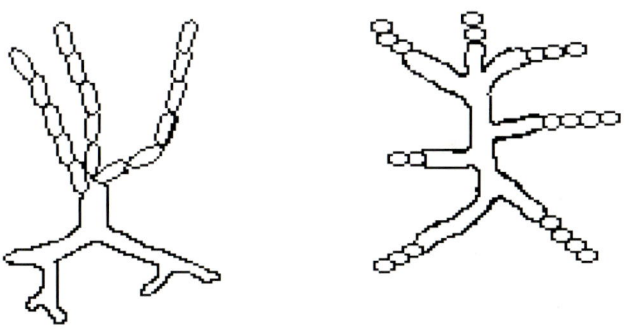

**Fig. 8.4.** Actinomycetes

Actinomycetes form long, thread-like branched filaments that look like grey spider webs stretching through the compost. These filaments are most commonly seen towards the end of the composting process, in the outer 10 to 15 centimetres of the pile. They sometimes appear as circular colonies that gradually expand in diameter.

## Fungi

Fungi include moulds and yeasts, and collectively they are responsible for the decomposition of many complex plant polymers in soil and compost. In compost, fungi are important because they break down tough debris, enabling bacteria to continue the decomposition process once most of the cellulose has been exhausted. They spread and grow vigorously by producing numerous cells and filaments, and they can attack organic residues that are too dry, acidic or low in nitrogen for bacterial decomposition.

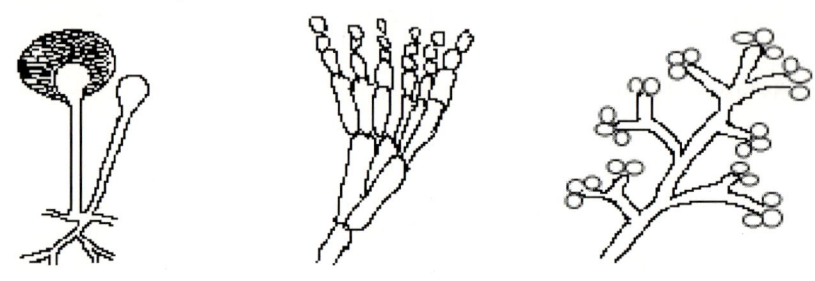

**Fig. 8.5.** Fungi

Most fungi are classified as saprophytes because they live on dead or dying material and obtain energy by breaking down organic matter in dead plants and animals. Fungal species are numerous during both the mesophilic and thermophilic phases of composting. Most fungi live in the outer layer of the compost when temperatures are high. Compost moulds are strict aerobes that grow both as unseen filaments and as grey or white fuzzy colonies on the compost surface.

## Protozoa

Protozoa are one-celled microscopic animals. They are found in water droplets in compost but play a relatively minor role in decomposition. Protozoa obtain their food from organic matter in the same way as bacteria, but they also act as secondary consumers ingesting bacteria and fungi.

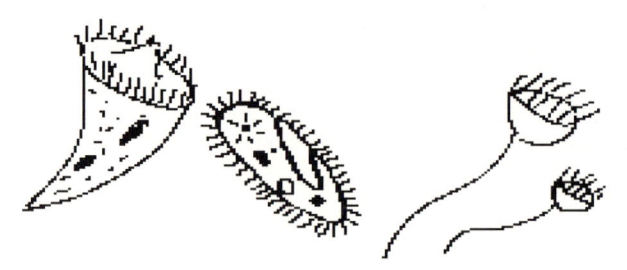

**Fig. 8.6.** Protozoa

## Rotifers

Rotifers are microscopic multicellular organisms also found in films of water in the compost. They feed on organic matter and also ingest bacteria and fungi.

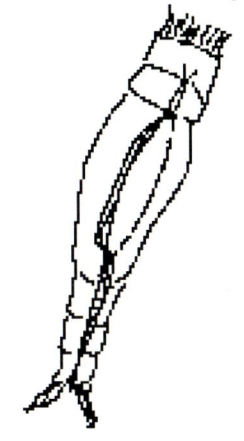

**Fig. 8.7.** Rotifers

Ref: http://www.cfe.cornell.edu/compost/microorg.html

### 8.2.4 Quality Criteria for Compost as a Product

#### Determination of Total Coliforms (TC) and Faecal Coliforms (FC)

The term total coliform organisms refers to a group of gram-negative aerobic and facultatively anaerobic, non-spore-forming rod-shaped bacteria which ferment lactose at 35°C. The term faecal coliform organisms refers to the thermo-tolerant forms of the total coliform group which ferment lactose at 44.5°C in 18-24 hours. Within this group, *Escherichia coli* and *Klebsiella pneumoniae* species are the interesting organisms. They accompany enteric pathogens like *Salmonella, Shigell* and *Vibrio cholerae*. High temperatures during the composting process are important to kill the pathogens. Freedom from pathogens is a quality requirement for the safe/hygienic use of compost. The TC bacteria colonies were grown on an Endo-type medium containing lactose and counted after 24 hrs as red colonies with a green-metallic sheen. The FC were grown on m-FC agar containing aniline blue.

#### pH

A pH value between 5.5 and 8.5 is optimal for compost micro-organisms. As bacteria and fungi digest organic matter, they release organic acids. In the early stages of composting, these acids often accumulate. The resulting drop in pH encourages the growth of fungi and the breakdown of lignin and cellulose. Usually, the organic acids are broken down further during the composting process.

### Salinity

The salinity was measured as a characteristic of conductivity. The content was calculated as potassium chloride. The optimum salinity for good compost is usually around 1-2% of the dry substance.

### Nutrients N, P and K:

Mature compost contains trace and essential elements, the most important of which are nitrogen (N), phosphorus (P) and potassium (K).These elements are available to soil and plants, depending on their initial concentrations in the raw compost material and on the degree of mineralisation.

## 8.3 Types of Toilets

### 8.3.1 Water Toilets

The most commonly used toilet type in Germany is the water-closet. As the name indicates, it is based on the principle of flushing the toilet bowl and diluting the faeces with water. This water is then centrally extracted and processed until it meets the quality criteria for drinking water. 6 to 7 litres of water are needed each time to flush and refill the toilet funnel in systems without a dosing mechanism. Even though moderate flushing can reduce the water requirement to about 1l, there is still a major imbalance between the volume of the faeces being flushed away and the amount of drinking water used to do so.

This toilet principle, which is commonly regarded as the most modern, is today characterised by centralised disposal via a rinsing sewage system with subsequent processing of the wastewater. This form of disposal became necessary as the water-closet grew in popularity and as manual disposal of the faeces in agriculture became economically unfeasible because of the vast increase in the amounts to be disposed of and because the water-soaked faeces-water mixture was unsuitable for composting into fertiliser.

One advantage of the water-closet is that users do not have to worry about the origin of material potentially added to the wastewater or its disposal. This job is taken care of, for a fee, by a water supply company, which assumes general responsibility. Consequently, precious drinking water is often wasted, and every year quite considerable amounts of highly toxic pollutants, such as varnish or aggressive cleaning agents, are discharged into the sewage system via toilet drainpipes.

The nutrient residues in the excrement, which were previously ingested via the users' food, are flushed away with precious drinking water. This means that the water circulation is "loaded" with nutrients, which are collected in the sewage system together with highly polluted industrial wastewater and can thus only be recovered by highly cost-intensive methods. As users generally have a fairly detached attitude towards both nutrients and drinking water, they rarely consider the consequences of their daily actions.

Apart from the above-described practical advantages of using water-closets, their current popularity can be attributed to their use being viewed as an "achievement of civilisation".

### 8.3.2 Waterless Toilets

There are currently four types of waterless toilets in use: 1) composting toilets, 2) incinerating toilets, 3) vault toilets and 4) pit toilets.

Composting Toilet: A system designed to store and compost (primarily by unsaturated, aerobic microbial digestion) human excrement (urine and faeces) into a stable soil-like material called "humus". Such systems are generally designed to accommodate faecal and urinary wastes (human excrement), toilet paper and small amounts of organic carbonaceous material added to assist their function. (Described in detail below).

Incinerating Toilet: A self-contained unit that reduces non-water-carried human excrement (urine and faeces) to ash and evaporates the liquid portion. Wastes are deposited directly into a combustion chamber and incinerated. The process is fuelled by LP or natural gas, fuel oil or electricity.

Vault Toilet: An on-site sewage system that incorporates 1) a structure enclosing a toilet above a water-tight (preventing liquid infiltration into the soil) storage chamber forhuman waste, 2) a sewage pumper/hauler and 3) the off-site treatment and disposal of the sewage generated. Portable chemical toilets are not included in this category.

Pit Toilets: An on-site sewage disposal unit consisting of a structure over a pit, not exceeding five feet in depth, in which human excrement (faeces and urine) is directly and permanently deposited in the ground. Owing to site and soil considerations, the application of pit toilets is very limited.

## 8.4 Composting Toilet Systems

**What is a Composting Toilet System?**
Composting toilet systems (sometimes called biological toilets, dry toilets or waterless toilets) contain and control the composting of moist organic solid matter, e.g. excrement, toilet paper, carbon additives and, in some cases, food wastes. Unlike a septic system, a composting toilet system relies on unsaturated conditions (material cannot be fully immersed in water), in which aerobic bacteria and fungi break down wastes, just as they do in a garden-waste composter. If properly dimensioned and operated, a composting toilet reduces waste to 10-30% of its original volume. The resulting end product is a stable soil-like material called "humus", which must by law be either buried or removed by a licensed seepage hauler in accordance with state and local regulations in the United States. In other countries, humus is used as a soil conditioner for edible crops.

The primary objective of the composting toilet system is to contain, immobilise or destroy organisms that cause human disease (pathogens), thereby reducing to acceptable levels the risk of human infection, without contaminating the immediate or distant environment or harming its inhabitants. This should be accomplished in a manner that:

- is consistent with good sanitation (minimising both human contact with unprocessed excrement and exposure to disease vectors such as flies)
- minimises odour and produces an inoffensive and reasonably dry end product that can be handled with minimum risk

### 8.4.1 Dimensioning Composting Toilets

There is no simple formula for determining the ideal size of a composting toilet system. Factors to be considered include the number of individuals using the system, the frequency and kind of use (e.g. residential or continuously used systems, day-use park systems, holiday-cottage or intermittently used systems) and the degree to which environmental factors are to be controlled (e.g. aeration, moisture content, temperature, carbon-nitrogen ratio and the presence of process controls) – factors with a significant impact on the speed and effectiveness of the composting process, and ultimately on the volume of the composting material in the composting chamber.

Under ideal composting conditions, the composting process can result in a significant volume reduction in a relatively short time. Under less ideal conditions, this would not be the case, however, and a larger storage capacity would be required. For instance, mouldering toilet systems (systems that support psychrophilic organisms, whose optimum temperature is above 5°C and below 20°C) are much larger in size than mesophilic composting systems (systems that support mesophilic organisms, whose optimum temperature is from 20°C to 44,5°C) to compensate for the reduced processing time (Enferadi 1980). A composter subjected to temperatures of 5°C or less will merely accumulate excrement, toilet paper and additives until the temperature rises. This is why composter manufacturers specify capacities for temperatures of 18,3°C (comfortable room temperature of an average human-occupied space) (Del Porto et al 1999).

Studies conducted in a community in northern Europe have shown that the average adult produces about 40.6 fluid ounces (1.2 litres) of urine and 20.3 fluid ounces (0.6 litres) of faeces daily. Performance-rating organisations like the National Sanitation Foundation base their calculations on "population equivalents" (p.e.), i.e. the average number of excrement events produced by an average adult person in a 24-hour period. For this standard, one p.e. is defined as 1.2 faecal events and four urine events per person per day. It is important to remember that the ratio of urine to faeces volume varies in different settings. In a day-use public facility, there will be a much higher ratio of urine to faeces (e.g. 10:1), but in a residential setting a ratio of between 3:1 and 4:1 is common (Del Porto et al 1999).

These factors should be considered when dimensioning a composting toilet system for specific needs.

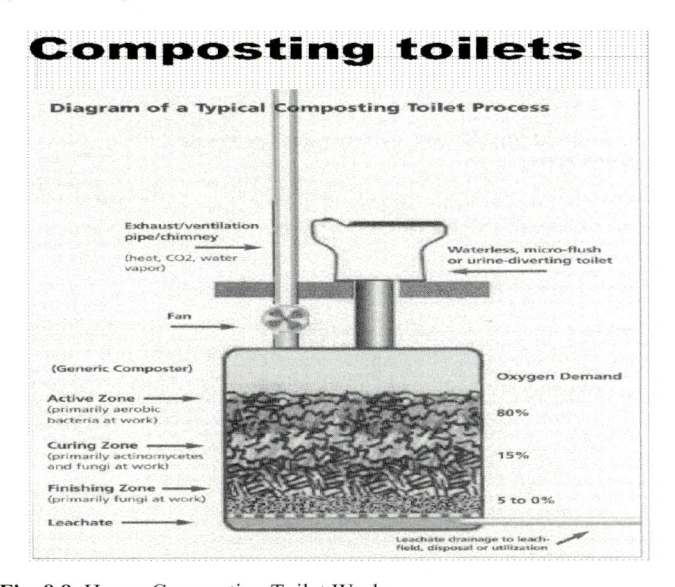

**Fig. 8.8.** How a Composting Toilet Works

**Dry Sanitation**
Dry sanitation is defined as the on-site disposal of human urine and faeces without the use of water as a carrier. This definition includes many of the most popular options for low-cost sanitation, including pit latrines, Ventilated Improved Pits, SanPlats, etc. There has always been interest in reusing human waste as a fertiliser, and much work has recently been done to develop composting and other processes enabling human waste to be reused.

## 8.4.2 Dry Sanitation with Reuse

### *Advantages of Dry Sanitation with Reuse*

Dry sanitation toilets with reuse are more susceptible to misuse than other sanitation systems. The advantages of well-functioning dry sanitation toilets are:

- Water requirements: no water is required to flush away the human waste, although basic cleaning of the toilet is necessary.
- Construction: when local materials are used, construction is simple and does not require skilled labour; when prefabricated, installation is quick and simple and maintenance is minimal; hard rock near the surface does not affect construction.
- Spread of disease: the faeces and urine are not accessible to animals; unlike conventional pit latrines, faecal material is isolated from the groundwater table by the walls of the composting or dehydration chamber; the end products contain only minimal concentrations of faecal pathogens; there are no smells; the system does not encourage fly breeding.
- Environmental contamination: raised chambers with concrete bases for storing the excreta prevents the contamination of soil, rivers and groundwater, especially where the water table is high.
- Environmental sustainability: the end product from the faeces pile can be used as a soil conditioner; the urine can be diluted and used as a source of nitrogen for plants.
- Community acceptance: if dry sanitation is introduced gradually and the donor spends sufficient time and effort on training and supporting users, dry sanitation is accepted by communities; children are less reluctant to use this system than a pit latrine.

### *The Disadvantages of Dry Sanitation Are*

**Usage**
- To use the system successfully, the operator must understand the basic principles of dehydration and decomposition.
- The system is more susceptible to misuse than other forms of sanitation.
- It can be difficult to keep the toilet basin above the chamber clean (which may encourage flies), since it is desirable to use only limited amounts of water for cleaning.
- Keeping the urine separator and pipe clean to avoid odours may be a problem. It is difficult to produce very smooth urine separators from cement mortar. This material is also slightly absorbent, leading to the retention of liquids and the risk of odours.

**Spread of disease**
- Incorrect use and maintenance can result in pathogens surviving in the end product from the faeces pile.
- If the storage time of the pile is too short, pathogens may still survive.
- Adding too little ash, soil or lime will affect the moisture content and pH of the pile and may provoke odours, fly breeding and the reduction of pathogen elimination.
- Seasonal lows in ambient temperature and increased humidity can result in reduced temperature and increased moisture in the storage vault and hence a reduction of pathogen elimination.
- Many designs recommend storage for 1 year. Where storage is for less than 1 year, sun-drying is sometimes suggested, which requires the handling of potentially infectious material.
- There is a small health risk from using the urine if it is contaminated with faeces or if the excreter has ascaris or schistosomiasis.

**Community acceptance**
- For urine diversion to work, men must sit to urinate, which is often unpopular in cultures where water is used for anal cleansing. Separate arrangements are required to keep the chamber dry.
- The constant addition of ash, earth or lime and the periodic mixing of the pile may be considered onerous tasks by the user.
- Users must acquire certain habits, which is only possible over a long period of time.
- Trial periods of several years in a community are necessary to demonstrate the advantages of dry sanitation.
- When defecating, women often preferred sitting (on the toilet) rather than squatting (open defecation).

The arguments presented above show why there is a quite intensive search for alternatives to the conventional water-closet. The aim of composting toilets is to utilise the nutrients and other valuable substances contained in faeces, rather than regarding them as waste to be disposed of. In composting toilets, the faeces are collected in separate containers – sealed to prevent infiltration of the groundwater – for conversion into a high-grade, soil-enhancing product that can be used in agriculture.

Apart from dealing with problems that may occur during the operation of composting toilets, such as unwanted odours, fly infestations or high production of leachate, it must be meticulously ensured that the material is rendered completely hygienic. Details of how traditional and currently used composting toilets are built and operated are given below.

### *Dehydration*

Toilets based on dehydration do not normally mix the faeces and urine. The urine is diverted and is either collected or flows into a soak-pit. The faeces are collected in one of two chambers below the toilet seat and dried by adding lime, ash or earth to the chamber after each defecation. Addition of these absorbents is also reported to reduce flies and eliminate bad odours. When the chamber is full, it is sealed and the other chamber is used. When the second chamber is full, the first chamber is opened. The contents of the first chamber are removed and used as soil conditioner, buried or composted (by home composting or at a local composting centre), depending on the recommendations of the toilet designer.

### *Decomposition (Composting)*

Toilets based on biological decomposition use bacteria, worms or other organisms to break down the faeces, producing compost. Many designs permit (or recommend) the addition of other organic matter such as vegetable scraps, straw, wood shavings or coconut husks. Temperature and airflow are carefully controlled in such designs to optimise the conditions for composting. It is important that airflow is sufficient to maintain aerobic conditions in the faeces pile. Urine is not normally diverted. The end product is a fine compost that can be used as soil conditioner. The additional liquid produced is either evaporated or allowed to flow into a soak-pit.

The Vietnamese double-bin toilet, the Mexican Dry Ecological Toilet, the Guatemalan DAFF, the South African urine-diversion dry toilet and the Ethiopian ECOSAN toilet are some examples of toilets working on the dehydration principle. Mexico is something of a "centre" or "focus" of work on dry sanitation with subsequent reuse. Mexican non-governmental organisations (NGOs) (e.g. Centro de Innovación Tecnológica, Espacio de Salud AC and Grupo de Technología Alternativa S.C) and government bodies (e.g. the National Water Commission and sections of both local and federal governments) have actively promoted dry sanitation with reuse, with differing degrees of success.

Dehydrating toilets are either built by the community using local materials or are prefabricated. Toilets based on the process of decomposition are usually prefabricated. Note that both systems involve construction of a self-contained chamber that isolates the faecal pile from the environment. Such a design offers greater protection from groundwater contamination than a conventional unlined pit-latrine, though presumably at greater cost.

### *Pathogen Reduction*

The isolation, reduction and/or elimination of pathogens (disease-causing organisms) is the primary objective of any sanitation system. There are, unfortunately, only a limited number of documented studies of pathogen die-off in dehydrating toilets, but these indicate consistent health aspects of dry sanitation with waste reuse findings. The two most crucial factors appear to be pH and residence time. Assuming appropriate addition of absorbent material, the reported storage time required before reuse of the faeces pile varies in the literature from 3-12 months. The Engelberg guidelines (< 1 nematode egg per kilogram wet weight and < 1,000 faecal coliforms/litre) are appropriate limits for compost applied to topsoil. Studies indicate that the content of both the Guatemalan DAFF and the South African urine-separation toilet fails to meet these standards. Other studies, however, show that the Vietnamese double-bin toilet can meet these criteria if the content is left undisturbed for 6 months. The only microbiological study of a composting toilet (the SIRDO model) indicates that its content would conform to the Engelberg guidelines. Unfortunately, most of the studies neglect to assess the health status of the families using the toilet. The absence of worm eggs in the faeces of families uninfected by worms is hardly surprising and says nothing about the efficiency of the process in destroying worm eggs.

### 8.4.3 Dehydration Toilets

#### Dry Ecological Toilet (Mexico)

A modified version of the Vietnamese double-chamber dry toilet is promoted by the Mexican NGO, Espacio de Salud AC (ESAC). It was designed by Cesar Anorve in Mexico (Sawyer 1998). The squatting slab is replaced by two toilet risers. A conventional-looking urine-separating toilet seat is placed on the toilet riser and the toilet is painted in attractive colours. Both these factors have increased acceptance of the toilet in rural areas. The urine-separating toilet seat has been modified as a result of feedback from users, and a domestic urinal has also been designed. The design has been adapted for use within homes as well. The total construction cost is about US$ 150, including labour and materials. ESAC reports that the dry ecological toilet has been successfully built in communities with a variety of climates, from humid and temperate to dry and tropical.

#### Vietnamese Dry Toilet

The Vietnamese dry toilet has two chambers built above the ground, made of concrete, stone or unbaked brick. There is a squatting slab with two holes above the chambers. The toilet cubicle over the two chambers is reached by two or three steps. Before a chamber is used, its floor is covered with a layer of ash, soil or lime. The faeces drop into one of the chambers; the urine is piped into a vessel or soak-pit. Paper used for anal cleaning is placed in a metal bucket and burned. After each use, additional ash, soil or lime is added to the chamber and the hole is covered with a wooden lid. When the chamber is nearly full, it is topped up with soil and the drop hole is sealed with mud. A process of anaerobic dehydration is then likely to begin. The second chamber is now used by the family in the same way as the first. When the second chamber is full, the first chamber is opened and emptied. This usually happens after a period of at least 2 months. The resulting humus is used as fertiliser. During the anaerobic dehydration process, the temperature in the chamber is usually 2-6°C higher than the ambient temperature, but has been reported to reach 50°C in summer when the ambient temperature is 28-32°C (McMichael 1978). Reports suggest that 85% of helminth eggs are destroyed after 7-8 weeks. Modifications of this design have been adopted in several countries because of its simple and low-cost construction.

## DAFF (Guatemala)

The DAFF is a dry-compost family latrine developed in Guatemala at CEMAT, (Centro Mesoameric. de Estudios sobre Technología Apropriada) (Chavez 1987). It is a modified version of the Vietnamese double-chamber dry toilet. It consists of two concrete-lined chambers, each with a hole at the top, on which the toilet seats are placed (rather than a squatting slab with two holes). The excreta and urine are collected separately. The excreta are deposited in one of two chambers. Ash, chalk or earth is added after each defecation to keep the excreta dry. The urine flows down a pipe into a pot. At the back of each chamber is a door through which the compost can be removed after 10-12 months. The compost is stored in sacks until it is used on the fields. The urine, a source of nitrogen, is diluted and used to water plants. This latrine can be built within the community by local unskilled labour using basic building materials. The total cost including labour is US$ 40-100, depending on the materials used for the superstructure. It is being promoted by the National Sanitation Programme in Guatemala. Although it proved possible to transfer the construction techniques, follow-up service to ensure correct usage and maintenance of latrines was often inadequate, resulting in low levels of usage (Strauss a. Blumenthal 1990). The DAFF has also been used in El Salvador, in such high-density urban squatter areas as Hermosa Provincia, in the centre of San Salvador The fact that the DAFF units were still functioning well after 6 years, with no odours or flies, was largely attributed to the high level of community participation. (Winblad 1996).

## Urine-Diversion Dry Toilet (South Africa)

The urine-diversion "dry-box" toilet is another modified version of the Vietnamese double-chamber dry toilet. The Council for Scientific and Industrial Research (CSIR) in South Africa saw urine diversion as a possible solution to many of the problems encountered with the VIP toilets (Holden 1999). A pilot project was conducted in 1997 with a blow-moulded plastic toilet seat. The chambers and the superstructure were constructed using locally available materials. The toilet is raised above the ground. There is a chamber below the seat with two containers. The faeces are collected in one container; when this container is full, it is sealed and the other container is used. The urine flows into a soak-pit (communities were not keen on the idea of collecting the urine and using it as fertiliser), with the option of converting to collection, should people be willing to try this later (Austin a. van Vuuren 1999). After each defecation, ash is sprinkled in the chamber. Most of the families participating in the pilot project were enthusiastic about the new technology, and the toilets caused no problems with odours or flies, despite the fact that there was no vent pipe (Austin a. van Vuuren 1999). It was concluded that, if properly implemented, urine-diversion sanitation works well (Austin a. van Vuuren 1999). Further pilot studies are now under way using the urine-separating toilet seat designed by Cesar Anorve in Mexico. This uses a fibre-glass mould and a cement-mortar mixture, which substantially reduces the cost of each toilet seat, from US$ 42 to US$ 10 (Holden 1999).

### ECOSAN Toilet (Ethiopia)

The ECOSAN toilet was developed by the Society for Urban Development in East Africa (SUDEA). It is a urine-diverting or non-mixing toilet that enables separate handling of faeces and urine (Terrefe a. Edström 1999a). The urine is collected in a special container. The faeces are also collected in special container and mixed with ashes, soil, leaves, grass or sawdust. The urine and the composted faeces can be used as fertiliser. It is recommended that neither urine nor composted faeces be spread on top of the soil, but be used instead under the topsoil. The toilet is constructed using locally available and appropriate materials. The total cost is about US$ 100 per toilet. SUDEA stresses the importance of a systematic approach to toilet installation and use, from initial contacts, construction and maintenance to recycling for urban or household agriculture.(Terrefe a. Edström 1999b).

### Single-Chamber Dehydrating Toilet (Yemen)

A single-chamber dehydrating toilet with urine diversion is used in the city of Sanaa, Yemen. The toilet is placed in a bathroom several floors above street level. The faeces drop down a vertical shaft inside the building, while the urine and water from anal cleaning are discharged through a vertical pipe on the outside of the building. Sanaa has a hot, dry climate. The faeces quickly dry out and are then collected and used as fuel. Most of the urine evaporates on the way down, and the rest drains into a soak-pit. (Winblad 1985).

### Tecpan Solar-Heated Toilet Prototype (El Salvador)

This design was developed in El Salvador (Gough 1997). It is basically the same as the DAFF, but the toilet has a single chamber and solar heating to minimise wetness. As with the DAFF design, after each use wood ash, soil or lime is added. The urine is piped away to a soak-pit near the toilet. Every 1-2 weeks, the pile is pushed to the back of the vault with a rake. Then, every 2 or 3 months, the dry and odour-free humus at the rear of the vault is removed and stored in a sack until used in the garden. The toilet costs US$ 164 including the chamber, the superstructure and the solar heater.

### Two-Chamber Solar-Heated Composting Toilet (Ecuador)

A two-chamber solar-heated composting toilet has been built in the Andes, in Ecuador. At this altitude, urine diversion is not necessary because natural evaporation eliminates any excess humidity. After defecation, a handful of sawdust and/or ash is added. Each chamber has a diagonally sloping lid, made of a wooden frame covered with thin galvanised iron, painted black. Each chamber has a ventilation pipe and the chamber lids have vents; both are covered with metal mesh. The toilet chambers and the superstructure are built from locally made sun-dried bricks, while the toilet seat, the lid of the toilet hole, the ventilation pipe and door are made of wood and are prefabricated (Esrey et al 1998). It is unclear, though, whether this design functions as a dehydration or composting toilet (Dudley 1993).

### *Ecological Sanitary Unit (Mexico)*

The Ecological Sanitary Unit is a prefabricated toilet constructed from high-impact recycled polyethylene. The design is based on the double-chamber Vietnamese dry toilet. It is promoted by the GTA. There are two versions available, the CA2 and the CA3. The CA2 consists of a pair of conical chambers made of recycled polyethylene. The CA3 also consists of two conical chambers, but each chamber comprises two conical pieces of recycled polyethylene which are screwed together on site to form the chamber. This version is designed for areas where donkeys are the only available means of transportation. In both the CA2 and the CA3, each chamber is covered with a polyethylene lid and a toilet seat which separates the urine and faecal material.

Ideally, the design requires the addition of 0.5kg per user per day of ash, lime and soil (in equal proportions) to the main chamber containing the faecal material. A mixture of three parts soil to one part lime is also reported to be effective. The urine drains away to a soak-pit of gravel or tezontle (a Mexican porous stone). Toilet paper may not be added to the chamber. The lime raises the pH value of the faecal material, which is reported to assist in the removal of pathogens. Owing to the high lime content, the humus produced is highly alkaline, making it unsuitable for use with many crops. If the ecological sanitary unit is correctly operated, the humus has < 100 total coliforms per 100ml (personal communication, Josefina Mena Abraham).

### 8.4.4 Decomposition Toilets

### *SIRDO (Mexico)*

The SIRDO (the Spanish acronym for Integrated System for Recycling Organic Waste) is a prefabricated solar-heated toilet developed in Mexico more than 15 years ago by the Alternative Technology Group (GTA). The SIRDO is promoted as a radical departure from the traditional pit-latrine, transforming faecal material into a biofertiliser that is free from pathogens There is initial anaerobic decomposition of the faecal sludge with rapid sedimentation (24-48 hours). After 48 hours, the sludge passes into the biological chamber, mixing with the organic waste from the house.

**Fig. 8.9.** Dry SIRDO

**Fig. 8.10.** The Dry SIRDO's Double Chamber for Compost and Organic Waste

The moisture content of the compost must be kept between 40 and 60%; excess water is removed by evaporation using a solar collector to generate heat. The solar collector is a black metal sheet in the roof of the biological chambers, inclined at 20° and facing south. Ducts in the roof ensure sufficient air flow. Temperatures of up to 70°C can be reached in the SIRDO.

The process of decomposition is carried out over a period of 6 months to guarantee pathogen elimination. The humus produced is reported to be high-quality compost, a fact soon recognised by owners of SIRDOs. As well as acting as a composter, the SIRDO can also treat "grey water", by passing it through a slow sand-filter, after which it can be used to irrigate plants. There are now six designs available: combinations of dehydrating or composting unit, with or without urine diversion and with one or two chambers. All units without urine separation had solar collectors to eliminate humidity. The basic design is based on a combination of the Vietnamese technology and the Mexican chinampa.

The polyethylene inclined chamber is divided into two vaults, with a baffle above the dividing wall. There is a window at the rear to collect solar heat. The design is based on a process of biological decomposition and is intended for use by a family. The SIRDO SECO "La Oruga" health aspects of dry sanitation with waste reuse (the caterpillar) is a modified version of the 6M.

**Fig. 8.11.** Prefabricated Dry SIRDO - Type 6M

The design is more efficient and cheaper and suitable for use by children in nursery, primary and secondary schools. Cost-benefit analyses indicate that, although some of the SIRDO models are more expensive than the basic dehydrating double-chamber dry toilet, the fertiliser value more than pays for the cost of system after several years (internal document, GTASC).

## Carousel Toilet (Pacific Islands)

A carousel toilet has a four-chamber fibre-glass waste-collection tank for batch/charging. This tank rotates on a pivot inside an outer chamber. An electric fan ensures adequate ventilation and aids evaporation. One chamber is used until full, and then the tank is rotated and the next chamber is filled. With recommended usage, the four chambers will be filled in about 2 years, by which time the humus from the first chamber can be removed and applied to the garden soil (del Porto 1996). To begin with, a bed of small stone gravel (3/4"), coconut husks and a scoop of garden soil is placed at the bottom of the chamber to allow drainage and provide soil bacteria to digest the waste. According to del Porto (1996), the minimum routine maintenance of adding a small handful of organic matter after each use (e.g. leaves or coconut husks) ensures aerobic digestion. The climate in the Pacific Islands enhanced the performance of the composting process. During cooler and wetter periods, humidity increased and an auxiliary heater was used until the excess liquid had evaporated (del Porto 1996).

### 8.4.5 Types of Composting Toilet Systems

## Single-Chamber Systems

**Algeria**
With this traditional system from the Algerian city of Ouargla, the lavatory is on the first floor. The solid excrement is conveyed through a hole in the ground to a collecting room on the ground floor, one wall of which borders the street. The liquid excrement is conveyed to the collecting room via a gutter. The floor of the collecting room is covered with palm leaves to support the decomposition process. The decomposing matter is removed and disposed of via a closable aperture in the wall. The compost is used as fertiliser in the palm plantations around the city. Since, with this system, compost and fresh excrement are close together, some hygienic problems may occur.

**India**
The dry highland area where this system is used belongs to India but has strong connections with Tibet. As with the systems in Algeria and Yemen, interior toilets are frequently situated in the upper storeys. The faeces are pushed, together with some earth, through a hole in the floor into the collecting room below. Toilet tissue is not used; the only additive sometimes used are hearth ashes. The decomposing matter is only removed in spring and autumn, via an outer door, and is then

used to enhance the thin humus layer on the fields. The problem of compost coming into contact with fresh faeces could be solved by having two holes in the ceiling, one for the summer and one for the winter. The system works well, however, and is helping to enhance the poor soil quality.

**Brazil**

The dry toilet used here, called a *bason*, was developed by Johan van Lengen, leading representative of the TIBA (a Portuguese acronym standing for Institute for Intuitive Technology and Bio-Architecture). In 1995, half a dozen toilets of this type were in use on the premises of this non-government-funded institute in the federal district of Rio de Janeiro. As the institute frequently offers coaching courses for groups, it has four of these systems installed in the main building to provide sufficient sanitary facilities. The entire decomposition process takes place in a inclined chamber. Only the upper part of the construction with the toilet seat is on the ground floor of the building, the longer bottom part of the container extending to the outside, where it can be warmed by the sun. The collecting container is divided into different compartments by two partition walls. There is no provision for urine or liquid separation. The mixture of excrement, toilet tissue and dried, crumbled leaves is regularly mixed using a manual crank made of bent structured steel. The incline of the container causes the decomposing matter to slide towards the extraction flap, the wooden lid of which has an air duct covered by wire mesh. From this air duct, the air stream flows over the decomposing heap to the upper end next to the toilet seat, where the ventilation is installed. The dark-painted pipes run along the outside wall of the building up to the roof.

The chamber, which consists of an upper and a lower part, and the partition walls are made of simple, reusable reinforced concrete elements. After the concrete has been shuttered and has set, the individual elements are put together and can used. Handling of the system is simple, and there are no problems with odours or excessive water in the decomposing matter, which is certainly due in part to the warm climate. Fly infestations are prevented by covering all apertures with small-mesh wire. Also, kitchen waste, which particularly attracts flies and may contain fly eggs, is not composted here. Because of the high temperatures – due in part to the solar effect – and the long dwelling time (approx.2 years) in the system, the compost can be assumed to be hygienically safe. Once the toilet is installed, no further financial investments are necessary.

Like the other TIBA projects, which can all be realised on a self-help, low-budget basis and which improve the living conditions of the poor rural population, the *bason* serves to improve the very poor sanitary conditions of those living in the *favelas* or in rural areas. It also prevents the pollution of rivers by the discharge of untreated wastewater from unofficial rinsing sewage systems or from overflowing three-chamber pits that are never emptied. At the institute, demonstrations of the construction and utilisation of the system are available to visiting groups or individual at any time. Also, construction workshops are regularly held with the active participation of the local population and of students from all over the country.

## Clamps or Multi-Chamber Systems

### China

In China, human excrement has traditionally been used for centuries, along with animal faeces, as fertiliser in agriculture. In a simple toilet system, the excrement is collected in a bucket or a shallow pit lined with stone slabs. This pit is designed such that the faeces and urine are separated by changing the incline. The faeces pits are emptied daily, their content being transported to a compost collection site. There, the excrement is fermented either in airtight tanks or in ventilated heaps.

Fermentation is sometimes achieved by high-temperature decomposition in specially designed clamps. For this purpose, the excrement is mixed in a 1:1 ratio with animal excrement, earth and street dust and is then stacked in layers. Optimal water content and reasonable ventilation are ensured by a specially designed channel. Under the local climatic conditions, the organic substances are mineralised in the summer (50-60°C) in 20 days, and in the winter in 60 days. The separately collected urine is often mixed with five parts of water and sprinkled on to vegetable plots. This highly labour-intensive, open method has a very high risk factor because of the direct contact with fresh faeces. Also, flies have unimpeded access to the excrement, which means a high risk of infection.

### Vietnam

This two-chamber compost toilet consists of two brick pits (each with a volume of approx. 0.32 $m^3$) which are both built above ground level and are used alternately. The urine is discharged via a gutter set into the ground and collected separately. The pit is fed only with faeces, ashes (to prevent the occurrence of insects), earth and toilet tissue. With an average of between 5 and 10 users, the pit has a storage capacity of about 2 months. During the next two-month period, while the second chamber is being used, the first chamber is filled up with dried, pulverised earth and all the apertures are filled with clay. Composting of the faeces takes place anaerobically, which means there is no self-heating comparable to that occurring during decomposition. The temperatures in the container are usually slightly above outside temperatures (an average air temperature of 30°C in the summer).

According to information from the local health authorities, after a dwelling time of approx. 1.5 months in the chamber the pathogenic germs have been killed and the organic substances have been mineralised. Practical analyses, however, show that this period is often not sufficient and that better results can be achieved with longer dwelling times. As there is no direct contact with the fresh faeces, the health hazard is lower than with the Chinese methods. The composted excrement is odourless at the time of extraction. Compared with the use of fresh faeces as manure, which is fairly common in Vietnam, using composted faeces increases the yield by 10-25%.

### Guatemala

A type of seat toilet based on the Vietnamese model is being used in Guatemala. The main difference from the model is the movable toilet seat, which is placed above the hole in collecting container that is currently being fed. The seat is equipped with an integrated device for urine separation. The brick chambers are not ventilated. The mixture of faeces, ashes (or earth) and toilet tissue in the chamber currently in use is stirred once a week with a stick. After an average of 10 months, the maximum filling level is reached. The container is then closed and the toilet seat placed over the neighbouring chamber. When the second chamber is also full, the first is emptied and then used again. The assessment of this toilet is similar to that of the Vietnamese model.

## India

Whereas in China and many other countries the human excrement collected in composting toilets is used as fertiliser, this is not the case in India, for cultural reasons. Two different models are in parallel use; in both cases the collecting containers are designed for a dwelling time of 6 months.

The *Gopuri double-chamber toilet* largely resembles the Vietnamese model, but with an additional vertical ventilation pipe. The movable toilet lid covers only one half of the collecting container; the other is covered by a metal plate. When the other side is to be fed, the toilet hut is moved. The odours produced escape through the ventilation pipe. No further constructive measures are needed to ventilate the compost heap. The heap is simply covered with dry earth, ashes, rice husks, dried and crumpled leaves, and straw. By contrast, the collecting containers of the *Sopa Danas toilet* are partitioned. They are not located immediately below the hut but are brick pits at the rear of the toilet. At ground level, they are covered with metal lids. The faeces slide down into the containers through an inclined pipe, which is then automatically closed with an lid, also preventing flies from entering the container.

Both types of toilet prevent the uncontrolled discharge of faeces and urine into the environment. It is not necessary to handle the fresh faeces. Flies breeding in the collecting containers created problems in the *Gopuri toilet*. By contrast, the decomposing heaps of the *Sopa Danas toilet* are warmed by the sun, inhibiting the development of fly eggs.

## Yemen

The narrow houses in the old parts of Sanaa are up to 9 storeys high. The bathrooms border on a vertical shaft leading into a ground-level collecting container. The produced excrement is often collected and dried in the sun to obtain fuel for heating the water in public baths. As these public baths are traditionally used by the inhabitants once a week, the people benefit directly from this recycling process. The hot, arid climate supports this system, which is optimally adapted to local conditions.

Urine and the washing water used for anal cleaning – toilet tissue is not used – are separated from the faeces. Through a hole in the wall, they flow outside through a gutter, where they evaporate on the hot façade. Environmental pollution is thus avoided in two ways: first, the water consumption is very low, and second, using the faeces as fuel saves firewood. The ash produced during heating of the public baths is used as garden fertiliser.

**Mexico**

This double-chamber toilet, which was developed for a low-cost settlement project, also features a partitioned brick collecting container. When regularly used by 6-8 persons, the containers have to be emptied once a year at the most. They are about 2 metres long, but only 75 centimetres wide. The cone formed below the toilet seat is regularly shovelled by hand to the end of the chamber. When the first chamber is full, a flap in the downpipe, which is manually adjustable, directs the excrement into the neighbouring chamber. There is, then, no need to move the toilet seat.

The collecting containers should ideally face south. Their metal lids are painted black and function as simple solar collectors. This increases evaporation of the trickling water and raises the temperature of the decomposing heap. Odours from the containers escape via a vertical ventilation pipe, the upper end of which is covered with wire mesh. This model is more expensive than the types discussed above, but it works very efficiently and has good handling characteristics. By designing the collecting container as a hot-air solar collector, it is possible to achieve such high temperatures that pathogenic germs should be exterminated after the usual dwelling time of the material (more than 1 year).

## 8.5 SIRDO

The **Dry SIRDO** is an alternative to the water-closet (WC). It eliminates the need to pipe sewage and hence the use of water to transport human excreta to a treatment plant. It prevents the contamination of rainwater, such as occurs with improvised latrines. It is often installed together with a small grey-water biological filter enabling the recycling of such liquid waste for irrigation. GTA has developed several dry SIRDO models: a double-chamber model to decompose human, kitchen and garden waste, and a single-chamber model to decompose human waste only. It uses a number of local materials (from adobe to bricks). The prefabricated and solar-heated composting units are made of fibreglass and polyethylene.

The **Wet SIRDO** is based on two techniques that were previously considered incompatible: a) aerobic decomposition producing high-quality biofertiliser in the form of dry soil, which traditionally requires the use of dry toilets; b) the use of conventional WCs with water flushing related to anaerobic processes. Both prove unsuitable for large sections of the urban (or semi-urban) population: the use of dry toilets often fails to meet the expectations of urban residents and the use of WCs presupposes a sewage system, usually conceived as a network connected to a large expensive treatment plant, which is beyond the means of the great majority of the Latin American population. Even where communities have sewage systems, these invariably discharge black water into rivers, lakes etc., contaminating water supplies. In many cases, water scarcity prevents the use of wet systems to transport human waste to such treatment plants.

## Sirdo Seco™ Solar-Heated Composting Toilet in Mexico

A prototype double-vault, solar-heated composting toilet was tested in Tanzania in the mid1970s. The idea was further developed in Mexico, where a prefabricated fibreglass unit has been in production for more than 15 years. Like the Vietnamese toilet, the Mexican type has a receptacle divided into two vaults. Above the dividing wall, there is a baffle (see figure below). The baffle directs the excreta into one vault. When the vault is full, the person in charge of the toilet turns a handle which causes the baffle to direct the excreta into the other vault. A ventilation pipe, which extends from the receptacle to above the roof, allows odours to escape. A screen on the top of the pipe prevents flies from entering. The two vaults have lids made of aluminium sheets painted black. The lids face the sun to collect as much solar heat as possible. This increases the evaporation from the vaults and raises the temperature on the surface of the compost pile. Each vault has a volume of 1.2 cubic metres. When the pile reaches the baffle, the person in charge of the toilet shifts the pile to the lower end of the receptacle. This means that the toilet only needs to be emptied once a year at the most. When properly managed, this toilet has a high capacity and works very well. The baffle makes it easy to change from one vault to the other.

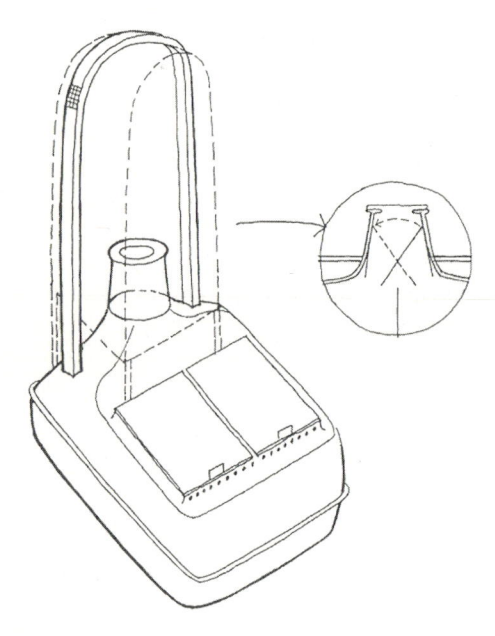

**Fig. 8.12.** The Sirdo Seco™ double-vault, solar-heated composting toilet in Mexico. The whole toilet, including the superstructure, is made of fibreglass (Esrey et al 1998)

In 1994, the cost of a prefabricated fibreglass toilet substructure was equivalent to US$ 445 (Mexican pesos 1,488). The cost of a prefabricated superstructure was US$ 109 (Mexican Pesos 360).The Sirdo Seco™ has been used with good results in Mexico for over 15 years. A particular advantage of this lightweight, prefabricated model is its mobility. People living in squatter settlements are often evicted at short notice. If this happens, they can arrange to have the toilet emptied and take it with them like a piece of furniture.

**Grupo de Technología Alternativa SC (GTASC)**
GTA is a 'for-profit' organisation that aims to develop alternative technologies for recycling domestic liquid and solid waste. GTA designed and now promotes the decomposing toilet system SIRDO (Integrated System for Recycling Organic Waste). Unlike many other dry sanitation technologies, GTA has worked not only in rural communities, but also in semi-urban and urban areas. Since GTA's founding in 1978, numerous sanitation projects have been undertaken and hundreds of SIRDOs have been installed in Mexico.

On the north-east side of the metropolitan area of Mexico City, near the Guadalupe Lake, over 70% of households dump their domestic waste in ravines and rivers. The lake was covered by a thick layer of irises. In 1994, GTA was commissioned to address this problem. It was established that most of the wastewater discharges, which were feeding the irises in the lake, came from an area in which most households had no sewerage facilities. In these areas, residents defecated in the open or had improvised latrines, which meant that the human waste went directly into the ravines and rivers and then entered the lake. GTA (with the financial support of SIDA) installed 24 dry toilet units made from local materials and 2 prefabricated fibreglass SIRDO SECO units to transform the human waste into biofertiliser (BF). All units without urine separation had solar collectors to control the moisture content of the compost.

A year later, 16 of the 24 dry toilets were functioning adequately, 6 families had chosen to be connected to a pipe discharging into the ravine, and 2 families had moved away. The two prefabricated units were in use, and the families had produced and sold biofertiliser. The urine separation, solar-heat collection and the (chemical or biological) process employed were assessed in terms of their effects on pathogen elimination. In 1996, with the help of private donations, further SIRDO SECO prefabricated units were installed. All had a solar collector, did not separate the urine and had a layer of earth containing the SIRDO innoculant.

Two years later, all the units were still in use. As well as promoting the SIRDO, GTA is currently installing 1,000 double-vault latrines per month in remote communities as part of the Mexican government programme "Fondo para la Paz". Once GTA has completed the initial installation and promotion of the SIRDO, follow-up service is provided by local NGOs (Non-Governmental-Organisations), with whom GTA has developed specific methodologies. GTA is currently working in Ciudad Juarez, in the north of Mexico, together with researchers from the Centre for Environmental Resource Management at the University of Texas in El Paso and the National Wildlife Federation. Installation of the prefabricated

SIRDO takes just 1½ hours; the rest of the time is spent explaining how to use the toilet and motivating the family.

## 8.5.1 Pathogens Elimination

### Dry Ecological Toilet (Mexico)

Preliminary studies of pathogen elimination in dry sanitation toilets in Mexico have been carried out by the Centre for Biotechnology Research (CEIB) at the University of Morelos. A wide variety of bacteria and parasites including Salmonella and *Ascaris lumbricoides* eggs were found in the stored faeces pile of dry sanitation toilets in San Juan Amecac, Puebla State (Ortiz 2000).

Another study at CEIB, using *Ascaris suum* eggs as a substitute for *A.lumbricoides* eggs, measured the viability of *A.suum* eggs in the dry sanitation toilets in San Juan Amecac, Puebla State. Twelve small "tea bags" made of polyamide cloth (of 2mm pore size) containing *A.suum* eggs were implanted in the middle of the full chamber of 10 dry sanitation toilets when the chamber was sealed. Each month, a bag was removed and the viability of the *A.suum* eggs determined. The pH, moisture and temperature were also determined at each sampling. Initially, 83% of eggs were viable. Results to date show that after five months between 12 and 57% of eggs were viable. Unfortunately, only five of the toilets were monitored for the full 12-month period. However the authors concluded that pH appeared to be the most important factor affecting the viability of the eggs: the higher the pH, the lower the percentage viability (Olvera-Velona a. Ortiz-Hernandez 2000).

Results of Dry Decomposition process (Mena Abraham 2000) Three sites were selected at which dry SIRDOS were already installed and in use for 3, 6 and 9 months, respectively. Dry compost from one of the samples was checked for its microbial population. It was found that there was a reduction in the number of *Faecal Coli* from several thousand to a few thousand, from the filling (wet) phase right through to the drying phase.

**Table 8.2.** Microbiological Quality of Compost (Dry Toilet) (Bockemühl a. Berger 1984)

| Total microbial population | Coliform Titer value | E.Coli Titer | Clostridium perfringens (per g) | Fungi | Psedumonas aeruginosa (per g) | Staphylococcus aureus (per g) | Enterococcus (per g) | Salmonella (per g) |
|---|---|---|---|---|---|---|---|---|
| 1.905 Million | 10 -4 | 0 | 2,500 | 1,050 | < 100 25g : + | < 100 | < 100 | Not found |
| 1.3 Million | 10 -3 | 0 | 7,000 | < 100 | 100 | < 100 | < 100 | Not found |
| 158,000 | 10 -2 | 0 | 400 | <100 | < 100 25 g : + | < 100 | < 100 | Not found |
| 745,000 | 10 -2 | 0 | < 100 | 1,700 | Not found | < 100 | < 100 | Not found |

Table 8.3. Microbiological Quality of Garden Soil (Bockemühl a. Berger 1984)

| Total microbial population | Coliform Titer value | E.Coli Titer | Clostridium perfringens (per g) | Fungi | Psedumonas aeruginosa (per g) | Staphylococcus aureus (per g) | Enterococcus (per g) | Salmonella (per g) |
|---|---|---|---|---|---|---|---|---|
| 4 Million | 10 -3 | 0 | < 100 | 200 | < 100 | < 100 | Not assayed | Not found |
| 2.2 Million | 10 -2 | 10 -1 | 600 | 1,100 | < 100 | < 100 | Not assayed | Not found |
| 2.6 Million | 10 -3 | 10 -1 | 400 | 300 | < 100 | < 100 | Not assayed | Not found |
| 3.2 Million | 10 -3 | 10 -1 | 300 | 400 | < 00 | < 100 | Not assayed | Not found |

The samples from the dry toilets showed a more or less absence of certain disease-causing micro-organisms, e.g. *E. Coli*, compared with the garden soil.

Aerobic static-pile composting of piggery solids was investigated on a 5m$^3$ scale. Sawdust was used as the bulking agent to provide additional carbon and to increase the porosity of the substrate. The temperature profiles indicated that the solid waste could be completely pasteurised. The nutrient analysis showed that 79% of initial nitrogen was conserved in the compost, while there was no significant change in the phosphorus concentration. The microbiological assays revealed that there was a decline in Enterococci counts by four orders of magnitude, while the MPN counts decreased by three orders of magnitude, suggesting that spore-forming bacteria may survive the composting process.

Table 8.4. Nutrient Analysis of Substrate and Compost

| Sample | TN mg/g | TP mg/g |
|---|---|---|
| Fresh Manure | 16.52 | 7.7 |
| Compost | 6.53 | 3.9 |

## *Microbial Counts*

The microbial counts of *Streptococci spec.* and MPN are presented in the figure below (Fig. 8.12.). Streptococci decreased by four orders of magnitude, the MPN by three orders of magnitude. The high temperatures of the pile for prolonged periods are expected to decrease the bacterial counts to levels lower than those observed.

From all four results, it was clearly evident that the dry toilets offer pathogen reduction (decrease in the microbial counts), while the nutrients are to some extent conserved. Some tests with dry SIRDOs were also attempted by *ifu* junior scientists, but the results could not be replicated and are thus not reported here. Rainy weather (low temperatures) at Suderburg prevented the desired thermophilic temperatures from being achieved; also, there was not enough raw material (excreta).

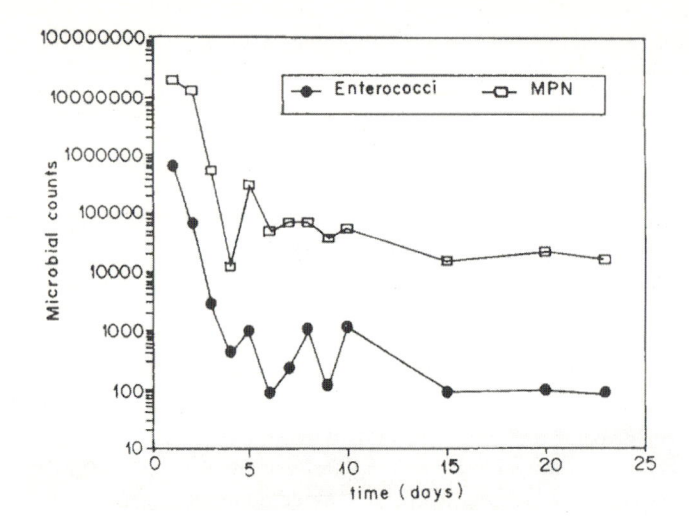

**Fig. 8.13.** Most Probable Number and Streptococci in the Composted Pig Waste (Rao Bhamidimarri a. Pandey 1996)

### 8.5.2 Social Evaluation

***USA and Mexico***

A research project on the effectiveness and safety of composting toilets was designed to determine the pathogenic content of the end product generated in the SIRDO and to report findings and make recommendations to the U.S. and Mexican authorities. The purpose of the study was to determine the United States Environmental Protection Agency (USEPA) classification rating of the composted material from a random sample of SIRDOs operating in Ciudad Juárez.

As defined in the standards laid down the "Disposal of Sewage Sludge Regulations" (Subsection D of 40 CFR 503), Class A compost, which contains safe and acceptable levels of pathogens, is considered safe for application to food and non-food plants. The end product can be considered a safe soil amendment (fertiliser) with a minimal health impact if it contains less than 1,000 MPN of faecal coliform per gram and less than one A. lumbricoides egg per 4 gram. The USEPA requirements for Class B compost state that the end product will be considered a safe soil amendment for ornamental plants if it contains less than 2,000,000 MPN of faecal coliforms per gram.

Project researchers collected samples from 90 households with installed SIRDOs randomly selected and stratified by community. Sample collection last three and six months. During the installation phase of the 300 SIRDOs, health promoters distributed a questionnaire to collect information on the number of users and the way in which latrines were used. Qualitative information on the use and maintenance of the new composting toilets was also collected for at least six months after installation. Although these alternative sanitation methods are nobly motivated, their acceptance is not an easy matter, being influenced by socio-economic and cultural factors.

## Suderburg

After installation of the composting toilet in Suderburg, it was observed that very few people used the facility. This became an important factor in the project findings because not enough material was compiled. The project group was therefore interested in determining which factors influenced the use of the composting toilet. A simple questionnaire was designed and distributed to *ifu* women junior scientists. A total of 39 questionnaires were returned. The results of the survey explaining the factors that influence the acceptance of composting toilets are given below. The questionnaire sought to ascertain:

- if junior scientists were familiar with the concept of composting toilets
- if they liked the concept of composting toilets
- how many people used the composting toilet
- reasons for not using the toilet
- what junior scientists liked or did not like about the toilet

## Conclusions Drawn by the Junior Scientists:

Based on the project findings, several conclusions were drawn. First, the SIRDO concept constitutes one alternative system of waste management. This is seen in the light of the fact that an increasing number of people in the world today lack proper sanitation facilities, making it difficult to safely dispose of human waste. The real significance of such a new, emerging approach to sanitation lies in its implications for health, ecology and economy.

Using composting toilets turns human waste into an important resource: biofertiliser. From our (preliminary work) results, it was observed that the biofertiliser produced by composting toilets was free of pathogens and very rich in nutrients. Such biofertiliser becomes a valuable resource that can be recycled. When returned to the soil, compost becomes a powerful natural soil amendment restoring and maintaining soil fertility as a result of its nutrient value. This is particularly important in view of the growing global need for food production. This, in turn, means that it is important to improve the nutrient and physical qualities of agricultural soils.

A significant aspect of the project was the acceptance of the composting toilets. Although most people said they liked the concept, they did not use the toilets. This is an indicator of the impact of socio-cultural factors on the acceptance of new ideas. Alternative sanitation systems should therefore take these factors in consideration if they are to be successfully adopted.

**Based on These Conclusions, the Junior Scientists came up with the Following Recommendations:**
- There is need for increased innovation, through both fundamental research and the widespread dissemination and adoption of such alternative systems. For instance, more innovation is needed in designing composting toilets that are socially acceptable for a wider population.
- There is a need for increased awareness about water issues and the education and training of people capable of bringing about the necessary changes. Once water is properly valued, users and producers will have greater incentive to conserve it and invest in innovation.
- Direct investment in alternative sanitation services should be encouraged in both the public and private sectors.
- It is also important to develop policies, regulations and institutional capacities to support alternative sanitation systems.
- Since sanitation issues at the household level are managed by women, it is imperative to focus on women-oriented approaches when developing alternative systems.

## *Summary*

Adopting a strictly medical approach to detecting and treating water- and excreta-related diseases will not enable their transmission to be completely stopped. Even if all those suffering from such diseases could be cured, without proper sanitation an endlessly recurring cycle of infection would continue. A world that is short of water is an unstable world. Chronic freshwater shortages are expected in various parts of the world as early as 15 years from now. Ecological sanitation therefore makes sense, if not for any other reason.

National and external funds must be made available to develop more holistic and sustainable solutions that eliminate the roots of the problems. Ecological sanitation projects are a wise investment for developing countries. Partnerships between research institutions, public and private organisations and, above all, community-based organisations are extremely important for the success of such projects. More specifically, we have learned that maintaining ideal conditions in a composting heap is a difficult task, that designing innovative systems for sanitation requires considerable co-ordination, and most important of all, that even the best ecological sanitation measures will fail after a few years unless appropriate participatory education and training is available and extensive monitoring visits are conducted.

## References

Austin LM, van Vuuren SJ (1999) Case Study: Urine Diversion Technology. Paper presented at the 25th WEDC Conference "Integrated Development for Water Supply and Sanitation", Addis Ababa, Ethiopia, 1999

Bockemühl J, Berger W (1984) Bakterlogische Untersuchung von Kompostproben aus Trockentoiletten. (Clivus Multrum), December.

Chavez A (1987) Diseño, Construcción y Estructura Convencional de LASF. Primero Seminario-Taller Nacional sobre Letrinas Aboneras Secas Familiares; Guatemala 22-26 de junio de 1987, CEMAT :21-26

Del Porto, Steinfeld D, Steinfeld C (1999) The Composting Toilet System Book, a Practical Guide to Choosing, Planning and Maintaining Composting Toilet Systems, a Water-Saving, Pollution-Preventing Alternative. The Center for Ecological Pollution Prevention, Concord, Massachusetts

Del Porto D (1996) Sustainable Strategies Provides Ecologically Engineered Solutions to Sewage Problems in Developing Island Nations of the Pacific. Unpublished document

Dudley E (1993) The Critical Villager - Beyond Community Participation. Routledge, London

Enferadi KM, Cooper RC, Goranson SC, Olivieri AW, Poorbaugh JH, Walker M, a. Wilson BA (1980) A Field Investigation of Biological Toilet Systems and Grey Water Treatment. State of California Department of Health Services, Berkeley, California. US Environmental Protection Agency, Wastewater Research Division, Municipal Environmental Research Laboratory, Cincinnati, Ohio, Grant Number R805942-01, EPA/600/2-86-069

Esrey SA, Gough J, Rapaport D, Sawyer R, Simpson-Herbert M, Vargas J a. Winblad U. (1998) Ecological Sanitation. Sida, Stockholm

Gough J (1997) El Salvador Experience with Dry Sanitation. In Ecological Alternatives in Sanitation, Proceedings of Sida Sanitation Workshop, August 1997: 6-9, Balingsholm, Sweden

Holden RD a. Austin LM (1999) Introduction of Urine Diversion in South Africa. Paper presented at the 25th WEDC Conference "Integrated Development for Water Supply and Sanitation", Addis Ababa, Ethiopia

http://oikos.com/library/compostingtoilet/html

http://helios.bto.ed.ac.uk/bto/microbes/thermo.htm

http://www.cfe.cornell.edu/compost/microorg.html

http://www.doh.wa.gov: "Water Conserving On-site Wastewater Treatment Systems", Recommended Standards and Guidance for Performance, Application, Design, Operation a. Maintenance, Washington State Department of Health: 8

McMichael J (1978) The Double Septic Bin in Vietnam. In: Sanitation in Developing Countries. Editor Arnold Pacey

Olvera-Velona A a. Ortiz-Hernandez ML (2000) Viabilidad de Huevos de Ascaris suum en Sanitarios Secos: Resultados Preliminarios. Centre for Biotechnological Research, University of Morelos, Internal document

Ortitz L (2000) Personal Communication. March 2000

Otterpohl R, Oldenburg M, Albold A a. Zimmermann J (2000) Kostengünstige, zukunftsfähige Abwasserkonzepte im ländlichen Raum. In: Schriftenreihe der Kommunalen Umwelt-AktioN U.A.N., Heft 37, Hannover 2000

Otterpohl R, Albold A a. Oldenburg M (1999) "Otter Wasser - The Most Important Tool for Source Control is the Toilet System".

Rao Bhamidimarri SM a. Pandey SP (1996) Aerobic Thermophilic Composting of Piggery Solid Wastes. Water Science and Technology Vol 33 No 8: 89-94

Sawyer R (1998) Meeting Demand for Dry Sanitation in Mexico. In: *Sanitation Promotion.* Eds: Simpson-Herbert M a. Wood S, World Health Organisation/Water Supply and Sanitation Collaborative Council (Working Group on Promotion of Sanitation), unpublished document WHO/EOS/98.5

Terrefe A a. Edström G (1999a) ECOSAN - Ecological Sanitation. Paper presented at the 25th WEDC Conference "Integrated Development for Water Supply and Sanitation", Addis Ababa, Ethiopia

Terrefe A a. Edström G (1999b) ECOSAN - An Integrated Sanitation Concept. Paper presented at the ECOSAN Conference, Harare, Zimbabwe

Winblad U (1985) *Sanitation without Water*. MacMillan, London

Winblad U (1996). Towards an Ecological Approach to Sanitation. Paper presented at the Japan Toilet Association Symposium in Toyama, Japan

# 9 River Development Planning

Andrea Töppe

## 9.1 River Protection for the Balance of Nature

Nature and the countryside (and rivers as an integral part of them) must be protected, cared for and developed in both inhabited and uninhabited areas to ensure that

- the efficiency of the balance of nature
- the utility of nature's resources
- the plant and animal kingdom
- the diversity, uniqueness and beauty of nature and the countryside

are sustainably protected as a basic prerequisite for their recuperation.

Impoverishment of the diversity of species and biocoenoses is also an indicator of damage to the balance of nature's efficiency, i.e. it indicates that the foundations of human life are threatened (Dahl a. Hullen 1989).

**The Evolution of River Systems**
If precipitation is (occasionally, regularly or irregularly) greater than evaporation and infiltration, runoff will occur, following the surface gradient. The force of gravity causes surface water to always take the shortest path downwards. It is collected in the deepest parts of valleys and flows further down the valleys in the form of small creeks or rivers. This gives rise to fluviatile erosion and denudation (surface levelling by wind and water).

Rivers are open dynamic systems that endeavour to attain equilibrium in terms of the transport of matter. Solid-matter transport can be subdivided into bed-load transport (bottom material) and suspended-load transport (suspended bed material and wash load). The causes of river sediments are weathering, erosion, surface runoff, channel erosion and sedimentation/resuspension.

The following variables of the catchment area and the river itself are responsible for differences in solid-matter transport: slope, water depth, flow velocity, precipitation, wind, temperature, type of soil, vegetation, soil utilisation, surface gradient, surface seal rate (buildings). A river forms an ecological unit from its source to its estuary and can therefore only be properly protected as a whole.

The aim of the project was to create a river development plan for the Stahlbach ("steel creek"), a first-order tributary in the German Elbe catchment area. The main objective was to survey the environmental situation of Stahlbach and develop improvement strategies to restore it to a natural or near-natural state.

The final survey included all inventory results along with proposals for individual restoration measures and a time schedule graded from topmost (as soon as possible) to long-term priority (up to 30 years). It was presented to the Subordinate Water Agency of Uelzen district for implementation.

## 9.2 Hydraulics and River Protection

### 9.2.1 Some River Characteristics

The longitudinal section of a large river can be divided into four main reaches:
1. source and headwater (upper reach), where usually erosion is dominant
2. middle reach, where the total amount of erosion is equal to the total amount of accumulation
3. downstream reach, where accumulation is dominant
4. estuary with tidal influence from the sea

The typical cross-sections are different within these reaches. In the headwater region, grooved deep valleys mostly occur, whereas "bottom valleys" with concave and opposite convex river banks are typical for the middle reaches, and at the lower river reach the cross-section appears with a wide deposited plane bottom. In the plan projection, the river line appears almost linearly at the upper reach, with furcations at the middle reach and meandering at the downstream reach. At the estuary, the river line will often be divided into several lines in the river delta.

The natural river morphology can therefore be seen very clearly in the longitudinal section, cross-section and plan. Some abiotic structure characteristics are discharge (increasing from spring to mouth), annual water temperature (increasing), flow velocity (decreasing), tractive force (decreasing), mud deposition (increasing) and oxygen concentration (decreasing).

Only artificial channels or rivers in which river works and construction have been carried out to a greater or lesser degree have regular, linear lines. Rivers made by nature do not have linear lines. Natural river lines are therefore more or less curved. Their curves can only be approximated by mathematical curves, such as circles, hyperbola or lines used for the construction of streets.

At every river turn, the water level is slightly inclined owing to the influence of centrifugal force on the current. The water surface is higher at the outer bank and lower at the inner bank. Since water tends to level its surface, a crosscurrent occurs which is directed downwards to the river bottom at the outer bank where it causes a scour, and upwards to the water level at the inner bank where it causes alluviation of the eroded material. The cross-sectional river profile will therefore always be asymmetrical.

In the main flow direction, the river turns to the left and the right alternately, being divided only by the turning points. Only at the turning points do symmetrical profiles with a levelled river bottom occur. This alternation of turns and asymmetrical profiles is the reason for the alternating of deeper scours in the river bottom at the apex of turns with fords (crossings, points of shallow depth between reversed curves in river course) at the turning points.

### 9.2.2 River Discharge

*Discharge Measurement*

River discharge can be measured using a variety of methods. What all these methods have in common is that they do not actually measure the discharge Q in m$^3$/s directly, but rather the mean flow velocity $v_m$ (m/s) of the cross-section's water area A (m$^2$). The discharge is subsequently computed by the rule of continuity:

$$Q = A \cdot v_m \qquad (9.1)$$

Owing to the presence of a free surface and to friction along the channel wall, the velocities in a channel are not uniformly distributed in the channel section. The maximum velocity will be at that point of the cross-section where there is the smallest friction influence, i.e. the point furthest away from the perimeter. Because friction between flowing water and air is less than between water and channel wall, the maximum velocity point is, vertically, not situated in the middle of the cross-section but shifted to the water surface. The measured maximum velocity in ordinary channels usually occurs below the free surface at a distance of 0.05 to 0.25 of the depth.

**Field Method**

Rough measurements of discharge can be made by timing the speed of floats. From many experiments it is known that a surface float travels at a velocity about 1.2 times the mean velocity. Objects floating with a greater submerged depth will travel at speeds closer to the mean velocity in the section. A float extending from the surface to mid-depth travels at a velocity about 1.1 times the mean velocity (Linsley et al 1982).

Given that

$$v = l / t \qquad (9.2)$$

only the river length (m) covered by the float and the time (s) taken to do so have to be measured. The mean flow velocity can then be approximated by

$$v_m \approx 0.8 \text{ to } 0.85 \cdot v \qquad (9.3)$$

If in addition the mean water area of the river stretch has been measured, the discharge can finally be computed by

$$Q = A \cdot v_m \qquad (9.4)$$

## Current Meter Measurement

The most common current meters are of the propeller type, employing a propeller turning about a horizontal axis. The vertical-axis meter is either a supporting shaft or cable. The relation between revolutions per second N of the meter cups and water velocity is given by a special calibration equation of the form

$$v = a + bN \tag{9.5}$$

by the manufacturer. Some differences in these constants a and b must be expected as a result of manufacturing variations, the effects of wear and accidental damage. Consequently, each meter should be periodically recalibrated by the manufacturer.

The channel cross-section is divided into vertical strips by a number of successive verticals. No section should include more than about 10 percent of the total flow. Velocity varies approximately as a parabola from zero at the channel bottom to a maximum near the surface. The variation for most channels is such that the average of the velocities at two-tenths and eight-tenths depth below the surface equals the mean velocity in the vertical.

The velocity at six-tenths depth below the surface also closely approximates the mean in the vertical. Mean velocities in verticals are therefore determined by measuring the velocity at either 0.6 of the depth in each vertical i, or, where more reliable results are required, by taking the average of the velocities at 0.2 and 0.8 of the depth

$$v_{mi} = (v_{0.2} + v_{0.8}) / 2 \tag{9.6}$$

or

$$v_{mi} = v_{0.6} \tag{9.7}$$

The average of the mean velocities in any two adjacent verticals multiplied by the area between the verticals gives the discharge through this vertical strip of the cross-section

$$Q_i = v_{mi} \cdot A_i \tag{9.8}$$

It is thus necessary to measure the water depth and the width of every vertical. The sum of discharges through all strips is the total discharge

$$Q = \Sigma Q_i \tag{9.9}$$

The mean velocity of the whole section is equal to the total discharge divided by the whole area

$$v_m = Q / A \tag{9.10}$$

## Computation of Discharge

The discharge may also be computed by using a flow formula (which computes $v_m$) and the rule of continuity. Nowadays, the most common flow formulas are those of Manning-Strickler (Chow 1985) and Darcy-Weisbach (DVWK 1991).

$$v_m = k_{St} \cdot r_{hy}^{2/3} \cdot I_E^{1/2} \quad \text{(Manning-Strickler)} \tag{9.11}$$

or:

$$v_m = \frac{1}{\sqrt{\lambda}} \cdot \sqrt{8 \cdot g \cdot r_{hy} \cdot I_E} \quad \text{(Darcy-Weisbach)} \tag{9.12}$$

Where $v_m$ = mean flow velocity in m/s
$k_{St}$ = Strickler's roughness coefficient in $m^{1/3}$/s (Table 9.1.)
$r_{hy}$ = hydraulic radius in m

$$r_{hy} = A/l_u \tag{9.13}$$

Where A = water area in $m^2$
$l_u$ = wetted perimeter in m
$I_E$ = slope of the energy line
$\lambda$ = resistance coefficient

$$\frac{1}{\sqrt{\lambda}} = -2 \cdot \lg\left(\frac{k_s / R}{14{,}84}\right) \tag{9.14}$$

Where $k_s$ = roughness in m (Table 9.2.)
g = acceleration of gravity in m/s² (g = 9.81 m/s²)

Natural streams rarely experience a strict uniform-flow condition. Despite this deviation from the rule, the uniform-flow condition is frequently assumed when computing flow in natural streams. The results offer a relatively simple and satisfactory solution to many practical problems. If there is steady uniform flow, the energy line, water surface and channel bottom are all parallel and their slopes are all equal

$$I_E = I_W = I_B \tag{9.15}$$

For these computations, measurements of the water area and the wetted perimeter (measure the cross-section), the representative slope for this cross-section (measure either water level slope or bottom slope) and an estimation of the bed roughness to determine either the roughness coefficient $k_{St}$ (Table 9.1.) or the roughness $k_s$ (Table 9.2.) are needed.

**Table 9.1.** Strickler Coefficient $k_{St}$ ($m^{1/3}/s$) for Rivers with Water Level with < 30m at Highest Water Level (DVWK 1990)

| Surface characteristics of riverbed / flooded forelands | $k_{St}$ ($m^{1/3}/s$) |
|---|---|
| Flat land | |
| Clean, straight, filled up to edges, no banks or scours | 30.0-40.0 |
| As above, but more stones and brushwood | 25.0-33.5 |
| Clean, meandering, some scours and shoals | 22.0-30.0 |
| As above, but some stones and brushwood | 20.0-28.5 |
| As above, but more stones | 16.5-22.0 |
| Clean, meandering, some stones and brushwood, small water depth | 18.0-25.0 |
| Stretch with sluggish flow, with brushwood, deep scours | 12.5-20.0 |
| Stretches with large amount of brushwood, deep scours | 6.5-13.5 |
| Flooding troughs with dense stock of trees and underwood | 6.5-13.5 |
| Mountainous | |
| No bottom vegetation, steep banks, flooded vegetation at banks: | |
| - Bottom of gravel, stones and some blocks | 20.0-33.5 |
| - Bottom of stones and big blocks | 14.5-25.0 |
| Forelands | |
| Willow without bushes, short grass | 28.5-40.0 |
| As above, but high grass | 20.0-33.5 |
| Acreage | 25.0-30.0 |
| Lined cultures | 22.0-40.0 |
| Areal cultures | 20.0-33.5 |
| Isolated brushwood, dense weed | 14.5-28.5 |
| Open brushwood and trees, in winter | 16.5-28.5 |
| As above, but in summer | 12.5-25.0 |
| Medium-dense to dense brushwood, in winter | 9.0-22.0 |
| As above, but in summer | 6.0-14.5 |
| Dense willows, in summer | 5.0-9.0 |
| Dense stock of trees, small underwood, water flowing below branches | 8.5-12.5 |
| As above, but flowing reaches branches | 6.0-10.0 |

**Table 9.2.** Single Roughness $k_s$ (m) (Wendehorst 1998)

| Material | Roughness $k_s$ (m) |
|---|---|
| Riverbed | |
| Muddy sand | 0.015-0.03 |
| Fine gravel | 0.035-0.05 |
| Sand with bigger stones | 0.07-0.11 |
| Gravel | ≈ 0.08 |
| Coarse gravel up to broken stones | 0.06-0.20 |
| Heavy riprap | 0.20-0.30 |
| Bottom pavement | 0.03-0.05 |
| Coarse stones and rock | 0.50-0.70 |
| Rock | ≈ 0.80 |
| Forelands | |
| Asphalt | 0.003 |
| Turf | 0.06 |
| Riprap 80/450 with grass | 0.3 |
| Grass | 0.10-0.35 |
| Grass and shrubs | 0.13-0.40 |
| Soil (acre) | 0.02-0.25 |
| Acre with cultures | 0.25-0.80 |
| Soil (wood) | 0.16-0.32 |

### Erosion and Sediment Transport

Every river tends to install a bottom slope which is able to transport the delivered bed load through this certain reach, under existing cross-sections and discharge variations. If the bed-load transport is disturbed, the river will react by changing its slope. For example, if the usual bed load of a reach is missing because of the upstream construction of a headwater that catches the material, the river will reduce its slope by eroding its bottom behind the headwater.

When water flows in a channel, a force is developed that acts in the direction of flow on the channel bed. This force, which is simply the pull of water in the wetted area, is known as the tractive force, also as the shear force or drag force. In a uniform flow, the tractive force is apparently equal to the effective component of the force of gravity acting on the body of water, parallel to the channel bottom. The average value of the tractive force per unit wetted area unit tractive force is given by:

$$\tau_0 = \rho_W \cdot g \cdot r_{hy} \cdot I \tag{9.16}$$

Where  $\tau_0$ = tractive force in N/m²
  $\rho_W$ = water density in kg/m³ (approx. 1,000kg/m³, depending on temperature and salinity)
  g = acceleration of gravity in m/s²
  $r_{hy}$ = hydraulic radius in m
  I = slope of the energy line (approx. $I_W$ or $I_B$)

The permissible tractive force is the maximum unit tractive force that will not cause serious erosion of the material forming the channel bed on a level surface. This value is known as the critical tractive force τcr (Table 9.3.). Thus, if

$$\tau_0 \leq \tau_{cr} \tag{9.17}$$

the riverbed will be stable and not erode (Chow 1985). The critical tractive force is applied to non-alluvial riverbeds and banks that do not consist of freshly sedimented material, but from older formations with an already polished surface, or an artificial protective layer.

In rivers with alluvial transport of non-cohesive loose sediments, the beginning of sediment motion can be estimated by the Shields curve, which is composed by:

D* ≤ 6:         $Fr^*_{cr} = 0.109 \cdot (D^*)^{-0.5}$
6 < D* ≤ 10:    $Fr^*_{cr} = 0.14 \cdot (D^*)^{-0.64}$
10 < D* ≤ 20:   $Fr^*_{cr} = 0.04 \cdot (D^*)^{-0.10}$
20 < D* ≤ 150:  $Fr^*_{cr} = 0.013 \cdot (D^*)^{0.29}$
D* ≥ 150:       $Fr^*_{cr} = 0.055$

$$D^* = d_{50} \cdot \left[ \frac{\left( \dfrac{\rho - \rho_W}{\rho_W} \right) \cdot g}{v^2} \right]^{1/3} \qquad (9.18)$$

Where  $d_{50}$ = mean sediment diameter in m
 $\rho S$ = density of sediments in kg/m$^3$ (approx. 2650kg/m$^3$ for quartz)
 g = acceleration of gravity in m/s$^2$
 v = kinematic viscosity coeff. in m$^2$/s (1.3 · 10$^{-6}$m$^2$/s at 10°C)

**Table 9.3.** Critical Tractive Force $\tau_{cr}$ (N/m$^2$) (Preißler a. Bollrich 1985)

| Surface characteristics | $\tau_{cr}$ (N/m$^2$) |
|---|---|
| Sandy, non-colloidal loam | 2.0 |
| Loamy deposits and alluvial mud (non-colloidal) | 2.5 |
| Common loam | 3.7 |
| Common quartz sand (0.2-0.4mm) | 1.8-2.0 |
| Common quartz sand (0.4-1.0mm) | 2.5-3.0 |
| Common quartz sand (up to 2.0mm) | 4.0 |
| Mixture of coarse sand | 6.0-7.0 |
| Compactly deposited sand and fine gravel | |
| - long-term stress | 8.0-9.0 |
| - short-term stress | 10-12 |
| Alluvial mud and stiff loam (very colloidal) | 12 |
| Round quartz gravel ( 5-15mm) | 12.5 |
| Loamy gravel (non-colloidal) | |
| - long-term stress | 15 |
| - short-term stress | 20 |
| Pure sandy loam | 11 |
| Coarse quartz débris (40-50mm) | ≤ 44 |
| Flat limestone bed load (10-20mm thick, 40-60mm long) | 50 |
| Turf | |
| - long-term stress | 15-18 |
| - short-term stress | 20-30 |

When the curve is plotted into a diagram, the part below the curve indicates no motion of sediments. Recent investigations have shown that the Shields curve, which used to be seen as the borderline for the beginning of motion, provides instead a reference value for which the risk of motion R is approximately 10% (Zanke 1990).

The amount of transported sediment can be computed by different transport formulas (e.g. van Rijn 1985). The common feature of all formulas is that they were drawn up for special conditions and cannot always be applied to other conditions, and that they are all still estimations of the highly complex natural processes. Their results will therefore often differ by a factor of about 2 from the actual sediment transport.

### 9.2.3 The River Protection System of Lower Saxony/Germany

The German National Nature Protection Law is designed to ensure long-term diversity of species by

- protecting species (protecting endangered species from direct persecution)
- protecting areas (protecting, caring for and developing the life-areas (habitats) of endangered species and their biocoenoses).

The protection of areas is regarded as more important than the protection of species. The aim of the German National Nature Protection Law, which sets out to offer long-term protection to flora and fauna, can only be achieved if the natural life-area diversity of rivers is restored.

Table 9.4. shows that species specifically adapted to rivers as their habitat are much more endangered than land-bound species (e.g. container plants) (Dahl a. Hullen 1989).

**Table 9.4.** Endangered Species (Percentage of the Total Species Figures)

| Species | Germany | | Lower Saxony | |
|---|---|---|---|---|
| | Total | Endangered (%) | Total | Endangered (%) |
| Container plants | 2,700 | 34.8 | 1,852 | 41.0 |
| Fish and round-mouth species | 70 | 70.0 | 46 | 57.0 |
| Types of dragonfly | 80 | 54.0 | 59 | 63.0 |
| Types of amphibian | 19 | 58.0 | 19 | 74.0 |
| Species of algae (Armleuchteralge) | 34 | 82.4 | 21 | 80.9 |

Rivers are open ecosystems and have a lineal form. The basic principle of rivers is that the current flows in one direction. From the spring to the estuary, there is a continuous transportation of nutrients and individuals. From the estuary to the spring, there is a continuous swimming, walking, crawling, creeping and flying of individuals of the different stretch of running-water species.

Individual sections of rivers are used by many species during different seasons or for different age-groups. From an ecological point of view, the following are part of a river

- the channel (riverbed and banks)
- the flood area (areas that are covered in water when the river bursts its banks)

The real flood area is generally larger than the theoretical flood area, which is laid down by law, made by people, and which serves for the discharge of high waters.

A river's bursting its banks is an important natural and highly impressive event. It causes parts of the river's life-area such as its dead arms to be periodically and episodically connected with the river. The river can move material, mainly oxygen-consuming mud, which is full of nutrients from the riverbed and deposit it on the foreshores. For this reason, a buffer zone should be included in the protection area, serving to reduce the charge of nutrients from the sides. The individual nature area has a decisive influence on

- the discharge regime
- the structure of the riverbed and the flood areas as well as
- the chemical composition

of the river and also its quality as a habitat for plant and animal species. Rivers of one nature area closely resemble each other in terms of their fundamental characteristics. There are generally basic differences between rivers of different nature areas. The characteristics of a river that flows through only one nature area are particularly pronounced. The characteristics of a river that flows through several regions are usually balanced and characteristic extrema do not occur.

The natural occurrence of species in a river is in any case dependent on the history of the river's development, which made the immigration of different species possible in the first place. The catchment areas of Lower Saxony's rivers assumed their present form after the ice age. Connections between current catchment areas were made during individual ice advances.

Depending on how long the catchment areas have already been separated from one another, within their own water regimes, rivers with the same nature areas but with different catchment areas have either fewer or more species occurring only in the rivers of certain catchment areas (Dahl a. Hullen 1989).

A river protection system is a continuous network of rivers providing habitats for all the native plant and animal species tied to them. Its main aims are:

- The representation of all types of rivers occurring in the affected area.
- The guarantee of the given range of species and biocoenoses in and on rivers (which is decided by conditions important to nature).
- The development of a continuous network of "close-to-nature" rivers among which basic biotic exchange is possible.
- The provision of ecological, generally intact cells within the network of a given region, from which the settling of other brooks and rivers can begin as soon as the living conditions have improved there.

Rivers also have different ecological functions (Dahl a. Hullen 1989):

**Connecting Rivers**
- flow through several nature regions or through one region for such a long distance that most of the existing small creeks have their outfalls into them
- connect all subordinate rivers with each other and
- provide continuity for living creatures from the sea to the headwater.

**Main Rivers**
- represent the natural characteristics of a river of one nature area
- are habitats for their natural biocoenoses within a catchment area
- are the central feature of the river protection system

**Side Rivers**
- are affluent of the main rivers
- should be made "close to nature"
- form ecological units with the main rivers
- are important reserves, i.e. starting points for resettlement in case a catastrophe occurs in the main river

**Demands of Main Rivers**
- The preservation, i.e. the re-establishment, of a water quality that can be expected under conditions close to nature (river quality grade II/I-II)
- The avoidance of (unnatural) pollution of the water quality, e.g. with chloride, heavy metals and waste heat
- The preservation, i.e. the re-establishment, of a close-to-nature current (avoidance of headwork and drawoffs)
- The preservation, i.e. the restoration, of a riverbed structure that is close to nature (close to natural longitudinal and cross-sections)
- The preservation, i.e. the extensive restoration, of a close-to-nature water meadow (close-to-nature flood areas in relation to size and utilisation)

### 9.2.4 Making an Inventory in Situ

Before any decision on the need to restore a river to a more natural condition can be taken, the degree of its non-natural state must be assessed. A study of the river's history and its construction works must therefore be followed by the carrying out of a detailed inventory (Sellheim 1996). The examination procedures applied in Lower Saxony are described by Rasper et al (1991).

*Bottom Structures (S-No.)*

All crossing structures of the river course are recorded. Important here are:

- weirs
- bottom falls
- bottom ramps, i.e. bottom slides
- build-up styles through dams and so forth, with pools, backwaters, etc.
- floodgate constructions

When recording the bottom falls and weirs, the following factors should be given particular consideration for the planned assessment:

- The height of the fall as a migrating restriction and interruption of thoroughfare, not only for small fish but also as a hindrance to invertebrate flowing-water organisms that swim upstream.
- The backwater effect and the substrata changes due to the diminished flow velocity, i.e. the sludgification of the gap system at the river bottom of the backwater. This gap system is an important habitat for invertebrate species typical of running waters.
- Concreted stretches with supercritical flow or smooth channels with a concrete bottom and supercritical flow, which also constitute migration hindrances, especially for invertebrates, and which mean a complete sealing of the riverbed.
- A decisive factor for bottom ramps and slides is the gradient, i.e. the height difference. This sort of construction is often too steep because of the excessive height differences and, from a nature-conservation viewpoint, cannot therefore be a satisfactory solution or alternative for bottom falls. They should only be constructed with small gradients and height differences. Where the overfall is high, several flat bottom slides should be built one after the other. The upstream and downstream existing gap system in the river's ground should be continued in the bottom slide.

## *Conduits (D-No.)*

The following is recorded:

- all bridges along the water (including small wooden footbridges)
- piping
- inverted siphons
- barriers, etc.

Besides general size information (length, breadth, height), it is very important to record bottom composition and side or bank reinforcements at all bridges. Possible bottom reinforcements are decisive for the evaluation. The streambed is sealed by either concrete or paved floors under bridges (e.g. box profiles).

This sort of smooth substratum can only be inhabited by a few specialised species and constitutes a hindrance to many migrating species. The embankment zone is interrupted by smooth, vertical bank reinforcements of bridges which stand directly in the water. A sufficiently wide conduit construction with a continuous bank patrol on both sides is necessary in order to ensure passability for migrating species.

The length of the conduit and its height are significant factors in the assessment. Very long conduits with little light can, for example, become hindrances to the migration of different organisms (tunnel effect). Piping (frequent in the headwater area) is life-threatening for almost all running-water species and is practically uninhabitable. Apart from this, it often constitutes an insurmountable migration hindrance for many water organisms because of the lack of light and the frequently very high flow velocity.

There is often a combination of different structures, especially conduits and bottom structures, along the river stretch, e.g. weirs under bridges, a lot of mills, floodgates or bottom falls at the end of pipings. As we are dealing here mostly with only one structure at a time, they are recorded and evaluated in double-numeration. More detailed information on this can be found in the tables and descriptions of the rivers.

### Extension Stretches (Bank and Bottom Reinforcements), (A-No.)

All bank and bottom reinforcements, made of a wide variety of materials, count here. The general condition of the extension of a river section of water (gradient, trapezium profile, etc.) is recorded in relation to the structure of the riverbed (close to nature, partially close to nature, etc.) and then presented in the respective river description and on a map.

Bank protection, i.e. bank shorings, can be positioned either on both sides or on just one and can be either continuous or just partial. Several smaller, shorter extension stretches with unsecured stretches between them can be subsumed under one A-Number. In addition to the length, the style and material of the embankment structure are important for evaluating the attachment; sheet pilings and similar materials constitute considerable impairments because of the absence of an amphibian zone, the complete separation of river and floodplain and the complete sealing of the bank areas.

On the other hand, shorter revetment stretches, old building-rubble stores and the like usually mean less impairment of the river and its biocoenosis for the following reason: migration of organisms from water to land and vice versa is usually possible and the different potential small habitats can be inhabited in the existing transition zones of the embankment area.

Close-to-nature bank reinforcements using living shores will not be classed as impairments within the framework of this examination, although they are not properly comparable with the natural condition. Bank protection and shoring that no longer fulfil their function (remains of old, overgrown stone heaps, broken wooden palisades and the like) do not normally need to be specifically recorded. They should be mentioned in the specific description of each river section.

Bank and floor attachments from former weirs will then always be treated as their own extension stretch (A-No.) when no threshold or bottom fall is available and it is only a question of the particular attachment remains of this sort of construction. Other than this, such attachments at weirs, bottom falls, etc (S-No.) are to be included.

### Discharges (E-No.)

Unlike with the above-mentioned constructions and structures, it will not be possible to make a complete, systematic record of the different discharges into rivers. Generally, data on the direct, obvious discharge is collected (e.g. from wastewater treatment plants, surface water drainage, etc.). As already mentioned, even these discharges cannot always be completely recorded during a single survey.

The evaluation of easy-to-overlook discharges and the description of the qualitative and quantitative composition of the discharged wastewater will be purely subjective, although exact measurements of the entries can be made. Equally, the diffuse entries and land drainage, etc cannot be taken into account when recording data. However, pipes that are dry at the time of data collection, discharge channels, etc should also be taken into account. For the above reasons, the confirmed discharges can, as a rule, only be partially evaluated, as opposed to other disturbing influences.

## *Evaluation of the Recorded Structures and Constructions along the River*

With the exception of discharges (E-No.), all established bottom (S-No.) and conduit (D-No.) structures, bank and bottom reinforcements (A-No.) should be evaluated according to the ecological criteria of running waters with regard to their possible impairing effect on the river and its biocoenosis. These comparable evaluation criteria encompass the following four levels of impairment for all structures:

- no recognisable impairment         (Table 9.5.)
- little impairment (!)              (Table 9.6.)
- strong impairment (!!)             (Table 9.7.)
- very strong impairment (!!!)       (Table 9.8.)

In the individual tables of the river descriptions, the appropriate evaluation is given using the exclamation marks (!, !!, !!!) defined above. As a rule, the studied structures can be properly assessed using these simple evaluation levels, ensuring comparability of impairments in the different rivers (Rasper et al 1991).

**Table 9.5.** Evaluation of Non-Impairments due to Structures and Constructions along the River Stretch

| Bottom structures (S-No.) (Weirs, bottom falls, bottom ramps, floodgates) | Conduit structures (D-No.) (Bridges, siphons, piping, flood barriers, etc) | Extension stretches (A-No.) (Bank and bottom attachments) |
|---|---|---|
| - | bridges without or with only small side attachments (stones, etc), or rather with passable, sufficiently wide bank stripes, bottom without reinforcement and passable | remains of old, non-effective bank reinforcements (not recorded as A-No.) |

**Table 9.6.** Evaluation of Small Impairments ! due to Structures and Constructions along the River Stretch

| Bottom structures (S-No.) (Weirs, bottom falls, bottom ramps, floodgates) | Conduit structures (D-No.) (Bridges, siphons, piping, flood barriers, etc) | Extension stretches (A-No.) (Bank and bottom attachments) |
|---|---|---|
| small swells mainly flowed around, passable bottom, no migration impairment | bridges with side attachments by smooth walls, sheet piling or similar in or adjacent to the water, missing bank stripes | shoring with revetments and building rubble (isolated or on short stretches), fascines or walls with gaps |
| flat bottom slides with appropriate structures (no backwater) | floor secured with revetment (dependent on landscape region) | bottom secured with revetment (dependent on landscape region) |
| secured bottom with revetment (no backwater) | bridge or short piping with sedimented and passable bottom | |
| | flood barriers with secured sides and partially secured bottom in tidal zone | |

**Table 9.7.** Evaluation of Strong Impairments !! due to Structures and Constructions along the River Stretch

| Bottom structures (S-No.) (Weirs, bottom falls, bottom ramps, floodgates) | Conduit structures (D-No.) (Bridges, siphons, piping, flood barriers, etc) | Extension stretches (A-No.) (Bank and bottom attachments) |
|---|---|---|
| swells, partially flowed around, bottom partially concreted or paved, or bottom attachment partially sedimented, no migration impairment | bridges with concrete or paved bottom, which are, however, partially destroyed or sedimented | bank protection with revetments (on longer stretches), palisade woods, wicker fences, synthetic fascine, sheet piling and smooth wall (isolated) |
| (= exception, generally with migration impairment: !!!, see above) | | shoring on inhabited river stretches with smooth wall, sheet piling or similar |
| | | paved or concreted bottom, which is, however, partially destroyed or sedimented |

**Table 9.8.** Evaluation of Very Strong Impairments **!!!** due to Structures and Constructions along the River Stretch

| Bottom structures (S-No.) (Weirs, bottom falls, bottom ramps, floodgates) | Conduit structures (D-No.) (Bridges, siphons, piping, flood barriers, etc) | Extension stretches (A-No.) (Bank and bottom attachments) |
|---|---|---|
| all bottom falls and weirs with migration and substrata impairment (backwater) (if passable: !!, see below) | bridges with concreted or paved bottom (if partially destroyed or sedimented: !!, see below) | Bank protections, outside inhabited areas, with smooth walls, sheet pilings or similar |
| steep bottom ramps with migration impairment and/or backwater | long conduits with tunnel effect | Concreted or paved bottom (if partially destroyed or sedimented: !!, see below) |
| floodgate constructions stopping tidal influences | pipings with concreted smooth bottom | |
| dams and similar constructions with interruption of the running water (ponds, outlets, etc) | inverted siphons | |

## 9.3 Stahlbach River Development Plan

River development projects are a valuable means for implementing Lower Saxony's River Protection Scheme in Germany. Over the past centuries, most German rivers and brooks have been hydraulically improved because of different utilisation requirements and are suffering from many human-made loads within the catchment area.

Now that Lower Saxony's Department for Ecology has developed a running-waters protection system and programme, it is the task of the districts to implement the program, for example, by creating single river development plans. This project originated from co-operation between the University of Applied Sciences in Suderburg and the Subordinate Water Agency of Uelzen district. The general idea was to choose a small river with high priority in the programme: the Stahlbach ("Steel Creek").

The aim of the project was to create a river development plan for the Stahlbach, a first-order tributary in the German Elbe catchment area. The main objective was to survey the environmental situation of the Stahlbach and develop improvement strategies in order to restore it to a natural or near-natural state.

## 9.3.1 The Elbe Catchment Area

The River Elbe has a total length of 1,091.47km from its spring in the Riesengebirge ("Giant Mountains") in the Czech Republic at a height of 1,384m above Normal Null (NN is the German datum for geodetic surveys, approximately sea level) to its mouth on the North Sea, near the city of Cuxhaven. Its total catchment area of 148,268km$^2$ is one of the biggest in Central Europe (Table 9.9.) (IKSE 1995).

**Table 9.9.** Comparison of Some Characteristic Data from big Catchment Areas in Europe

| River | Length (km) | Catchment area (km$^2$) | Mean annual discharge (m$^3$/s) | Population in catchment area (in millions) |
|---|---|---|---|---|
| Danube | 2,857 | 817,000 | 6,550 | 82.1 |
| Rhine | 1,326 | 183,800 | 2,300 | 50.0 |
| Elbe | 1,091 | 148,268 | 877 | 24.7 |
| Loire (France) | 1,012 | 115,000 | 400 | Not known |
| Oder | 866 | 119,149 | 539 | 16.9 |
| Po (Italy) | 676 | 75,000 | 1,500 | Not known |
| Weser | 432 | 46,136 | 402 | 7.7 |

The Elbe catchment area covers 4 different countries (Germany, Czech Republic, Austria and Poland). 27.2% of the total area of Germany (356,954km$^2$) is within the Elbe catchment area. In the Czech Republic, it covers 63.6% of the total area (78,864km$^2$). Ten of Germany's federal states have their territory partly or completely within the Elbe catchment area.

The biggest catchment areas of the main tributaries are: Moldau (28,090km$^2$), Havel (24,096km$^2$), Saale (24,079km$^2$), Mulde (7,400km$^2$), Schwarze Elster (5,541km$^2$) and Eger (5,164km$^2$). On the border between the Czech Republic and Germany the catchment area covers a total of 51,393.5km$^2$. The mean discharge over many years has been 313.8m$^3$/s (9.9 billion m$^3$/year). The mean discharge over many years at the mouth on the North Sea is 833.3m$^3$/s (27.7 billion m$^3$/year).

Given its discharge parameters and regime coefficients, the Elbe belongs to the streams of the rain-and-snow type. This means that the discharge behaviour is shaped mainly by winter and spring floods. The seasonal distribution of floods at the Dresden gauge (catchment area = 53,096km$^2$) from 1800 until 1994 shows that 81% of all floods occurred in winter.

Analysis of floods at the Barby gauge also shows that 86% of floods occur between December and April, with the highest frequency in March. Different behaviour is exhibited by streams of an alpine/mountainous nature, e.g. the River Rhine, where winter precipitation is mainly retained as snow and only gradually forms part of the spring and summer discharge. Together with the summer precipitation, then, an alpine discharge regime with summer floods exists.

In the Czech Republic, there are two different stationings for the River Elbe, one for water-resources management purposes and another for shipping. The stationing for water resources management purposes by the water resources management offices starts at the border with Germany (right bank) at km 0.0 and counts up to km 396.92 at the spring.

In Germany, there is only one stationing (the one for shipping purposes) from the border with the Czech Republic (left bank, km 0.0) to the North Sea (km 727.70). As some lengthening and shortening stretches have to be taken into consideration, the real length on German territory is only 726.96km. Also, a common length of 3.43km, representing the border between Germany and the Czech Republic, has to be subtracted. The total length is therefore 1,091.47km.

After a decision taken at the 5$^{th}$ meeting of the International Conference for the Protection of the Elbe (IKSE) on 21/22 September 1992, the following subdivision of the Elbe was made:

- Upper Elbe: From the spring in the Riesengebirge to the point of transition to the North German flatland (German Elbe station km 96.0)
- Middle Elbe: From km 96.0 to the weir at Geesthacht (German Elbe station km 585.9)
- Lower Elbe: From the weir at Geesthacht (km 585.9) to the seaward boundary at Cuxhaven (km 727.7).

Unlike comparable European rivers, the River Elbe and its floodplains have, in terms of their structure, numerous stretches that are largely "near to nature". Typical floodplain habitats have been largely preserved, with connections between the floodplains, flatwater areas and floodplain forests. They offer an unparalleled habitat for a wide variety of animal and plant species, all of which are threatened. Here, they can still live in representative populations.

The Elbe and its floodplains have supraregional importance as a rest, repose and migration area for numerous bird species. This is why many areas within the Elbe's floodplains have been placed under legal protection over the past decades. The Middle Elbe area is characterised by a very wide valley with river meanders, floodplain meadows, oxbow lakes, flood channels, floodplain forests, areas of dry grass on inland dunes and well-water regions behind the dikes. This mosaic of varied habitats offers an ideal environment for unparalleled animal and plant associations. The occurrence of 50 threatened species of water and land-bound plants demonstrates the botanical value of this floodplain landscape.

The largest connected floodplain forests of the middle Elbe, with an area of 117.4km$^2$, are in the UNESCO "Middle Elbe" biosphere reserve, between the affluence of Mulde and Saale. This varied and extensive floodplain landscape, which has been largely conserved in a near-to-nature condition, forms an ideal basis for a unique variety of flora and fauna, along with the floodplain meadows and the diverse water areas in the floodplains.

226 species of bird live here – more than the half of the species regularly occurring in Europe. 150 species of bird brood in this reservation, many of them on the German "Red List of Birds". In addition, the water areas are important as a resting-place for passing of water-bound bird species or as a hibernation area (if free of ice). Over a period of 15 years, an average of 29,000 water-bound birds belonging to 43 species were counted every winter.

Special emphasis should be given to the significance of the Elbe floodplains for the white stork population. The existing perennial meadows and certain hydrological conditions in the Elbe depression offer the minimum ecological requirements for the white stork. More than a fifth of Germany's total stork population lives in the Elbe floodplain areas.

The "Middle Elbe" nature reserve is also a stock population area of a characteristic animal of the Elbe floodplains, the Elbe beaver. The centre of its distribution is Saxony-Anhalt, with 1,500 individuals of a current total population of 2,850. The many water areas within the Elbe floodplains are of great importance as habitats and spawning-grounds of amphibious species on the German "Red List of Amphibious Species" and as habitats for certain types of beetle that are threatened with extinction. The mosaic of dry and humid areas offers a habitat for 200 threatened species of butterflies, bees and dragonflies. More than 700 species of large dragonflies have been observed there (IKSE 1995).

## 9.3.2 The Stahlbach River

The River Elbe connects various nature areas in northern Lower Saxony. For nature area "Lüneburg Heath", the Elbe tributary Ilmenau was chosen as the main first-priority river (RASPER et al., 1991). Of the three headwaters, the River Hardau, which flows through Suderburg, represents the headwater course of the Ilmenau. The small creek Stahlbach is an affluent of the River Hardau and therefore an important tributary in the overall Ilmenau catchment area.

The Stahlbach's spring is west of the village Bahnsen; the creek then flows to the east, through the villages of Böddenstedt, Hamerstorf and Holxen where it discharges into the Hardau. All the villages are situated in the immediate vicinity of Suderburg. The total length of the river is about 7km. A stationing of the Stahlbach exists only from the downstream side of the street bridge in Bahnsen (km 0+000) to the point where it discharges into the Hardau at Holxen mill (km 5+830). Here, the Stahlbach is a second-order river. Upstream from the Bahnsen street bridge to the spring, the Stahlbach is a third-order river and it is not the maintenance association but private land owners who are responsible for its maintenance.

The total catchment area of the Stahlbach is 28.65km$^2$. It can be divided into three parts: from the spring to Bahnsen bridge $A_{Eo}$ = 8.01km$^2$; from Bahnsen bridge to Hamerstorf street bridge $A_{Eo}$ = 11.87km$^2$; and from Hamerstorf street bridge to the Hardau $A_{Eo}$ = 8.77km$^2$.

## 9.3.3 River Stahlbach Development Project (Project Modules)

The overall inventory consisted first in making a study of existing regional planning and the information collected by local agencies on the Stahlbach, and second in carrying out practical field work. Junior scientists made a survey of the approx. 7km-long river from its spring to its mouth.

They examined the utilisation of the floodplain area; collected and identified samples of plants; identified the river fauna to calculate the saprobia value; measured water temperature, pH value, electrical conductivity, nitrate and phosphate content; made a stationing of the upstream river course; collected samples of the river substrate and analysed the grain size distribution in the laboratory; measured flow profile and flow velocities and calculated the discharge; noted all visible impairments such as conduit constructions, bottom constructions, bank and bottom protections and discharges into the river and measured their dimensions; documented all locations on maps and all work with numerous photographs as well as a video.

The final report included all inventory results together with proposals for individual restoration measures and a time schedule graded from topmost (as soon as possible) to long-term priority (up to 30 years). It was presented to the Subordinate Water Agency of Uelzen district for implementation.

## *Excursions*

In addition, the junior scientists went on excursions to gain practical knowledge relating to their development work. The excursion were to:

- the exhibition at "Elbtalhaus Bleckede" and the weir at Geesthacht/Elbe with its fishway
- the river maintenance works at Bad Bodenteich and sugar-beet factory at Uelzen
- the Ilmenau barrage at its mouth to the Elbe, the discharge measurement station at Bienenbüttel and the Ilmenau weir at Uelzen
- the old mill at Holxe.

The Elbtalhaus Bleckede is situated on the downstream part of the Middle Elbe. The exhibition there gives a detailed impression of the ecological situation, including flora, fauna and protection of this part of the catchment area. The weir at Geesthacht was the only existing migration barrier of the Elbe in Germany (the only other weirs are in the Czech Republic). For this reason, a new and bigger fishway has been constructed. The function of the weir and fishway were explained to the junior scientists by experts from the responsible authority. One part of the weir was out of order owing to damage, and junior scientists had the unique opportunity to go down the steps into the pier and have a look from inside. This excursion gave junior scientists an idea of the dimension and peculiarities of the River Elbe and its catchment area.

The visit to Bad Bodenteich was designed to familiarise junior scientists with the sort of maintenance work generally carried out on small rivers and the equipment used. Most junior scientists were astonished at the volume of this work and said that no work of this kind was done in their own countries, even on bigger rivers.

At the Uelzen sugar-beet factory, junior scientists witnessed sugar production from beet. They learned something about the crop, which is very common in this region, and that extensive irrigation is needed to grow it on the poor sandy soil.

There was also a presentation of an actual project involving a large reservoir for production wastewater that is to be treated and used for irrigation, thus saving the groundwater currently used.

The third excursion took junior scientists to various locations on the River Ilmenau. The storm-surge barrier at the mouth to the Elbe is situated upstream from the weir at Geesthacht, but storm surges originating in the North Sea can run over the weir and these caused severe flooding before the barrier was constructed. It normally leaves the flow profile open and is only closed for a short time during storm surges. The Ilmenau weir at Uelzen is now complemented by a fishway; junior scientists were able to see for themselves that there are no longer any migration hindrances on the river system downstream. This conveyed to them the importance of restoring even headwater rivers like the Stahlbach to a near-to-nature condition.

The old Holxen mill on the River Hardau, downstream from the mouth of the Stahlbach, has been restored by its owner and is open to the public. It is part of the Lower Saxony "mill route" established for tourists and including several different mills that can be visited on a single trip by car. The Holxen mill includes the barrage construction (wooden vertical lift gate), waterwheel and millstones for the corn. Some such barrages still exist on the Ilmenau tributaries. The water agencies plan to replace these migration hindrances by riverbed ramps and slides wherever possible, or at least to install bypass creeks or fishways.

### *Project Junior Scientists*

The home countries of the junior scientists in this project were: Austria, Nicaragua, Germany, India, Nigeria, Romania, Russia, South Korea, Sudan, Syria, Turkmenistan and Vietnam.

The junior scientists had bachelor's or master's degrees/diplomas or a Ph.D. in following disciplines: biochemistry, biology, environmental engineering, environmental sciences, fishery/fish microbiology, geography, hydrology, physics and water resources management. One junior scientist specialising in environmental sciences also had a degree in mathematics.

### 9.3.4  Results

The project fully met the goals set. The final report was presented to the Subordinate Water Agency of Uelzen district and will form the basis for future planning and development measures on the Stahlbach. The proposed measures will be implemented in co-operation with the local river maintenance agency, the owners and the local nature-protection associations. The promised additional planting of 40 new alder trees at Stahlbach in spring 2001 is proof of the Subordinate Water Agency's gratitude to the junior scientists for their highly efficient work.

It was the firm intention of the group to complete the inventory of the whole river and incorporate all the results in the final report because some of the junior scientists' common knowledge base would have been lost after their departure.

When writing the report and compiling the results, teamwork and the contribution of each junior scientist's specialised knowledge were extremely important. The main results are outlined below.

## Literature Study

Historical maps of Stahlbach dating from 1777 were compared with current ones. Approximately 23 decades ago, Stahlbach was in its most natural state. The riverbed had its natural undulations and curves allowing aquatic animals to shelter and breed in the shallow waters along the banks. These undulations formed natural nurseries for the young animals. The brook flowed through the valley and the floodplains had their natural flora and fauna, forming several food chains in the ecosystem. The current maps show that the course of the Stahlbach has been altered in many areas. For example, near Bergwiese the brook undergoes four 90° diversions. Again, close to the dam, there are two such 90° diversions. Moreover, in some areas, the riverbed has lost its natural undulations and curves because of the artificial tubes, concrete constructions, bridges, etc.

The Suderburg community covers an area with a population density of up to 50 inhabitants per $km^2$. The total population of region is 7,001, accounting for 1.43% of whole Upper Elbe region (Niedersächsisches Umweltministerium 1992).

The floodplain of the River Stahlbach is largely composed of clay fluvial deposits and loamy soil based on two different geological divisions: sandy floodplain and glacial valleys, and end moraines. Salt layer covers the river stretch from the spring up to the village of Böddenstedt. The area's annual precipitation is 600-700mm (1931-1960). The groundwater lies at a depth of between 10m and 7.5m, declining from source to mouth (Niedersächsisches Umweltministerium 1992).

All the areas around the Stahlbach creek are arable land, constituting an "area with special significance for agriculture", or are used for forestry, especially the area between Bahnsen and Böddenstedt. They also have a special significance in terms of nature and the landscape. The spring marshland has been designated as a priority area for leisure and recreation in nature and the countryside. Between the village of Bahnsen and Hamerstorf, the floodplain on right hand is a designated area for improvement of landscape structures (Landkreis Uelzen 1993).

The main part of river stretch belongs to communities or part of communities with a public water supply; the area of the river source belongs to communities with a part public and part private water supply. Water pipelines with diameters of up to 300mm cover the river stretch and are connected to a container with a capacity of 1,000-5,000$m^3$. All villages located on the Stahlbach – Bahnsen, Böddenstedt, Hamerstorf and Holxen – are linked to the wastewater treatment plant at Suderburg, which is of key importance (Niedersächsisches Umweltministerium 1992).

Approximately ten wells have been drilled for irrigation purposes in the Stahlbach floodplain. All of them are authorised and controlled by the district of Uelzen. It is the policy not to allow extraction to affect the water balance of the Stahlbach.

All the area covered by the Stahlbach creek is protected by the Lower Saxony Law for the Conservation of Nature (Section 26 of NNatG); only in the village of Bahnsen does Section 24 of NNatG also apply. Excluded from government protection are private properties in Böddenstedt and the railway bridge near Hamerstorf (Landkreis Uelzen 1988).

Apart from the private property in Böddenstedt, there are no buildings in the villages at a minimum distance of 40-100m from the Stahlbach (Borough of Suderburg 1999). Stahlbach is included in the area with little protection potential (Niedersächsisches Umweltministerium 1992).

A fishery association of nine people have rented the Stahlbach and the River Hardau downstream from the Suderburg community border. They treat the Stahlbach as a sort of nursery in which fish are cultivated and protected and do only very little fishing. At the watermill at Holxen, they put in about 500-800 trout every year which were bred in a special pond.

As a result of earlier measures to straighten the Stahlbach, many alder trees have vanished, which means the fish are unable to find enough places to shelter nowadays. The Stahlbach contains mainly brook trout and eel, and at one time grayling (Thymallus thymallus). The Stahlbach has a better fish population than the Hardau. Trout grow better in the Stahlbach and are healthier than Hardau trout. The Stahlbach appears to have enough nutrients for trout. Nowadays, fish are found mostly after bridges and conduits, which are the natural deeper parts of the river. If they are disturbed, they will immediately retreat up- or downstream (Brammer 2000).

Downstream from Suderburg, there are *Phoxinus phoxinus*, *Gasterosteus aculeatus*, *Gobio gobio*, *Rhodeus sericeus amarus* and *Lampetra planeri*. From the ponds come *Rutilus rutilus*, *Persa fluviatilis*, *Cottus gobio* and *Thymallus thymallus*. Pike and carp occur only rarely (Düver 2000).

Administrative documents from the Landkreis Uelzen (2000) show that only nine fish ponds are permitted. Four of them are on the stretch between Bahnsen and Böddenstedt and the other five around the village of Holxen. Existing rights end during the period 2002-2009. There are restrictions on their withdrawal from and discharge to the Stahlbach. Extensive use is forbidden.

No report is available on the current water quality of the Stahlbach. However, a 1970 study of the Ilmenau showed that it was in water-quality class II, little pollution (betamesosaprobe). In 1984, the water quality of the Hardau and Ilmenau was between classes II and III, critical pollution between betamesosaprobe and alphamesosaprobe (Niedersächsisches Umweltministerium 1992). The water quality was the same in 1996 (StAWA Lüneburg 1995) as in 1970, which indicates that the water quality is improving from the critical pollution in 1984.

The maintenance framework plan includes the kind of maintenance works that are done downstream from Bahnsen bridge (maintenance and cleaning of side, mowing and cleaning of slopes, mowing and cleaning of river bottom, removal of all flow obstructions, trimming of plants), the equipment used (front mower, slope mower, dredger with mowing basket, handwork) and the remark that maintenance and development of river strips are measurements of the maintenance of nature and the landscape from Böddenstedt mill to the mouth at the River Hardau

(Unterhaltungsverband Gerdau 2000). Actually, the maintenance association does not carry out all possible maintenance regularly, but only when it is deemed necessary. This ensures that disturbance to flora and fauna is kept to a minimum (Hilmer 2000).

## *Survey*

The nature of the substrate is of paramount importance because most lotic or running-water invertebrates are benthic. The substrate provides habitat space for a variety of activities such as movement and rest, reproduction, for rooting or fixing to, and as refuge from predators and flow. It also provides food directly (organic particles) or surfaces on which food aggregates (e.g. algae, coarse and fine detrital particles).

The substrate itself comprises a wide variety of inorganic and organic materials. The inorganic material (ranging in size from silt, sand and gravel to pebbles, cobbles, boulders and bedrock) is usually eroded from the river basin slopes, river channel and banks and modified by the current. The organic materials vary from organic fragments and leaves to fallen trees derived from the surrounding catchment and upstream habitats, as well as aquatic plants such as algae, moss and macrophytes (rooted aquatic plants) (Giller a. Malmqvist 1998).

Junior scientists were able to identify six different types of substrate in the Stahlbach: sand, sand and gravel, sand and pebble, gravel and sand, sand and silt, and silt. The visual results were detailed in maps of scale 1:5000. Furthermore, some samples of the river bottom were taken and their grain-size distribution analysed. The group also described various artificial devices that were installed in the riverbed and thus changed passage conditions for vertebrates and invertebrates. There are also concrete pipes, one pontoon and waterfalls. Most of these pipes are not filled with sediment and have such a slippery surface that invertebrates cannot pass through them. In some cases, the pipes are put very profound into the riverbed and this has a negative effect on light levels inside them.

There are a number of reasons for bank erosion. The structure of the soil is of prime importance for assessing bank erosion. When soil is not stable, this means it contains a high percentage of sand or silt. In this case, the bank is very easily eroded by the water current. Current velocity also has an impact on bank erosion. The greater the current velocity, the larger the particles that can be moved. Current velocity and substrate type are related, then. Some banks are eroded by animals going to the river to drink.

The junior scientists noted some measures that had been taken in the past to protect banks from erosion. Tree branches ('fascine') were installed in the bank slopes underwater. Also, big stones were put in the water along the banks to reduce the current velocity. These 'larger' particles, then, can 'protect' smaller ones from being entrained in the current and carried away. The group documented the embanking of river slopes near one bridge. In most cases, embanking has been undertaken to provide protection against erosion damage. In the long term, embanking could result in the disappearance of communities characteristic of the pioneering and intermediate stages in ecological successions (Petts a. Amoros 1996).

The map of soil indicates that the floodplain of the Stahlbach stretches between the main roads of the area. The average width of both strips along the river is about 225m. The upper part of the Stahlbach is swampy, a general characteristic of most springs in Lower Saxony. The whole area of the spring is covered by swamp plants, which are an indication of poor soil conditions. The floodplain along the whole of the river's length is generally flat with slight slopes at some locations. The elevation above sea level is 75-50m for the first half of the river and 50-37.5m for the second half.

Pastures and meadows dominate the floodplain of the Stahlbach. More than 30% of these pastures are used for animal breeding. Forests occupy only about 10% of the floodplain area. Most of these forests, however, do not represent the natural condition of the stream. Most of them are anthropogenically affected by forestry. Despite the fact that there is a great deal of human intervention in the floodplain of the Stahlbach, there is still considerable diversity of plant and animal species. The floodplain is not deemed worthy of designation as a protected area in the Study of the River Elbe (Niedersächsisches Umweltministerium 1992).

As mentioned before, pastures and meadows dominate the floodplain areas, which are used mainly for animal breeding. Generally, farming activities affect the river's ecology in many ways. These can be summarised as follows:

- Increase in nutrients and toxic agents through the use of fertilisers, pesticides, veterinary medicines and hygienic agents
- Lack of sufficient water retention and occurrence of extreme water flooding
- Increase in nutrients through drainage, resulting in severe mineralization
- Increase in the mobility of heavy metals through drainage and oxidation of soil
- Decline of flora and fauna
- Impairment of the river's the self-purification potential

According to the results of the chemical analysis conducted by one team, no excess of nutrients (like phosphate and nitrate) was found in the Stahlbach. This might be due to the fact that most of the floodplain is used as pasture which is not normally treated with fertilisers. Soil melioration which occurs mainly with drainage in the floodplain of the Stahlbach also has an ecologically negative impact. First, the stream should be deepened and excavated in order to discharge the drainage water properly – which in turn increases the velocity, reduces sedimentation, promotes erosion and affects the groundwater level. Secondly, the drainage water usually contains high levels of iron, which damages the river's water quality and promotes mineralization. Pastures and meadows, which have to be mowed regularly, contribute to the decline in fauna and flora in the floodplain. It has been mentioned before that more than 60% of the floodplains are pastures that need mowing, which destroys animals' natural habitats.

The use of the river by animals as a drinking-water resource could cause very severe damage to the soil, in turn leading to erosion and an increase in nutrients and sediments in the river. In the case of the Stahlbach, such damage was observable at some locations, especially along the stretch from the spring to Bahnsen. Most of the grazing pastures, however, are equipped with a kind of small pump that the animals are trained to use. Such a measure can alleviate the effects of animal breeding in the floodplain. Another negative impact of animal breeding is the deterioration in water quality. Disposing of faecal waste near the river contaminates the water and leads to an increase in the number of coliform bacteria in the river. No bacteriological analysis has been made by our team, but waste disposal of this kind was observed at many locations.

Along the Stahlbach there are more than 18 fish ponds, which vary in size and stretches. For example, the area of one of the biggest fish ponds is more than 18,000m$^2$; the small ones are about 1,200m$^2$. Almost all of these ponds discharge their wastewater directly into the Stahlbach. The wastewater of the fish ponds contains very high nutrient loads that can strongly affect the ecological conditions in the river.

Maintenance of the Stahlbach is done in two phases. Excavating machines and handwork are used for mowing and cleaning the riverbed and banks (Unterhaltungsverband Gerdau 2000). This causes eradication of almost all organisms and destroys the habitat's structure.

43 velocity measurements (in m/s) were made along the river, and the discharges (m$^3$/s) were computed for those locations where the flow profile had been measured. The velocity measurements carried out during the survey in the Stahlbach were either done using a current meter or floating pieces of wood, a measuring tape and a stopwatch. All locations were shown on the inventory maps.

At 24 locations along the river course, also shown on the inventory maps, chemical and physical water parameters were investigated. In the spring region, the pH value showed an acidic influence, originating in the surrounding soil, which is a moor region. It then changed to nearly neutral or slightly alkaline, between pH 6.92 and 7.57. The temperature ranged between 11 and 14.6°C, owing to the mostly shadowed areas. Junior scientists expected an influence from the salty underground layer on the Stahlbach. Though the conductivity was very low (31.9-87.1µS/cm), there is no dependence.

Nutrients can be considered relatively low for the whole length of the creek. The nitrate content rose from 0 mg/l at the spring to stationing 0+000, to 11mg/l between Bahnsen and Böddenstedt. This region may be affected by additional nutrients through drainage pipes from arable land. Phosphate content decreased from spring to mouth, where the junior scientists found nearly no phosphate at all. The pH range and the nitrate content are comparable with that of other small rivers of the region (Voss a. Preschel 1996), but the conductivity was much lower. The phosphate levels of both Stahlbach and Hardau are higher than that of other tributaries. Compared to levels measured in the Jeetzel and Elbe in 1985, the Stahlbach has a higher nitrate and phosphate content in some places (Niedersächsisches Umweltministerium, 1992).

At 15 floodplain locations, flora samples were collected and identified (Haslam 1978). The locations were again noted on the inventory maps. The results can be summarised as follows: only some of the studied areas are overgrazed. In these areas, the vegetation is poor and undeveloped. In areas without overgrazing,

9 River Development Planning    247

the grass is developed and there are some species of *Poaceae*. *Alnus glutinosa* is the main tree species on the banks. Ungrazed banks have tall herbs, while grazed banks have short grasses that exhibit damage through trampling by cattle.

Some of the stems are shaded and the banks are stabilised by alder (*Alnus glutinosa*) and therefore require very little maintenance. Alder (*Alnus glutinosa*) roots growing down a stream bank stabilise the channel. On the banks, a developed vegetation (even luxuriant vegetation) was observed in some areas, with a few species such as *Urtica dioica* and some herbs of the *Poaceae* family owing to abiotic factors (pH, moisture).

At three locations, the saprobia index was investigated as a biological parameter for water quality (Croft 1986). The frequency of the species was estimated, the saprobia values were calculated for 8-12 species of the river. All locations had saprobia values around 2.1, which indicates class II water quality (between 1.8 and < 2.3). Tables with detailed results of all investigations and analyses were included in the project group's final report. The inventory of impairments from the spring to the final inflow into the River Hardau showed a total of 44 conduit constructions, 7 bottom constructions, various protected bank stretches and 60 discharges. For better comprehension and to facilitate the development of a measurements catalogue, the river was divided into 4 stretches:

- Spring to Bahnsen bridge (Stretch I)
- Bahnsen bridge to Böddenstedt bridge (Stretch II)
- Böddenstedt bridge to Hamerstorf bridge (Stretch III)
- Hamerstorf bridge to inflow into River Hardau (Stretch IV)

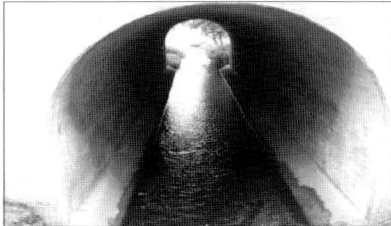

**Fig. 9.1.** One River – Two Views

See Tables 9.10-9.14: these contain exemplary classifications of the individually numbered structures and their corresponding measurements. The evaluation of the impairments of the Stahlbach is shown here:

## Stretch I
Impairments by conduit constructions (D-No.):
1. The area from the spring to the first impairment D1 is predominantly marshland, so that the first conduit reported on the topographic map (1:5000) could not be found at the time of the field inventory. The map showing the actual situation of the impairment structures does not therefore include this particular conduit construction.
2. The first four conduit constructions, D1, D2, D3 and D4, were judged to be very strong impairments as they could present a hindrance for many migrating species. They consist of concrete pipes with smooth substrata without sediments, ranging in length from 4 to 6 metres; D3 causes a fall in the bottom level. The conduit construction D5 is a set of two parallel concrete pipes that do have sediment substrata and could allow migration; D5 was thus judged to be a small impairment.
3. D6 causes no impairment, as it is a black pipe above the water level.
4. The bridge at Bahnsen was evaluated as a small impairment as the bridge structure is supported on the banks and therefore not directly in the bed sediment structure.

**Table 9.10.** Total Impairments: *Example Conduit Constructions* D-No.

| Conduit constructions (pipings and bridges) | D-No. | 1 | 2 | ... | n |
|---|---|---|---|---|---|
| Pipe | Diameter (cm) | C | C | | |
|  | Material | 5 | 4 | | |
| Length (m) | | - | - | | |
| Width (m) | | - | - | | |
| Height (m) | | - | - | | |
| Other structure | | - | - | | |
| Bank protection | Wall, smooth | - | - | | |
|  | Sheet piling | - | - | | |
|  | Wood | - | - | | |
|  | Riprap | - | - | | |
|  | Other | - | - | | |
| Bottom protection | Concrete/paving | - | - | | |
|  | Riprap | - | - | | |
|  | Other | - | - | | |
| Increased flow velocity | | - | - | | |
| Total impairment | | !!! | !!! | | |

## 9 River Development Planning 249

Impairments by bottom constructions (S-No.):
1. The bottom fall (S1) of 1 metre (two steps) constructed of a plastic rubber-like material in this river stretch is evaluated as a severe impairment because of the change in flow level which constitutes a migration restriction. Although this was evaluated as a very strong impairment, it is important to note that the creek is narrow at this point and lacks accompanying bank protection. Measures to change this construction are therefore very easy to carry out and inexpensive.

**Table 9.11.** Total Impairments: *Example Bottom Constructions* S-No.

| Bottom constructions | S-No. | 1 | 2 | ... | n |
|---|---|---|---|---|---|
| Bottom fall (cm) | | - | - | | |
| Weir | | - | - | | |
| Bottom ramp | | - | - | | |
| Other | | - | D | | |
| Height of fall (m) | | 1 | - | | |
| Width (m) | | - | - | | |
| Backwater (m) | | - | - | | |
| Material | Sheet piling | - | - | | |
| | Wood | - | - | | |
| | Stone/concrete | - | - | | |
| | Other | Ru | - | | |
| Bank protection | Wall, smooth | - | - | | |
| | Sheet piling | - | - | | |
| | Wood | - | - | | |
| | Riprap | - | - | | |
| | Other | - | - | | |
| | Length left (m) | - | - | | |
| | Length right (m) | - | - | | |
| Bottom protection | Concrete/paving | - | - | | |
| | Riprap | - | - | | |
| | Other | - | - | | |
| | Length (m) | - | - | | |
| Impairment of substrate | | - | - | | |
| Fishway | | - | - | | |
| Migration impairment | | - | - | | |
| Total impairment | | !!! | - | | |

Impairments by bank protection (A-No.):
1. Concrete bank protection on both sides of the bank, A1, was evaluated as a small impairment as the length is only 0.75 metres and its removal presents no great difficulty.

Discharges (E-No.):
1. The 60 discharges indicated on the map and found in all stretches were not evaluated individually as impairments; considering the water quality data, the discharges do not seem to cause changes in the Stahlbach's nutrient concentrations.
2. It is important to note that inlet creeks and discharges were in many cases found to be different from the topographic map (1:5000). Some discharges were missing and others were found that were not indicated on the map.

**Table 9.12.** Total Impairments: *Example Protected Stretches* A-No.

| Protected stretches | A-No. | 1 | 2 | ... | n |
|---|---|---|---|---|---|
| (only bank and bottom protections) |  | 0,75 | - |  |  |
| Bank protection | Length left (m) | 0,75 | - |  |  |
|  | Length right (m) | + | - |  |  |
|  | Wall, smooth | - | - |  |  |
|  | Riprap | - | - |  |  |
|  | Sheet piling | - | - |  |  |
|  | Wood palisade | - | - |  |  |
|  | Wattle/bandle | - | - |  |  |
|  | Other | - | - |  |  |
| Bottom protection | Length (m) | - | - |  |  |
|  | Width (m) | - | - |  |  |
|  | Concrete/paving | - | - |  |  |
|  | Riprap | - | - |  |  |
| Total impairment |  | ! | - |  |  |

**Table 9.13.** Total Impairments: *Example Discharges* E-No.

| Discharges | E-No. | 1 | 2 | ... | n |
|---|---|---|---|---|---|
|  |  | - | - |  |  |
| Wastewater treatment plant |  | - | - |  |  |
| Fish pond |  | - | - |  |  |
| Other discharges |  | - | - |  |  |
| Pipe | Diameter (cm) | - | - |  |  |
|  | Material | - | - |  |  |
|  | Side | left right |  |  |  |
| Clouding / smell |  | - | - |  |  |
| Total impairment |  | - | - |  |  |

## Stretch II
Impairments by conduit constructions (D-No.):
1. This stretch has five conduit structures (D8-D12). D8 and D10 are located before the fish pond, have a length of 4.4 and 2 metres, respectively, and were evaluated as a very strong impairment because of their smooth concrete substrata. D9 is close by, but is a well structure beside the riverbed used to protect some drainage inlets and has no effect on the main riverbed structure.
2. The Böddenstedt bridge can also be evaluated as a small impairment as the sediment structures are not strongly disturbed and the supports are on the river bank.

Impairments by bottom constructions (S-No.):
1. S2 is a siphon, which causes no impairment to the riverbed since it is located under the bottom level.
2. S3 is a small fall evaluated as small impairment only caused by accumulation of rocks, which could be avoided by better maintenance.
3. The Böddenstedt mill fall, S4, has a height of 2.32m. This fall causes a severe change in the flow level and constitutes a strong migration restriction. This bottom construction is therefore evaluated as a very strong impairment.

Impairments by bank protection (A-No.):
1. There were no bank-protection constructions found on this stretch.

Discharges (E-No.):
See Stretch I

## Stretch III
Impairments by conduit constructions (D-No.):
1. This stretch of the Stahlbach has a large number of conduit constructions, 13 in all (D13-D25). The first 12 conduit structures (D13-D24) are all within a distance of 650m. They all consist of concrete piping, have a length of 5m, a width of 2m and a diameter of 80cm. They were evaluated as a small impairment since they contain sediment bed structures and to this extent would not seem to be hindrance to migration, though their concentration over a short stretch could cause an accumulated effect on certain migration phenomena.
2. The bridge at Hamerstorf, D25, has gravel sediments and was therefore evaluated as a small impairment.

Impairments by bottom constructions (S-No.):
1. There were no bottom constructions found on this stretch.

Impairments by bank protection (A-No.):
1. The bank protection of A2 is a very strong impairment as both banks have 4.4 metres of concrete bank protection. They therefore seal off a considerable length of the bank areas.
2. A5 is a concrete bank protection of 1 metre concrete on both sides of the bank, considered to be a small impairment and easily removable.
3. A3 and A4 merely mark the beginning of areas of wood-palisade bank enforcement's, which are not considered an impairment.

Discharges (E-No.):
See Stretch I

**Stretch IV**
Impairments by conduit constructions (D-No.):
1. This section of the river has the most structures and constructions, with a total of 18 conduit pipe bridges spread out over the overall length.
2. The first conduit construction here, D26, is the most severe impairment of all river stretches. This construction is a tunnel-like conduit crossing the railway where the river widens, causing a very shallow water level. It is completely embanked on both sides and has a concrete bed, thus constituting a clear hindrance to migration of all types. It is therefore judged to be a very strong impairment.
3. D27 is a wooden bridge, anchored on the riverbank, causing no impairment.
4. D28 is a conduit structure which forces an artificial change of flow direction and widening of the river. This also results in a reduced water level. Taking into consideration these factors, it was evaluated as very strong impairment.
5. D29 is conduit piping, evaluated as a small impairment because the bed structure continues in the piping.
6. D30 is a wooden bridge with a concrete construction underneath. This is a new construction and must be evaluated as a very strong impairment because of the vertical concrete walls and the narrow width causing higher flow velocities and – in the case of high water levels – a backwater stretch and possible flooding.
7. The following nine conduit pipes, D31 to D39, are made of concrete and have a smooth bottom with no sediment substrates; they are therefore considered a very strong impairment. Also, some of the pipes have a step from the bottom, creating a change in the flow stream.
8. The rest of the conduit pipes, D40 to D42, are small impairments as they have some sediment substrate inside and cause no change in the flow dynamics.
9. The Holxen bridge, D43, was also evaluated as a small impairment.

Impairments by bottom constructions (S-No.):
1. The two bottom falls, S4 and S5, are small impairments, S4 being caused by an accumulation of stones and S5 by a discharge pipe. These can be easily changed by normal maintenance work.
2. S6 does not represent an impairment as it is a protection well in the river bottom.
3. S7 is a siphon and causes no impairment as it is located under the bottom level.

Impairments by bank protection (A-No.):
See Stretch III

Discharges (E-No.):
See Stretch I

It is important to mention that there are a large number of structures and constructions for a relatively small river of only 7km flow. This is especially true of the conduit structures, which are concentrated in some areas, such as on Stretches III and IV.

The railway tunnel (D26), the Böddenstedt mill fall and some of the conduit pipes, which hinder migration owing to the absence of sediments, are indicated as the strongest impairments and disturbing factors and ones which must be changed in order to start the Stahlbach's renaturation process.

## *Proposals/Conclusions*

A wide range of restoration measures can be undertaken for the Stahlbach creek on the basis of the inventory conducted there. The junior scientists were divided into three main groups to generate ideas for restoration, which can then be forwarded to landowners and the authority responsible for river restoration. The studies and inventory conducted can be divided into the following subgroups and proposals made for each of them:

- Inventory of impairments such as all conduits, weirs, bank and bottom constructions and all discharges into the river.
- Detailed study of the riverbed characteristics and floodplain biota (fish, birds, invertebrates, plants).
- River hydraulics such as discharge and velocity measurements and observations with respect to flooding and riverbed erosion.
- Floodplain and land-use-pattern assessment.
- A water-quality survey, looking at the nutrient content of the water by chemical analysis such as pH, temperature, conductivity, nitrate and phosphate tests.

## Time Schedule
There is necessary to draw up a time schedule for achieving the objectives or individual measures within the time-frame indicated. For this purpose, the junior scientists listed and categorised the measures as follows:

Topmost Priority (TP): Measures that can be carried out in the immediate future fall into this category. Such measures can be taken without the need for much investment or cost and without many lifestyle changes or agreements having to be made with farmers and landowners. Small to medium impairments are likely to fall into this category.

Second Priority (SP): These measures can be achieved within a period of 1-5 years. The investment required for such measures may be low to medium and may involve talks and agreements with the community. Strong impairments might fall into this category.

Long-term Priority (LP): These measures can be achieved in the long term and may also require long-term planning, monitoring and implementation. They could be carried out in the distant future, within a period of 6-30 years, and might involve high investment and changes in the lifestyle of the community as well as long discussions, dialogues and agreements with landowners.

## Measures to Be Undertaken
All impairments along the creek were measured and mapped to enable proposals to be made for improving the creek and restoring the river to a natural or near-natural state. All bottom ramps, falls, conduits and bank constructions that constitute a hindrance to the natural flow of the water are impairments. They range from small to strong to very strong impairments, and accordingly special measures must be adopted to improve each of these and also to reduce channel erosion and siltation.

Individual measures refer to specific river locations. Junior scientists proposed the following individual measures:

- Sedimenting would involve sedimentation and introducing bottom substrates underneath the pipes or conduits that are smooth and slippery and those causing problems for the migration of fish and invertebrates. The stretches or the number of the pipes where such a measure must be implemented are D1, D2, D3, D4, D8, D10, D31, D32, D33 and D39, with Topmost Priority.
- Elimination would involve removing some of the conduit constructions that were categorised as a strong impairment and where they are most concentrated (Stretches III and IV). The junior scientists proposed removing alternate pipes within these stretches, according to the owners wishes, with Second Priority.
- Introducing bigger tubes will involve introducing tubes with bigger diameters so that the problems of flooding are reduced; also some bottom substrate can be introduced. The bridges or pipes where such a measure must be undertaken are D1, D2, D3, D4, D7, D8, D10 and D30, with Second Priority.

- Remeandering and narrowing down the river profile and increasing water depths are a measure that can be taken at the large bridges with a wider channel cross-section and thus small water depths, which are not conducive to fish migration. The stretches or bridge numbers where such a measure needs to be adopted are D12, D25, D26, D28 and D43, with Topmost Priority.
- Removal of bank concrete is a measure that can be taken at places where the bank concrete may impair the discharge profile of the river. It is also a measure that can be adopted to restore the river to a more nature-like state. It may involve introducing clay or silt and removing the concrete walls. When taking such measures, the cost factor must also be kept in mind. If they involve high costs, such measures can be postponed until a later date or put on the long-term planning and implementation agenda (A1, A2 and A5, with a Long-term Priority).
- It is essential to remove two falls as Topmost Priority as they constitute a serious barrier to migration. S1 can easily be removed because the Stahlbach is narrow at this point. The fall upstream from the Bahnsen bridge (S1) needs to be smoothened and the rubber or plastic sheet removed because this constitutes a hindrance to the migration of invertebrates and fish; also, there is a sudden change in the water level which may cause further erosion of the riverbed and subsequent siltation. This is of topmost priority and can be done immediately. S4 is more complex, but its elimination is one of the most important measures required to restore the creek to its natural state. All other falls (S3, S5 and S6) can be removed by improved maintenance performance.

The creation of a narrow pedestrian path for anglers, nature-lovers, scientists, etc. along one bank will also help maintenance work. All artificial impairments can be replaced by natural devices like wooden bridges. Junior scientists proposed planting trees like alder, which are water-loving trees, to protect the banks and to maintain the ecological balance of the river ecosystem. There is a need to re-introduce aquatic animals like fish (eels, grayling) and semi-aquatic animals like salamanders into the Stahlbach. It would be better if farmers kept their cattle in an particular area for shorter periods and then moved them to another area. This would reduce the effects of overgrazing.

In order to improve conditions in the floodplain and restore it to a more nature-like state, implementation of the following measures is suggested:

Planting alder trees along the riverbanks: Such measures could prevent the excessive growth of water plants, balance the water temperature and fix the river bank. Moreover, the roots of the trees provide a refuge for the water organisms.

Setting up fences along each side of the river about 4m from the bank: This measure would prevent animals from approaching the river, alleviate erosion of the river banks and prevent bacteriological contamination.

Controlling logging in the forests along the river: There are many small tributaries that collect the runoff from the whole catchment area and ultimately drain into the Stahlbach. These drainage channels are not a problem because the water quality does not deteriorate as a result of this natural runoff. Since the water quality of the Stahlbach is good, as noted earlier, no real measures need to be taken in this regard.

It is important to mention that there are already some nature-like areas on the Stahlbach, especially the forest areas which should be protected from any further negative influences, logging, etc.

## 9.4 Summary

By sharing knowledge about rivers and their problems and dealing with these in different countries and climates provides junior scientists with suggestions and ideas for solving problems in their own countries. Catchment-area management, for example, includes knowledge of topography, soil, vegetation and animals, land use (forestry, agriculture, surface sealing by buildings and roads) and erosion, surface runoff and river systems, river hydraulics, limnology, nutrients and destructive substances, administration (laws and their application), river maintenance, and much more.

The Stahlbach development project group was an ideal combination of experts from different fields. The botanist contributed investigations of flora; the limnologist collected and analysed invertebrates and computed the saprobia values; the fish microbiologist provided information on the river's fauna and biological water quality; another biologist conducted most of the chemical field tests; the geography teacher accompanied the different work teams with a video camera and conducted interviews; environmental scientists and engineers did most of the impairment inventory and measurements; and the floodplain teams were also responsible for the photographic documentation.

Although certain teams were formed from the start of the inventory, some teams or individual junior scientists changed to different working groups or sometimes had to accompany them. This enabled every junior scientist to gain insights into the work of the others and learn about the whole project. Since junior scientists were theoretically and practically familiarised with the different tasks during inventory, no special prior knowledge was required.

However, if there had been no chemists, botanists or limnologists participating in the project, much of their work could not have been done by others and would have had to be omitted altogether. The interdisciplinary project group gained a detailed idea of the river's state and, with its river development plan, produced results that could not have been achieved by experts from only one discipline. This provided a clear indication of how future river development plans should be undertaken.

The internationality of the project group was a great help for junior scientists who had no such programmes in their own countries. They will be able to share what they learned about the protection of small rivers in Germany. The statement of an Indian junior scientist perhaps best sums up the potential for knowledge transfer:

"This project has helped me in better understanding the hazardous consequences of overharnessing rivers and has given me a deep insight into such problems and the necessity of restoring rivers to their natural state. Inventory mapping and surveying of the rivers has enabled me to acquire a deeper understanding of river hydraulics and the consequences of each impairment. The interpretation of survey results and the measures subsequently suggested by us were based on visual observation and our own perceptions of river dynamics and river problems. This project has certainly made me more experienced in terms of carrying out inventory mapping and analysis and interpretation of the results, which I might be able to apply in my home country where such problems exist for big rivers and the big river-valley projects. We certainly do not have such problems with small creeks and inlets, but this has been an eye-opener in terms of the future problems of rivers in our country if appropriate planning and monitoring is not done right from the start."

Other junior scientists came from arid countries where small rivers dry out when there is no precipitation and where there is no inventory or information programme, for small or large river systems. They regarded it a matter of urgent necessity to initiate a regular programme for monitoring water quality and an inventory process in order to draft an improvement plan.

During the five weeks of the team's work, it became apparent that the group's internationality was more important for most of the foreign junior scientists because in their home countries water-resources and catchment-area management rarely take into account ecological aspects and problems of small rivers, their importance for a functioning river system from spring to sea often being totally neglected.

The European Union has already recognised how important it is that catchment-area management should not be disrupted by national borders and different laws. It has consequently drafted general water directives in the form of a framework to be fleshed out according to the specific conditions in the different member states. International commissions already exist for the protection of some big rivers stretching over several countries, e.g. the International Commission for Protection of the Rhine (IKSR) or the similar body for the protection of the Elbe (IKSE).

The Federal Ministry of Transport is responsible for all German rivers used for shipping purposes, but rivers that are not waterways are the responsibility of the individual federal states and their administrative hierarchies, from the environment ministries down to the district subordinate water agencies. This administrative structure does not match the EU's general water directives and German experts are still developing the missing organisations.

Water is the basis for all life on earth:

- evolution began in water
- two-thirds of the human body consists of water
- water is the most important food (humans can survive without nourishment for longer periods, but only a few days without drinking)

The protection of water is, then, the responsibility of every human being, especially women who have to provide food for their families and take care of their family's health. Women should therefore be involved in all decision-making processes concerning water resources. They should not be content with being only consumers, but should ask questions and demand answers about insufficient supplies and/or the poor quality of water and find solutions that may differ from those obtained by traditional (male-dominated) approaches. The growing awareness of global ecological problems begins with the realisation that something is wrong on your own doorstep and that sometimes individual solutions are needed. Knowledge from all other countries is a great help here. It should not be simply transferred, but rather used as basis for developing new solutions.

## References

Brammer O (2000) Telephonic Information. 12th September 2000
Chow V te (1985) Open Channel Hydraulics. International Student Edition, McGraw-Hill
Croft PS (1986) A Key to the Major Groups of British Freshwater Invertebrates. Field Studies 6: 531-579
Dahl HJ, Hullen M (1989) Studie über die Möglichkeiten zur Entwicklung eines naturnahen Fließgewässerschutzsystems in Niedersachsen (Fließgewässerschutzsystem Niedersachsen). Naturschutz Landschaftspfl. Niedersachs. 18: 5-95
Düver F (2000) Telephonic Information. 12th September 2000
DVWK (Deutscher Verband für Wasserwirtschaft und Kulturbau e.V.) (1990) Hydraulische Methoden zur Erfassung von Rauheiten. Schriftenreihe des Deutschen Verbandes für Wasserwirtschaft und Kulturbau e.V. 92
DVWK (Deutscher Verband für Wasserwirtschaft und Kulturbau e.V.) (1991) Hydraulische Berechnung von Fließgewässern. DVWK Merkblätter zur Wasserwirtschaft 220
Giller PS, Malmqvist B (1998) The Biology of Streams and Rivers. Oxford University Press Inc. New York 5: 32-43
Haslam SM (1978) River plants, The Macrophytic Vegetation of Watercourses. Cambridge University Press
Hilmer H (2000) Personal Information. 28th September 2000 (Kreisverband der Wasser- und Bodenverbände Uelzen)
IKSE (Internationale Kommission zum Schutz der Elbe) (1995) Bestandsaufnahme von bedeutenden punktuellen kommunalen und industriellen Einleitungen von prioritären Stoffen im Einzugsgebiet der Elbe. Magdeburg
Landkreis Uelzen (1988) Landschaftsrahmenplan
Landkreis Uelzen (1993) Regionales Raumordnungsprogramm (RROP)
Landkreis Uelzen (2000) Wasserbuch (Auszüge)
Linsley RK Jr., Kohler MA, Paulhus JLH (1982) Hydrology for Engineers. International Student Edition, McGraw-Hill
Niedersächsisches Umweltministerium (1992) Wasserwirtschaftlicher Rahmenplan Obere Elbe. Hannover
Petts GE, Amoros C (1996) Fluvial Hydrosystems. Chapman & Hall, Konolon: 245-248

Preißler G, Bollrich G (1985) Technische Hydromechanik. Band 1, VEB Verlag für Bauwesen, Berlin

Rasper M, Sellheim P, Steinhardt B (1991) Das Niedersächsische Fließgewässerschutzsystem - Grundlagen für ein Schutzprogramm (unter Mitarbeit von Blanke D, Kairies E). Naturschutz Landschaftspflege Niedersachsen, 25: 1-4, Hannover

Rasper M (1996) Charakterisierung naturnaher Fließgewässerlandschaften in Niedersachsen, Typische Merkmale für die einzelnen naturräumlichen Regionen. Inform. d. Naturschutz Niedersachs., 5: 177-197

Rijn van LC (1985) Sediment Transport. Delft Hydraulics Laboratory, publication no. 334

Samtgemeinde Suderburg (1999) Flächennutzungsplan mit 1. bis 17. Änderung

Sellheim P (1996) Hinweise für die Erstellung eines Gewässerentwicklungsplanes (GEPl), Gliederung und Leistungsverzeichnis. Inform.d. Naturschutz Niedersachs., 5: 198-201

StAWA Lüneburg (Staatliches Amt für Wasser und Abfall Lüneburg) (1995) Gewässergütekarte 1996

Unterhaltungsverband Gerdau (2000) Unterhaltungsrahmenplan

Voss JH, Preschel C (1996) Schwermetallbelastung der Sedimente von Fließgewässern im Landkreis Uelzen. Institut für Angewandte Abfallwirtschaft an der Fachhochschule in Suderburg (IFAAS), Suderburg (unpublished)

Wendehorst R (1998) Bautechnische Zahlentafeln. Teubner, Stuttgart Leipzig

Zanke U (1990) Der Beginn der Sedimentbewegung als Wahrscheinlichkeitsproblem. Wasser und Boden

# 10 Water and Soil
# Towards Sustainable Land Use

Brigitte Urban

## 10.1 Overview

Following the project idea of the International Women's University, the project area Water focused on environmental topics. The goal was to create awareness of the problems faced and seek solutions cutting across scientific disciplines and national borders. Selection of the topics in this project area was based on resolutions adopted at international conferences such as UNCED – United Nations Conference on Environments and Development, Rio de Janeiro 1992 (in line with the Agenda 21). The idea was to pursue a holistic approach to the subject-matter with a view to developing the women junior scientists' knowledge and skills.

At the beginning of the project, an introductory study was conducted covering a variety of fields and disciplines including biology, chemistry, biochemistry, geology, geography, physics, agronomy economics, agricultural sciences, environmental sciences, civil engineering and educational science. This is briefly summarised in the following chapters. And to provide guidelines for further environmental studies, a manual for the analysis of soils and related materials has been compiled.

### 10.1.1 Soils

**What Is Soil?**
Soil is by definition the uppermost layer of the earth's crust, a weathering product of the parent material or accumulated organic matter. It is mainly composed of mineral particles, has an organic content and pores filled with air and water. Its most important aspect for human beings is its fertility, which is due mainly to the organisms living in the uppermost layer of soil. These organisms help to decompose and recycle organic and inorganic components. Soils therefore constitute the main habitat of all kinds of terrestrial plants and are in particular the basis for growing crops.

## How Does Soil Form and What Are Its Characteristics?

Five soil-forming factors interact by controlling the physical and chemical processes that transform the parent material into soil:

**Table 10.1.** : Soil Formation

| Soil formation factors: | Processes of soil formation: |
| --- | --- |
| Parent material | Physical weathering |
| Climate | Chemical weathering |
| Topography | Translocations |
| Biota | Clay formation |
| Time | |
| People as soil formers | |

All kinds of geological material, such as hard rock, windblown sand or silt (loess), glacial and river or lake deposits, form the basis (parent material) from which soil genesis starts. Precipitation and temperature, the two main climatic factors, affect soil formation.

Water is the main agent in the weathering of the parent material. The amount of precipitation and its distribution over the year influences the kind of soil that is formed and its rate of formation. By contrast, the amount of water entering the soil depends on how much water falls as precipitation and how much runs off the surface of the soil as a result of its texture and structure. Climate determines which soil-forming processes dominate, forming a soil profile. Without water, the chemical processes of hydrolysis, hydration, acidification, oxidation and reduction, chelation, solution and the physical processes of transport and deposition cannot take place.

There is a strong correlation between landscape and soil formation. The landscape, mainly the factor topography, influences the amount of water that enters the parent material or runs off. The length and degree of slope strongly affects the balance between soil creation and erosion that destroys the soil. Both plants and animals are included in the biotic factor. Plants are extremely significant as a soil-forming factor. They bind atmospheric carbon and add it to the soil as organic matter. Root exudates are also sources of nutrients and energy for micro-organisms, which in turn form part of the biotic component and are, perhaps, its major element.

$CO_2$, which is released by roots and micro-organisms' respiration, contributes to the pH of the soil solution. Roots remove ions from the soil solution and add ions during growth. The uptake and return of ions is called nutrient recycling. Leaves and wood that fall to the ground are decayed. Pores and structure in the soil are created by roots and soil animals of all sizes, from microscopic unicellular organisms to hamsters or even wombats, large soil mammals living in Australia. Such animals play a role in the decomposition of organic matter and in the biological mixing of surface organic matter into the subsoil and of subsoil horizons with the surface (Singer a. Munns 1999).

Organic matter and biological mixing are important factors in soil genesis. Vegetation cover contributes significantly to the stabilisation of the land surface.

The vegetation cover is particularly important in landscapes with soft parent material occurring on slopes, supporting as it does landscape stability and helping to prevent erosion and soil loss.

Soils have been described ever since people began using them. Very old terms like "heavy" and "soft" soil reflect the difficulties and advantages of cultivation. The oldest areas of land used for growing crops on the different continents are mainly situated in the continental and semi-arid climatic regions, e.g. the Near East, North Africa, Central Asia, China, western South and Central America and Central and Eastern Europe. In most of these areas, we also find the oldest urban civilisations, which are linked to agriculture and herding (Williams et al 1998).

As precipitation in these regions is limited and/or unevenly distributed during the growing period, certain physical and chemical conditions of the soil, responsible for cultivating and growing crops, had to be taken into account. In addition, soil formation is time-dependent, the duration of soil genesis varying and depending on several factors. In the temperate regions, for example, where the tremendous climatic deterioration during the quaternary geological period changed the earth's surface, soil genesis began again about 10,000 years ago.

Agricultural land use in different parts of the world dates back more than 6,000 years. The cumulative effect of all the different kinds of land use and interventions in the environment since the proto- and early neolithic periods has resulted in major problems in terms of the status of the world's soil resources. Soil and water, environmental protection and sustainability of the use of resources are relatively new terms and issues. For example, until 1999 there were no regulations on soil protection in Germany. It is questionable whether, prior to this, soils were sufficiently protected by, for example, water- and nature protection regulations.

**Human-Induced Soil Degradation**

The scale of soil degradation throughout the world is alarming, and yet exploitation still goes on. The main types of soil degradation, shown on the World Map on the Status of Human-Induced Soil Degradation (United Nations Environmental Program and ISRIC) produced as part of the GLASOD programme (Global Assessment of Soil Degradation, Oldeman et al 1991), are listed below:

<u>Water Erosion</u>
Wt      loss of topsoil
Wd     terrain deformation/mass movement

<u>Wind Erosion</u>
Et       loss of topsoil
Ed      terrain deformation
Eo      overblowing

<u>Chemical Deterioration</u>
Cn     loss of nutrients/organic matter
Cs      salinisation
Ca      acidification
Cp      pollution

Physical Deterioration
Pc    compacting/crusting
Pw   waterlogging
Ps    subsidence of organic soil

The causative factors are:
F    deforestation and removal of the natural vegetation
G   overgrazing
A    agricultural activities
E    overexploitation of vegetation for domestic use

(bio)industrial activities (leading to degradation type 'Cp pollution')

There seems to be an obvious difference as regards the major causative factors of soil degradation between the industrialised and the so-called developing nations. While factor I has in modern times played an important role in most of the industrialised world, factors F, G, A and E have a long and well-documented history, but most developing countries lack the sort of management strategies that have already been established to deal with these factors in the industrialised nations.

## 10.1.2 Soil and Water

As already stated, water plays an important role in soils. Water and water-soluble compounds are among the main biota taken up by plants and therefore essential to their life. Depending on their texture, i.e. the grain size of their mineral components – important here is the amount of clay – and on their organic matter content, soils are able to hold water and water-soluble components against the force of gravity. Again, depending on soil texture and organic components, only a certain portion of the stored water is available to plants. Plants growing in clayey soil – which has about the same water content as sandy soil – might die owing to the former's lower level of available moisture.

The texture and water-retention capacity of an unsaturated soil are therefore integral parts that are useful for assessing the threat to groundwater by pollutants. Some ecological properties of soil, such as its water-retention capacity and fertility, were recognised by people in very early times, irrigation and rainwater harvesting already being established techniques by the neolithic period, during which river flood plains were also inhabited. If we look at the current global distribution of loess (a wind-blown calcareous silt), we find that it often occurs in areas with the oldest land use.

Sequences of loess soil profiles (catena) of arable land in the hilly areas of Northern Germany (Kuntze et al 1994) show a complete loss of the original soil cover along the slopes due to water erosion and accumulation of soil material in the valleys forming the so-called colluvium. This process has been going on since the early neolithic period. Other areas of Europe show a strong degradation of soils, caused mainly by clearing and growing annual crops or overgrazing. These areas share the same features with other temperate zones with a history of land use dating back several thousand years.

In the subtropics and tropics, water erosion is not as time-dependent as it is in the temperate zone. It is therefore known to constitute a severe problem, depending on soil properties. The current yearly loss of topsoils due to water erosion is highest in the tropics, with maximum levels of between 10 and 100t/ha · a ($10^3$ t/ha · a) in South-East Asia, where soft volcanic soils occurring on steep slopes are easily eroded by rainfall. The Universal Soil Loss Equation, established by Wischmeier and Smith (1978) and adapted for many different countries, is based on empirical research and statistical analysis of field experiments. It was originally developed to calculate water erosion on individual agricultural plots (Schäfer 1999).

The Universal Soil Loss Equation (USLE) is generally written in this form:

Soil Loss  A: R* K* L* S* C* P (tons/ha/year)

| | |
|---|---|
| A | is the long-term average annual soil loss for a location |
| R | is the long-term average rainfall-runoff erosivity factor |
| K | is a soil erodibility index |
| S | is a slope angle factor |
| L | is a slope length factor |
| C | is a soil cover factor |
| P | is an erosion control practice factor |

Water erosion is mainly dependent on soil texture and aggregate stability, organic matter and plant cover (Bergsma 1996). Hence highly infiltrating soils are well protected against water erosion, but might be affected by wind erosion if they are not or only loosely covered by vegetation. On the other hand, high infiltration rates can cause leaching of soil components, with nutrients or pollutants being transported to the groundwater. Soil and water pollutants affect human beings and biota. Soil contamination occurs when sufficient quantities or concentrations of harmful substances accumulate, in excess of their natural or background levels in soils. Sources of soil contamination include industrial waste, sewage sludge, landfills and the use and misuse of agricultural chemicals.

Contamination may present a risk in terms of chemical toxicity, which reduces crop yields and the usefulness of soils for agricultural land use. It is or can become dangerous to humans and animals through direct contact, through drinking contaminated water or through the food chain, by eating products grown on contaminated soils. Soil contamination often affects groundwater quality, recharged seepage water leaching the solum (lat. soil). Such contaminants include chemical elements such as lead (Pb), mercury (Hg), cadmium (Cd), copper (Cu), chromium (Cr), arsenic (As) and their compounds as well as organic chemicals such as pesticides, solvents, oils and chlorinated hydrocarbons from industrial production, refuse incinerators, etc.

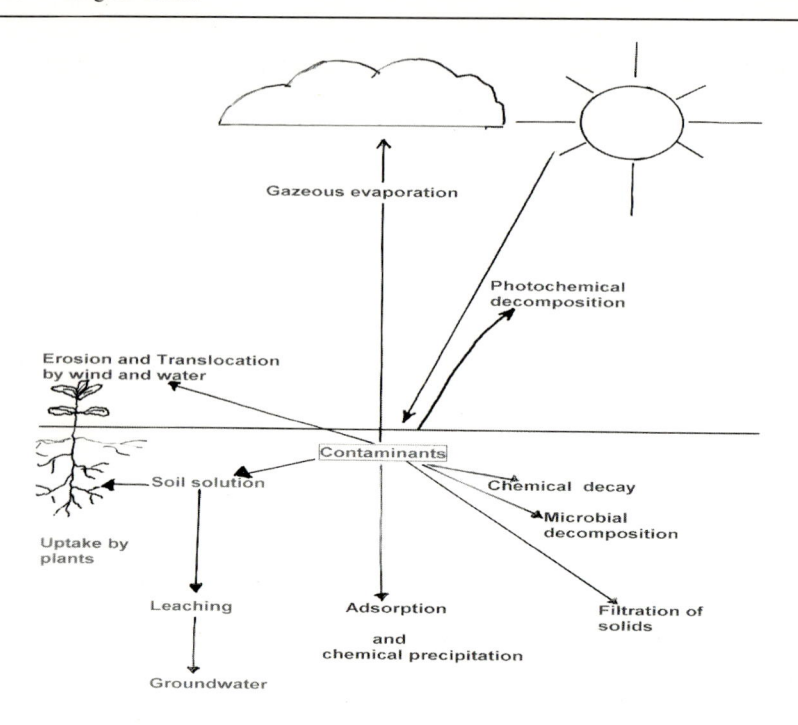

**Fig. 10.1.** Paths of Soil Contamination

Radioactive materials, sodium, salts and acids as well as asbestos and other hazardous materials – all carry environmental risks, as do nutrients such as nitrate and ammonium. Alkalinity, sodium and salts originate in mineral weathering. Saline soils contain large amounts of soluble salts, most commonly Na, Ca and Mg with chloride, sulphate and bicarbonate. An accepted measure of and criterion for salinity is the electrical conductivity (EC) of the soil's paste or saturation extract.

Conventionally, soils are considered saline if their EC exceeds 4 milliSiemens/cm. Many plants suffer at this level. Sodic soils contain Na (sodium) as a significant proportion of their total amount of exchangeable cations (>10%). If a sodic soil is irrigated or receives rain, the structure of the soil is destroyed and the pores clog. In nonsodic but saline soils, soil structure persists when water infiltrates. Ions can be transported to the groundwater, which is often used for irrigation, increasing the problem of salt accumulation and limiting plant growth. Another human health hazard is the use of saline drinking water, which, for example, causes diseases due to high blood pressure.

## 10.1.3 Global Significance

Many countries have adopted environmental laws or specific soil- and water protection regulations, e.g. the German Federal Soil Protection Ordinance, which only recently came into force (BBodSchV 1999). The 9$^{th}$ Conference of the International Soil Conservation Organisation (ISCO 1998) was held under the motto "Towards Sustainable Land Use" and presented several examples of Furthering Co-operation Between People and Institutions to reinforce the international community in its continued commitment to supporting programmes for improving land quality.

Financial and economic support by national and regional governments for baseline studies is called for, and prevention is proposed as a key principle rather than rehabilitation, which is extremely difficult and often impossible because of severe soil degradation. Another of the options outlined in the conference proceedings (ISCO 1998) is that transboundary eco-regional land conservation and basin-wide watershed development can be facilitated under the auspices of international conventions (e.g. AGENDA 21), action programmes or regional frameworks.

The European Union recently (EU-WRRL 2000) established guidelines for the protection of water aimed at transboundary

- protection and improvement of aquatic ecosystems
- support of sustainable use of water resources

River Catchment Management plans (basin-wide watershed management) must be developed at the national level to reduce water pollution and minimise the ecological impact of flooding and droughts. The analysis and classification of River Catchment Areas should provide the basis for deriving and enforcing measures to improve the present situation. Such watershed management plans will based on the strategies of the International River Catchment Commissions.

## 10.2 Project Water and Soil

### 10.2.1 Skills and Aims

The project water and soil covered interactions between land use, soil properties and seepage-water and groundwater quality, taking into account the above-mentioned integrative aspect of sustainable watershed management.

Water and Soil – Towards Sustainable Land Use, was a joint project involving women junior scientists from 14 different countries. They came from Nigeria, Romania, Madagascar, Ukraine, Germany, Georgia, India, Sudan, Russia, Philippines, Korea, Kenya, USA, Myanmar.

Though they had backgrounds in fields as diverse as biology, chemistry, biochemistry, geology, geography, physics, agronomy economics, agricultural sciences, environmental sciences, civil engineering and educational sciences, the junior scientists created a common sense of purpose within the project and contributed their varied skills to the group's joint endeavour.

The first phase of the project included field trips and excursions, e.g. to the Nature Conservation Centre at Bleckede/Elbe, Lower Saxony, to study the results of recent research in the form of field and laboratory investigations on the floodplains of the River Elbe and examine the projects for future management of arable lands. In the second phase of the project, studies, field and laboratory work covering the physical, chemical and biological properties of soils and plants were conducted in close co-operation with the Stahlbach Creek Project (cf. chapter 9).

Landscape analysis, soil mapping and analyses of soil material from different locations in the village of Hamerstorf (District of Uelzen/Lower Saxony, Northern Germany) along the Stahlbach Creek served as an information base for data relating to the filtering and transformation capacity of soils and their corresponding susceptibility to degradation, and generally encouraged the adoption of a more holistic view of sustainable land and water use.

### 10.2.2 Stahlbach Creek Project

Following the pioneering River Elbe Ecology Project, the Stahlbach Creek Project aimed to examine relations and interactions between managed soils in the floodplain of the Stahlbach creek and adjacent areas. Plant communities – including pasture, lower plants in fallow and crops – as well as seepage(soil)-water and groundwater quality were identified and analysed at three locations in the village of Hamerstorf, near Suderburg in the district of Uelzen/Lower Saxony in Northern Germany between 4 September and 5 October 2000.

The impact of land use and interactions between soil and water (surface water, seepage water, groundwater) were studied in soil sequences along managed lands in the floodplains. Field and laboratory tests to quantify and classify the content of nutrients and heavy metal pollutants in soils, plants and water in order to predict their threat to water quality, crop production and the ecosystem, as well as to human beings, were also conducted. Preliminary conclusions were drawn and measures for soil, plant and water protection were derived from the topics studied and initial data.

The results of the work done in this project could serve as a basis for future investigations on water and soil in relation to sustainable land use. Various places were visited in order to gain a better understanding of the impact of landuse and management on soil, on water and on groundwater. They included: the River Elbe and the Nature Conservation Centre at Bleckede (Elbtal Haus); the Elbe Ecology and Research Project at Radegast; the weir and fishway at Geesthacht, River Elbe; Hösseringen Agricultural Museum in Hösseringen/Suderburg; wastewater irrigation in Wolfsburg; the River Banks Maintenance Department/District of Uelzen;

the sugar beet factory in Uelzen; and the Uelzen Composting Plant (landfill and waste dump).

## 10.2.3 Methodology

The practical project work was preceded by a one-week introductory course on how a soil is formed, its components, soil units and the processes relating to groundwater recharge and quality as well as a study of regional geological and soil maps. In view of the large number of participants, three groups were formed to investigate three different management locations. These then began field work including mapping and sampling of soil, water and plants at their respective locations in Suderburg/Lower Saxony, Germany.

The three selected land sites were: arable land used for growing carrots and onions, close to an old reclaimed waste dump (Arable Land); a meadow used to graze Galloway cattle under a programme for the extension of grassland (The Meadows); and a fallow whose only management for more than 10 years has consisted in mowing the grass once a year, leaving the cuttings on the soil to decay (The Fallows). All three locations belong to the catchment area of the Stahlbach Creek, which is a contributor to the River Hardau, which in turn drains to the Ilmenau, a direct contributor to the River Elbe.

**Field Work and Laboratory Analysis**
Collection of Soil Samples:
Soil samples were collected by the three groups, using Edelman soil augers, from depths of 0-15cm and 15-30cm from the ground surface. The samples from each depth were mixed thoroughly, quartered and taken in approximately 1kg quantities for laboratory analysis. Soil profiles were checked in exposures and disturbed and undisturbed (steel rings) samples were collected from the different soil profiles for laboratory analysis.

Collection of Plant Specimens:
Carrot and onion samples and their leaves were collected from the arable land and stored in the freezer for further analysis. Plant material was also collected from the other two sites for identification and preparation of a herbarium. These samples were oven-dried at 60°C for two days.

Water Sampling:
Soil water samplers were installed in the fields using a Pürckhauer soil auger. Water samples from the creek were also taken for analysis.

Field Tests:
Field tests for the presence of ammonium, nitrate and potassium in the soil and water samples were done using field kits (Agro Quant 14602) and the results obtained were subsequently compared with the laboratory data.

Laboratory Analysis: Well-mixed soil samples were prepared for laboratory analysis. The tests were intended to determine (detailed procedures are described in chapter 13):
- moisture content and dry weight
- organic matter
- pH value
- soils' salinity (electrical conductivity, EC)
- total amount of micro-organisms in solids (microbial count)
- soil respiration, biological oxygen demand (BOD), respiration activity of compost
- carbon content
- nitrogen (Kjeldahl procedure)
- C/N and C/P ratio
- carbonates
  (determination of plant-available phosphorus and potassium (calcium acetate-lactate method, CAL)
- Nmin (NO$_3$ and NO$_2$)
- Nmin (NH$_4$)
- exchangeable cations (cation exchange capacity, CEC)
- nitrohydrochlorid acid disintegration (total heavy-metal content)
- soil moisture-retention capacity, pF value
- soil texture
- grain-size distribution
- X-ray fluorescence spectroscopy

Creek and soil water samples were analysed for
- pH, EC, chloride, nitrate, sulphate, ammonium

The status of plant nutrition and soil parameters were assessed after Fink (1989, 1982), Blume (1992) and references quoted in chapter 13.

## 10.3 Results of The Project

### 10.3.1 Case Study: River Elbe Ecology Project

One of the main goals of the project Water and Soil - Towards Sustainable Land Use conducted in the project area Water was to impart knowledge about integrative, interdisciplinary and practicable tools and techniques for identifying and describing the impact of different types of land use systems on the environment, and to develop strategies for the participatory development of appropriate measures. Care was therefore taken to select techniques and methodologies that are applicable to other climates, phyto-geographical, geo-morphological and edaphic conditions, taking into account different socio-cultural backgrounds.

Studies carried out in river catchment areas are therefore suitable for providing more comprehensive information on the relevant issues. The case study of the River Elbe Ecology Project served as the baseline project here. The results of this project were discussed and helped in understanding the different aspects of land and water use and the interactions between environmental factors. The structure, procedures and main results of the case study, which constituted the preliminary objectives of the project Water and Soil, are described below.

To enhance knowledge about the diversity of biotic and abiotic components, about river hydrology and the sustainability of land use systems, the German Federal Ministry for Education and Research launched the "Elbe Ecology" project. The subproject "Aims and models of nature conservation and properties of their realisation in co-operation with (agricultural) land users in the Elbe Valley, Lower Saxony" which ran from 1997 to 2000 was supervised by the Alfred Toepfer Academy for Nature Conservation (NNA), Lower Saxony and comprised a number of contributing disciplines and components.

This was reported at the session "Water Characteristics and River Catchment Management (17 August 2000, *ifu*: "Aims of nature conservation and its realisation in terms of agriculture aims, instruments and costs of a sustainable agriculture in the valley of the River Elbe in Lower Saxony"), focusing on integrated development of the region, strategies and aims (NNA Elbe-Projekt 2000). The participation of farmers in the development process was one of the fundamental requirements to achieve the goals set. The introduction to the project promoted understanding and acceptance of the interdisciplinary and mediated method of applied research in the field of environmental sciences by *ifu* junior scientists. The project, which is summarised in the following chapter, can thus be regarded as a pioneering example of an integrated research project on land use, the status and role of soil properties, and water quality.

## Brief Description of The River Elbe and Protection Situation

The River Elbe has its source in the Czech Republic, then flows mainly through German territory to the North Sea basin. It has a length of 1,091.47km. The Elbe forms the axis of an important broad regional biocorridor for aquatic biological communities. On one or both sides of the river, established zones of protection are looked after by the International Elbe Protection Commission (IKSE 1997).

On German Territory, The River Runs Through
5 national parks
5 UNESCO biosphere reserves (mainly on the coastal mud flats)
113 nature reserves
32 landscape protection areas

On Czech Territory, The River Runs Through
2 large landscape protection areas
2 national nature reserves

By connecting biotopes along the river, the maintenance and restoration of landscape interactions are guaranteed. The protection programmes and international agreements reflect concern for improving and protecting river-water and soil quality, in addition to ecological considerations.

## Management of Arable Land in the Elbe Floodplains

### Germany
Several options exist for regulating land management:
Farming in nature reserves is based on contracts between government and land users. When the state or nature-protection associations are the land owners, they work out regulations for suitable land management.

### Czech Republic
Regulations on the application of fertilisers and pesticides exist only for protected water supply zones. Cattle grazing, which is appropriate to the nature of the landscape in the Czech Elbe floodplains, is not an environmental problem.

### Example
- reduced management of pasture in the floodplains
- no cattle grazing along the river bank
- farmers would receive compensation
- or/and offered land in exchange (IKSE 1997)

### Furthermore
- Regulations on management of the river forelands are aimed at increasing water retention in those areas.
- No landfill sites and removal of inherited waste dumps along the floodplain.

## The River Elbe Ecology Project

This subproject is concerned with the analysis of biotic and abiotic features, land use and agronomy, and is being conducted in co-operation with seven land users (farmers) in an area covering 570km$^2$. The project will develop scenarios predicting future development with reduced, increased or without human impact and provide tools for sustainable land use. The area under investigation spans several nature reserves and is a planned biosphere reserve.

## The Subproject Component Water and Soil

The subproject component Water and Soil ran between April 1999 and October 2000 at the Department of Civil Engineering (Water and Environmental Management), University of Applied Sciences, Suderburg. The aim of this project is the ecological characterisation of selected land units, relations and interactions between managed fluvisols, plant communities and groundwater, groundwater dynamics and quality.

Fluvisols (ISSS, ISRIC, FAO 1999, Urban et al 2001) in the area are mainly composed of river clays with a low to medium content of sand and receive floated material during flooding, in some cases more than twice a year. The Elbe transports nutrients and has transported pollutants contained in the floated material.

## *Research*

The research conducted at about 30 locations in the floodplain included: geological and pedological mapping of the locations by drilling to a depth of 4 metres below the surface; describing the profiles using field methods (texture, colour, groundwater level, stagnation horizons, oxides, etc.); and sampling the top metre of the sequence for laboratory analyses. In addition, surface-soil samples were taken (upper 30cm), twice a year on arable land and four times on pasture.

**Physical Properties**
Soil water-retention capacity, texture, electrical conductivity and pH value provide information on soil filter capacity and its retention capacity with respect to nutrients and pollutants.

**Chemical Properties**
Cation-exchange capacity, potassium, calcium, magnesium, phosphorus, nitrogen, carbon, total organic matter, C/N ratio, nitrate and ammonium and total element analysis provide information on cation fixation and cation exchange and retention capacity and its total amount of chemical elements as well as ecologically relevant nutrients and pollutants.

**Biological Properties**
Denitrification capacity provides information on the trophic status, microbiological activity and tools for management recommendations.

**Analysis of Soil Water (Monthly)**
At 10 sites, we installed soil water samplers that pump seepage water out of the soils from depths of 110cm and 60cm below the surface.

**Physical Analyses**
pH, electrical conductivity

**Chemical Analyses**
DOC, dissolved organic carbon, P, Ca, K, Cl, sulphate, nitrate, ammonium

Groundwater wells were installed at 12 sites, 10 of them adjacent to the soil water sampling sites. The analysed parameters are identical to those measured in the soil water. The amount and chemical composition of rainfall was measured at each site and the relationship with soil moisture assessed.

The results are scaled against certain soil features, management types and plant communities or land use systems. The final results on the status of soil and water quality and soil properties are presented on 1:5,000 scaled maps and will be used for modelling scenarios .

## Prediction and Strategies for Future Development of Landscape and Biotopes in Relation to (Agricultural) Land-Use Systems

**Table 10.2.** Regional Models

| Environmental Quality Standard | Environmental Quality Goals |
|---|---|
| Model A<br>The ecological model: | natural development of the area without any further human impact |
| Model B<br>The biodiversity model: | achieving highest level of biodiversity |
| Model C<br>The abiotic resources protection model: | soil and water protection by sustainable land use |

Scenarios for all models and mixes between model patterns will be acquired and presented on different scales, predicting landscape development over the next 30 years (Alfred Toepfer Akademie für Naturschutz, NNA 2001). Within the subproject component Water and Soil, studies were conducted to investigate the present status of water and soils in the area of the Lower Middle Elbe river floodplains in order to characterise sites located between the main dyke and the river (external dyke) as compared with the areas behind the dyke (inner dyke).

Most of the land use systems in the external dyke are grazing lands exposed to flooding up to three or four times a year. Arable lands predominate in the inner dyke where there is little risk of flooding. Reports from ARGE (ARGE-elbe.de) show that the suspended materials in the River Elbe contain heavy metals and nutrients, which are accumulated in the floodplain soils during flooding. Since the sources of the contamination of the floodplain soils are the suspended particles and sediments in the River Elbe, dividing the Elbe's water quality into different classes, could help in predicting the level of contamination in the soils. The results of total element analysis of topsoil samples are summarised by way of an example.

Because of regular flooding, the investigated locations in the external dyke area contain higher amounts of heavy metals than those of the inner dyke locations (Gebrenegus Beyene 2001). The amount of mobile fraction of heavy metals (bio-available) is even more important than the total content from an ecological point of view. To assess the biotoxicity of these metals, analysis was done using ammonium nitrate extraction and comparisons made with existing threshold levels (BundesBodenschutzGesetz und Altlastenverordnung 1999). Depending on the plant species and other factors, the transfer of bio-available metals to plants varies, and in this study the transfer coefficient was calculated and compared with some ranges determined in other research and the existing Federal Foodstuffs Ordinance (Futtermittelverordnung 1992).

To determine the mobility of heavy metals in soils, their toxicity and to take countermeasures to remove the threat they may represent to human beings, the soil factors influencing the availability of metals must be assessed. In our research, the key soil factors – texture, pH value (CaCl$_2$), humus content and cation-exchange capacity – were correlated with the mobility quotient (concentration of the mobile fraction by ammonium nitrate extraction/total content from nitrohydrochlorid acid disintegration). As regards the main objective of identifying pollutants in the fluvisols using different detection methods, the following conclusions can be drawn:

- Nitrohydrochlorid acid disintegration dissolves only environmentally mobile metals held on surface coatings and in oxides and sulphides, but does not dissolve crystalline silicate structures. Hence the amounts measured using atomic absorption spectrometry or inductively coupled plasma spectroscopy (ICP-AES, AAS) are lower than those obtained using the X-ray fluorescence spectroscopy (XRFS) method. As for the second method, the analysis of heavy metals is done directly by X-Ray fluorescence spectroscopy (XRFS) without any special preparation of the samples, except for drying and graining. Both methods showed an excellent correlation for Ni, Cu, Zn and Pb. The total amounts of Hg and As are too small to be detected by XRFS and so no comparison was made for either of them.
- The results from both methods of total metal analysis showed that the locations in the external dyke (sites between the river and the dyke) have a much higher total content of heavy metals than the areas at the back of the dyke. The main reason is that the sites in the external dyke are subjected to direct inundation during flooding, and the pollutant-bearing sediments that are transported by the river are deposited there, thus enriching the heavy-metal content. The evaluation of heavy-metal contamination under existing regulations (Federal Soil Protection Ordinance 1999) showed that most of the sample areas in the external dyke contained a total amount of Zn, Pb and Cd (nitrohydrochlorid acid disintegration) above the recommended limits. At most of the sites, the amount of metals obtained was a great deal lower than the action values, with the exception of sites of one locality for As and Hg.
- Plant samples from grassland were collected and their heavy metal concentration was correlated with the existing Foodstuffs Ordinance. Maximum permissible levels exist for Cd (1.1mg/kg) and Pb (45mg/kg) only, elements with extremely broad ranges of mobility. With the exception of two samples for Cd, all other plant material showed values lower than the permissible level of Cd and Pb. In most samples, the concentration of elements in plants increases in this order: Zn > Cr > Cu > Ni > Pb > Cd. The concentration of metals in plants was poorly correlated with the mobile fraction in soil.

The probable reason is that the plant samples were not washed before analysis, which means that the analysed elements were not necessarily only the ones by roots; they may have included metals, which remain adsorbed on the external part of plants (pollutant pathway). It remains difficult, then, to predict the correlation between the mobile fraction of metals in the fluvisols and the concentration in plants. However, the transfer coefficient (total concentration in plants/total content in soils) was taken to assess at least the probable risk.

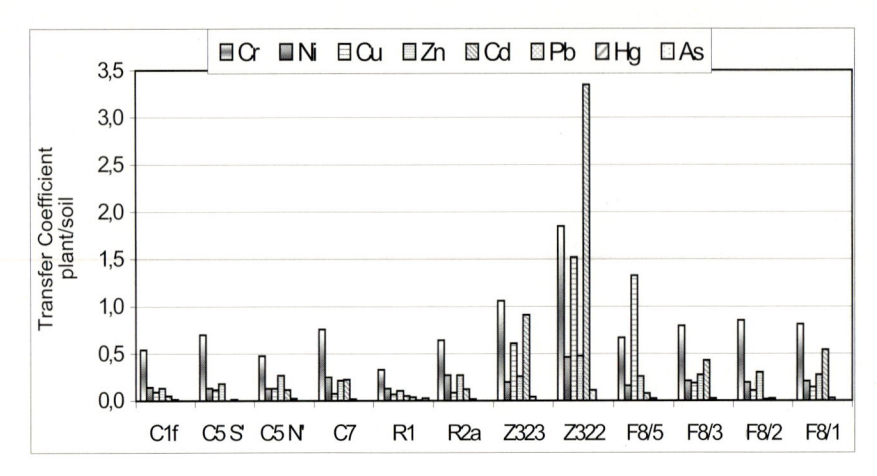

Fig. 10.2. Transfer Ccoefficient (Total Concentration of Heavy Metals in Plants / Total Content in Soils) of Grasslands Situated on the External Dyke (Except Z 323 and Z 322) in Flood Plains of River Elbe (Urban et al 2001)

The transfer-coefficient values were compared with values obtained in other research (Kahl et al 1994). The values lie within the normal ranges, with the exception of Cr in two samples. Although one of these sites is located on the inner dyke, it showed a higher transfer coefficient for most elements (Z322). It can be deduced that the prevalent type of plant (Alopecurus spec.) can absorb large amounts of heavy metals entering the soil through the use of fertilisers, which contribute to the pollution.

Geological and pedological mapping of the sites provided important information for deducing and monitoring the potential threat to groundwater by the leaching of nitrate, pesticides and heavy metals. Owing to the top clayey-silty soil texture at most of the locations, moisture-retention capacity and plant-available soil-moisture content is quite high, and the risk of nitrate leaching at most of the locations is considered to be very low to medium. However, the nitrate levels of the investigated groundwater wells and soil water samplers of arable land, situated on both sides of the dyke, showed in many cases mean annual concentrations (12 and more tests/a) exceeding the Federal and EU Ordinance for Drinking Water (50mg/l; 30mg/l). For pasture and grassland, none of the 21 locations had mean groundwater concentrations of nitrate above the mentioned threshold values.

This reflects soil properties – low sorption capacities of locations with sandy topsoils – and the type of land use – highly fertilised arable land. To summarise: an extensive land use type can be recommended to meet the environmental goal of soil and water protection. This would mean no arable land use at external dyke locations, low or no fertilisation of grassland (containing endangered species) and adapted methods of management, as proposed by the biodiversity working group (NNA 2001).

**Table 10.3.** Measures within Sustainable Development

| | |
|---|---|
| To maintain grassland in the floodplains: | -no fertilisation of soils<br>-mowing twice a year (2$^{nd}$ cut after 1 Sept)<br>-no ploughing up of grassland<br>-no draining, no soil melioration<br>-no late grazing |
| To support development of grassland in the floodplains: | -no fertilisation of soils<br>-mowing twice a year<br>-to reduce nutrients, even mowing 3 times a year<br>-no ploughing up of grassland<br>-no draining<br>-no soil melioration<br>-no late grazing |

## 10.3.2 Description and Results of the Three Project Sites

The results of the multiple analyses conducted by the three subgroups were compiled. Preliminary conclusions drawn from the results and analyses are presented in this section.

### Arable Land

**Table 10.4.** Arable land

| | |
|---|---|
| Parent material: | Cover sand overlying glaciofluvial deposits of the Saalian period (Quaternary) (NLfB Quartärgeologische Übersichtskarte 500) - (sand / sandy loam) |
| Soil type: | Cambisol (Braunerde) (BÜK 50, L3128 Uelzen) |
| Groundwater depth: | >200cm |

The soil profile, taken in the onion field, revealed the existence of three distinct horizons. The upper layer was 30cm thick, consisting of the root zone for the crops. The second horizon, measuring 60cm in thickness, was characterised by sand and stones. The third horizon was made up of light sand of the parent material. The groundwater was not reached in the 1-metre-deep pit dug for soil profiling. The analysis of physical parameters revealed the presence of medium to fine sand grains, a normal soil pH and a low CEC, which were typical of the poor sandy parent material and of leaching.

Most biological activity was found in the topsoil layers, as shown in the BOD and microbial-counts data. Little microbial activity was observed in the lower horizons. The chemical parameters investigated provided a plethora of data, the most notable being the nutrient data and the heavy-metal analysis data. The C:N ratio was optimal for the topsoil samples, but the C:P ratio indicated a lack of phosphorus. The heavy metals found in the soil were all below regulation limits. A sample from the former waste-dump site provided some variation in this data. The levels of Cd and Zn in the top remediated layers covering the dump (> 30cm below surface) exceed regulation limits.

**Table 10.5.** Soil Content of Total Heavy Metals (mg/kg DS), Arable Land

|  | Al | Cd | Cr | Cu | Fe | Hg | Ni | Pb | Zn |
|---|---|---|---|---|---|---|---|---|---|
| C 0-30 | 2890,5 | 0,175 | 9,867 | 3,7245 | 3789,5 |  | 3,0665 | 11,495 | 24,38 |
| Dump | 4260 | 1,574 | 10,59 | 28,28 | 6446 |  | 6,59 | 60,1 | 198,5 |
| O 70-100 | 4984 | 0,255 | 14,13 | 10,73 | 14610 |  | 7,979 | 25 | 80,49 |
| Limit |  | 0,4 | 40 | 20 |  | 0,1 | 15 | 40 | 60 |

**Table 10.6.** Plant Content of Total Heavy Metals (ppm), Arable Land

|  |  | Al | Cd | Cr | Cu | Fe | Ni | Pb | Zn |
|---|---|---|---|---|---|---|---|---|---|
| Leaves: (shoot) | Carrots | 371,9 | 0,8135 | 2,0255 | 4,9995 | 519,75 | 2,607 | 2,0025 | 27,86 |
| Ranges |  |  |  |  | 7-15 |  |  |  | 30-80 |
|  | Onions | 71,265 | 0,2405 | 0,4825 | 4,0645 | 85,36 | 0,2995 | 0,309 | 9,128 |
| Ranges mg/kg DS |  |  |  |  | 7-15 |  |  |  | 20-70 |
| Fruits: (roots) | Carrots | 79,695 | 0,639 | 0,9495 | 6,24 | 78,73 | 2,8915 | 0,722 | 21,975 |
|  | Onions | 42,405 | 0,3325 | 0,3825 | 5,948 | 59,04 | 0,7795 | 0,584 | 31,81 |

**Table 10.7.** Nutrients and Trace Elements of Crops (ppm), Arable Land

|  | N % | P | K | Ca | Mg | Na | S |
|---|---|---|---|---|---|---|---|
| Onions | 1,11 | 284,55 | 581,45 | 245 | 86,08 | 47,69 | 182,85 |
| Carrots | 1,02 | 343 | 1639 | 1931,55 | 312,5 | 102.51 | 385,3 |
| Carrots leaves | 1,455 | 475,6 | 1467 | 265,3 | 95,76 | 132,2 | 130,6 |
|  | B | Mo | Cu | Mn | Zn | Co | Fe |
| Onions | -2,847 | -0,135 | 1,6535 | 1,908 | 2,272 | -0,003 | 4,4895 |
| Carrots | -0,121 | 0,0645 | 1,8975 | 13,55 | 3,2355 | 0,054 | 52,31 |
| Carrots leaves | -0,916 | -0,144 | 1,665 | 2,146 | 3,166 | 0,06 | 7,601 |
| Ranges mg/kg DS | 20-100 | 0.2-5 | 5-15 | 20-200 | 10-100 | 0.03-0.3 | 50-200 |

The concentration of heavy metals in both the carrots and onions does not exceed the limits for either cadmium or lead (1,1mg/kg for Cd and 45mg for Pb) set in the German Foodstuffs Ordinance (1992). The level of nutrients and trace elements in the crops is generally below the lower mean value of such elements in crops, except for Fe and Co.

**Table 10.8.** Final Results, Arable Land, Carrots Field

| | Biological Properties | | Physical Properties | | |
|---|---|---|---|---|---|
| Depth (cm) | Microbial Count ($10^6$/g DS) | BOD1 (mgO$_2$/kg DS) | Texture | Field Capacity (pF) | Class. Field Capacity |
| 0-15 | 4,63 | 18,37 | S-fm | | |
| 15-30 | 1,29 | 17,74 | S-fm | 20,73 | 13-26 |
| Dump (>30) | 0,87 | 0 | S-fm | 22,54 | (low) |

| | Chemical Properties | | | | | | | |
|---|---|---|---|---|---|---|---|---|
| Depth (cm) | pH CaCl$_2$ | EC (mS/cm) | Ignition Loss % | C % | Organic Matter | N % | C:N | Class. C:N |
| 0-15 | 5,885 | 0,015 | 2,233 | | | | | |
| 15-30 | 6,015 | 0,018 | 1,857 | 1,104 | 1,898 | 0,075 | 14,62 | optimal |
| Dump (>30) | 6,07 | 0,054 | 3,268 | 1,796 | 3,089 | 0,08 | 22,45 | wide |

| Depth (cm) | N min (kg/ha) | P (kg/ha) | C:P | Class. C:P | K (kg/ha) | Class. K |
|---|---|---|---|---|---|---|
| 0-15 | 3,211 | 272,7 | | | 294 | high |
| 15-30 | 6,528 | 260,3 | 89.72:1 | narrow | 260,4 | high |
| Dump (>30) | | 216,7 | 179.3:1 | optimal | 268,8 | high |

## The Meadows

**Table 10.9.** The Meadows

| Parent material: | (peat and) fluvial deposits, Holocene (Quaternary, (NLfB Quartärgeologische Übersichtskarte 500) |
|---|---|
| Soil type: | Gleysol (BÜK 50, L3128 Uelzen) |
| Groundwater depth: | 70cm |

There are three distinct zones comprising the soil profile of the meadows along the Stahlbach creek. These include the topsoil, the transition zone and a reduction horizon. The topsoil is 30cm thick and consists of the root zone. The transition zone is only 15cm thick and is characterised by the presence of mottling of iron. The reduction horizon is observed to be influenced by the occasional presence of groundwater at this depth. The groundwater table was reached at 85cm.

Grain sizes in the Meadows were found to be in the range of very fine sand. The colour (10 YR) of the soil indicates the presence of iron oxides. The soil pH ranges from 5 to 7 (field measurements) for the topsoil and shows the expected increase from top to bottom in the soil-profile laboratory analysis.

The field capacity (soil water-retention capacity) is low, which is characteristic of the medium-to-coarse sandy texture of the alluvial layers. The largest amount of water can be retained by the Ah horizon at 0-30cm below the surface, which contains about 2% organic matter.

**Table 10.10.** Description of Samples, Meadows

| Sample No. | Descriptions | |
|---|---|---|
| 1 | 0-15 cm Stahlbach creek | Single sample of topsoil / river bank |
| 2 | 0-15 cm Meadow | Soil pit |
| 3 | 15-30 cm Meadow | Soil pit |
| 4 | Transition Zone | Soil pit |
| 5 | Gr Horizon | Soil pit |

**Table 10.11.** Heavy Metal Content of Soil Horizons (mg/kg DS), Meadows

| sample | Al | Cd | Cr | Cu | Hg | Ni | Pb | Zn |
|---|---|---|---|---|---|---|---|---|
| 1 | 26930 | 0,251 | 915 | 7,5 | <2,0 | <1,0 | 14,6 | 16,5 |
| 2+3 | 28050 | 0,185 | 878 | 10,9 | <2,0 | <1,0 | 18,1 | 23,9 |
| Limit | | 1,5 | 100 | 60 | 1 | 50 | 100 | 200 |

The CEC is low owing to the absence of clay and silt fractions and the low levels of organic matter. The microbial counts show high activity in the Ah horizon, the same being true of soil respiration, which indicates a higher level of microbial activity in the topsoil and no activity in the reduction horizon. Optimal values for the C:N ratio were obtained, balanced by low $NH_4$ and high $NO_3$ levels. The phosphorus content is higher in the second layer (15-30cm) compared with the top part of the Ah horizon. Higher amounts of potassium occurred in the deeper layers, indicating leaching and groundwater influence. Heavy metals in the soil are well below regulation limits. Chromium must be excluded because of analytical problems.

Plants were collected at the three sites and identified up to the species level. The absence of flowering branches made identification difficult, especially that of the grass species. Nevertheless, fifteen plants were identified, as shown in the table (Appendix), representing locations close to the creek. Most of the members belong to the families Asteraceae, Plantaginaceae, Urticaceae, Fabaceae, Polygonaceae, Poaceae and Apiaceae. Each plant was described based on habit (life form), habitat (occurrence), and its presence as an indicator of certain properties of soil and water. Three other plants were identified as well: *Festuca* (Poaceae), *Senecio* (Asteraceae) and *Sphagnum* (Musci).

**Table 10.12.** Plant Indicators, Meadows

| Species | Indicator |
|---|---|
| Family Asteraceae: | |
| Taraxacum officinale | -fertile soil, moist, clay and silt soil, neutral-mild pH |
| Taraxacum laevigatum | -loose sand, loess and loam soil, carbonate- containing soil |
| Cirsium arvense | -fertile soil, moist, rocky, sandy loam, loam and nitrogen indicator, pioneer species |
| Chamomilla recutita | -fertile soil, moist, sandy-clay and sandy-silt soil, loamy soil, neutral-slightly acidic pH |
| Achillea millefolium | -fertile soil, rocky, sandy loamy soil, mild-slightly acidic pH, nutrient indicator, sensitive to moist places, soil stabiliser, fodder plant |
| Matricaria spec. | |
| Family Plantaginaceae: | |
| Plantago major | -fertile soil, moist, compacted clay and silt soil, pioneer species, salt tolerant |
| Plantago lanceolata | -fertile soil, moist, sandy loam |
| Family Urticaceae: | |
| Urtica dioica | -fertile soil, humus, moist-damp, loose clay and loam soil, neutral-mild pH, nitrogen indicator, as humid indicator in forested areas |
| Urtica urens | -fertile soil (ammonia), humus, moist, loose clay and loam soil, neutral-mild pH, nitrogen indicator |
| Family Fabaceae: | |
| Trifolium repens | -fertile soil, humus, moist, dense loam an clay soil, mild-slightly acidic pH, nitrogen indicator, on arable land moist indicator, valuable fodder plant |
| Family Polygonaceae: | |
| Rumex acetosa | -Fertile soil, humus, moist-damp, loose loam and clay soil, also on peat soil, nitrogen indicator, mild-slightly acidic pH |
| Rumex crispus | -Groundwater, fertile soil, dense loamy and clay soil, nitrogen and moisture indicator, indicates dense soil |
| Family Poaceae: | |
| Avenochloa pubescens | -moderate fertile soil, humus, loose clay and loamy soil, neutral pH, fodder plant |
| Family Apiaceae: | |
| Daucus carota | -fertile soil, clayey and silty soil, pioneer species, slightly acidic-mild pH |

Basically, most of the plants collected indicated that the soil is fertile and moist. Specifically, *Urtica dioica*, *Trifolium repens* and Rumex species have been reported as good indicators of moist soils. Members of Asteraceae and *Plantago lanceolata* grow typically in soil conditions that are either rocky, loose sandy, sandy clay, sandy loam or sandy silt.

The rest of the species collected thrive well in loose or compacted clay, clay loam and in humus-rich soils. Characteristically, soil pH ranges from neutral to mild to slightly acidic. On the other hand, *Taraxacum laevigatum* is a good indicator of base saturated and also sandy soils. *Cirsium arvense, Urtica urens, Trifolium repens* and *Rumex* species indicate nitrogen-enriched locations close to the creek.

Table 10.13. Final Results Table, Meadows, Soil Profile (pit)

| Depth (cm) | Biological Properties | | Physical Properties | | |
|---|---|---|---|---|---|
| | Microbial Count ($10^6$/g DS) | BOD1 (mgO$_2$/kg DS) | Texture | Field Capacity (pF) | Classification Field Capacity |
| 0-15 | 5,51 | 40,68 | medium sand | 26,32 | low |
| 15-30 | 27,90 | 32,64 | medium sand | 23,32 | low |
| 30-45 | | | fine sand, silty | 19,12 | low |
| >45 | 0,0037 | | fine sand, silty | 13 | low |

| Depth (cm) | Chemical Properties | | | | | | | |
|---|---|---|---|---|---|---|---|---|
| | PH CaCl$_2$ | EC (mS/cm) | Ignition Loss % | C % | Organic Matter | N % | C:N | Class. C:N |
| 0-15 | 5,62 | 0,03 | 2,73 | 1,1 | 1,892 | 0,12 | 9 | narrow |
| 15-30 | 5,92 | 0,035 | 2,67 | 1,11 | 1,909 | 0,11 | 10 | optimal |
| 30-45 | 5,99 | 0,026 | 0,56 | 0,24 | 0,413 | 0,02 | 12 | optimal |
| >45 | 6,01 | 0,026 | 0,32 | 0,1 | 0,172 | 0,01 | 10 | optimal |

| Depth (cm) | N min (kg/ha) | P (kg/ha) | K (kg/ha) | Class. K | CEC (cmol/kg) | Class. CEC |
|---|---|---|---|---|---|---|
| 0-15 | 13,358 | 84,872 | 82,32 | medium | 3,86 | very low |
| 15-30 | 14,936 | 167,130 | 88,20 | medium | 3,86 | very low |
| 30-45 | 3,857 | 57,389 | 99,96 | medium | 1,22 | very low |
| >45 | 3,284 | 19,992 | 141,12 | high | 0,58 | very low |

## The Fallow Region

Table 10.14. The Fallow

| | |
|---|---|
| Parent material: | Peat and fluvial deposits (BÜK 50, L3128 Uelzen) |
| Soil type: | Gleysol (BÜK 50, L3128 Uelzen) |
| Groundwater depth: | 170cm |

The Fallow Region site was broken down into three sites: Site A (close to the road, under trees, oak), Site B (representing best the Meadows/Fallow) and Site C (along the river bank, Stahlbach). Soil mapping of Sites B and C reveals the occurrence of a Late Holocene peat layer at different depths below the surface, which is covered by sandy material. This cover sand was most probably added to the topsoil by people and the beginning of the last century, when land melioration techniques such as meadow irrigation were applied to the poor areas of the Lüneburg Heath.

The general soil profile of this site, which was dug at the transition between Sites B and C, covers four horizons: the topsoil, sand, sand with mottling, and peat. The topsoil has a thickness of 40cm, the sandy horizon is only 15cm, and the sand with a mottled horizon is 20cm thick. The underlying peat extends to the groundwater at 170cm, which was the lowest part of the pit.

Grain sizes of the topsoil samples from the three sites of the Fallow Region are in the range of medium to fine sand. The soil pH of the topsoil and down the profile is slightly to strongly acidic. The CEC is low in the top samples (0-15cm and 15-30cm) of Site B, and the amount of retained water at the same soil depth (23.12Vol%) is low. The microbial counts reveal high activity, as is clearly shown by the rates of soil respiration.

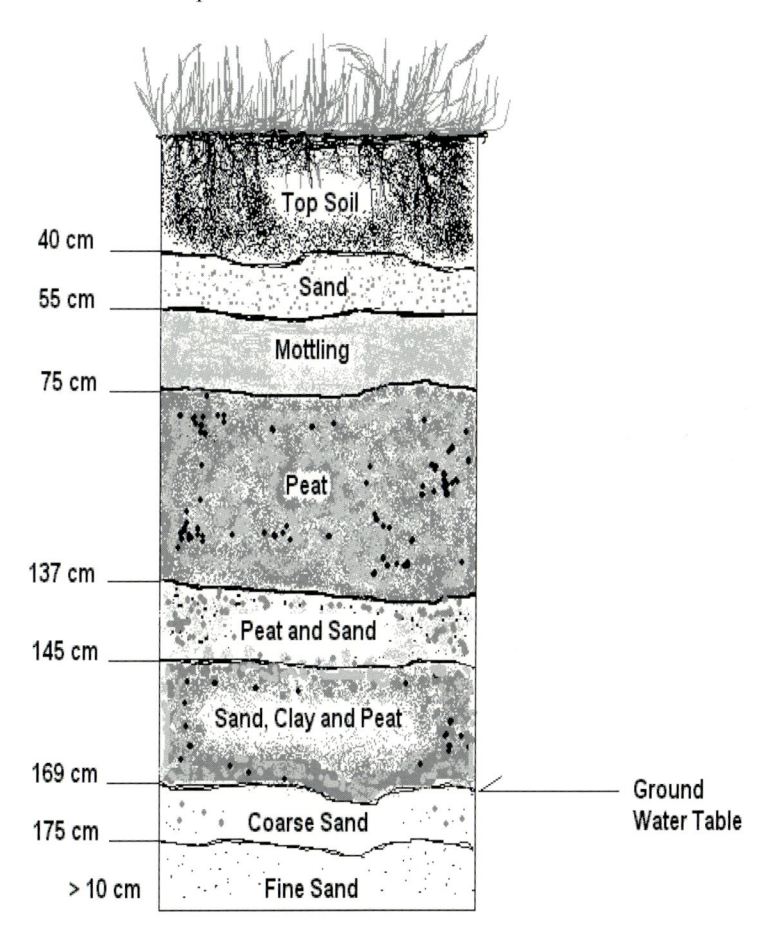

**Fig. 10.3.** Typical Soil Profile Under The Fallows

**Table 10.15.** Soil Content of Total Heavy Metals (mg/kg DS), The Fallows

| sample | Al | Cd | Cr | Cu | Hg | Ni | Pb | Zn |
|---|---|---|---|---|---|---|---|---|
| 0-15cm (6+7) | 27260 | <1,0 | 1040 | 11,7 | <2,0 | 5,5 | 20,2 | 29,7 |
| 0-15cm (8+9) | 31220 | <1,0 | 828 | 12,9 | <2,0 | <1,0 | 18,3 | 25,9 |
| Limit | | 1,5 | 100 | 60 | 1 | 50 | 100 | 200 |

Even an increase in the BOD values down the profile is found, indicating oxygen supply of the lower layers due to soil texture. Values for the C:N ratio indicate a different quality of organic matter depending on the location: high amounts of organic nitrogen (N-Kjeldahl) are found under the trees, whereas the C:N ratio of the topsoil at the Stahlbach creek varies widely, and because of the high organic carbon content of the decomposed peat a very widely varying C/N ratio can be stated. The amounts of potassium and phosphorus in the topsoil samples from sites B and C are indicative of management, as there is no harvesting of the grasses. Heavy metals in the soil of two samples taken from a depth of 0-15cm were well below regulation limits.

**Table 10.16.** Description of Soil Profile, Fallow Region, Study Site B (Soil Pit), Colors and Structure after Earth Colors, Soil Color Book (1997)

| Soil | Horizon | | | |
|---|---|---|---|---|
| Depth | 0-61 | 61-70 | 70-75 | 75-100 |
| Texture | Medium coarse sand with organic matter | Increasing amounts of sand, less organic matter | Medium sand without organic matter | Wet, highly decomposed peat material with fine sand |
| Structure | 2-5mm and 5-10mm | | 5-10mm | |
| Colour | Very dark greyish brown 10YR 3/2-3/3 | Dark yellowish brown 10YR 4/6 | Dark brown 10YR 3/3 | Dark brown 10YR 2/2 |
| Mottling | | | Present | Present up to 85cm |
| Prominence of Roots | Present | | | Peat, wet, roots of plants still visible, flint stones |
| pH | pH: 5 | pH: 5 | pH: 5 | pH: <5 |

**Table 10.17.** Description of Soil Profile, Fallow Region, Study Site C, Colors and Structure after EarthColors, Soil Color Book (1997)

| *Soil* | *Horizon* | | | | | |
|---|---|---|---|---|---|---|
| *Depth* | 0-45 | 45-60 | 60-64 | 64-82 | 82-90 | 90-100 |
| *Texture* | Medium and coarse sand, highly decomposed peat | Sand decomposed peat | Moister, less peat material feels loamy | Sandy and peaty layers, alternat. | Sandy, quite, wet | Middle sand, organic material |
| *Structure* | 2-5mm and 5-5mm aggregates | | | 1-2mm aggregates | | |
| *Colour* | Greyish brown 10YR 3/2 | Dark brown 7.5YR 3/3 | Dark brown 7.5YR 3/2 | Dark reddish brown 5YR 2.5/2 | Dark reddish brown 5YR 2.5/1 | Reddish 5YR 3/2-3/3 |
| *Special features: Stony, low pH* | pH:<5 leaching | pH:≤ 5 stoniness | pH:<5 | pH:4 | pH: ≤ 5 | pH:5 |

**Table 10.18.** Final Results Table, Fallow Region

| | **Biological Properties** | | **Physical Properties** | | |
|---|---|---|---|---|---|
| *Depth (cm)* | *Microbial Count ($10^6$/g DS)* | *BOD1 (mgO$_2$/kg DS)* | *Texture* | *Field Capacity (pF)* | *Classification Field Capacity* |
| 0-15cm Stahlbach | | 72,98 | medium sand | | |
| 15-30cm Stahlbach | | 72,98 | medium sand | | |
| 0-15cm Meadow | 15,80 | 73,29 | medium sand | 23,12 | low |
| 15-30cm Meadow | 10,22 | 73,29 | medium sand | | |
| 0-15cm Under Trees | | | medium sand | | |
| 15-30cm Under Trees | | | medium sand | | |
| 85-95cm Peat | 3,2937 | | | 69,85 | very high |
| 30-70cm Sand | 0,1657 | 127,01 | medium sand | | |
| 70-90cm Sand/Peat | | | medium sand | | |
| 90-100cm Peat | | 64,26 | | | |

**Table 10.19.** Final Results Table, Fallow Region

| Depth (cm) | pH CaCl₂ | EC (mS/cm) | Ignition Loss % | C % | Organic Matter | N % | C:N | Class. C:N |
|---|---|---|---|---|---|---|---|---|
| 0-15cm Stahlbach | 4,97 | 0,022 | 5,89 | 2,29 | 9,9 | 0,08 | 29 | wide |
| 15-30cm Stahlbach | 4,96 | 0,018 | 4,72 | 1,98 | 3,4 | 0,08 | 25 | wide |
| 0-15cm Meadow | 5,10 | 0,019 | 4,67 | 1,74 | 3 | 0,09 | 19 | Opt. wide |
| 15-30cm Meadow | 5,12 | 0,016 | 4,38 | 1,81 | 3,1 | 0,08 | 23 | wide |
| 0-15cm Under Trees | 4,53 | 0,009 | 5,4 | 0,82 | 2,1 | 0,22 | 4 | narr. |
| 15-30cm Under Trees | 4,57 | 0,009 | 1,88 | 0,78 | 1,34 | 0,08 | 10 | narr. |
| 85-95cm Peat | 4,86 | 0,06 | 38,29 | 16,08 | 27,6 | 0,08 | 201 | very wide |
| 30-70cm Sand | 5,52 | 0,01 | 3,91 | 0,14 | 0,2 | 0,92 | 0,15 | very narr. |
| 70-90cm Sand/Peat | 5,44 | 0,012 | 4,46 | 1,56 | 2,7 | 0,09 | 17 | opt. |
| 90-100cm Peat | 4,68 | 0,041 | 40,46 | 11,48 | 19,7 | 0,12 | 96 | very wide |

| Depth (cm) | N min (kg/ha) | P (kg/ha) | K (kg/ha) | Class. K | CEC (cmol/kg) | Class. CEC |
|---|---|---|---|---|---|---|
| 0-15cm Stahlbach | | 161,5 | 310,8 | ex. high | | |
| 15-30cm Stahlbach | | 203,3 | 176,4 | high | | |
| 0-15cm Meadow | 9,9 | 189,2 | 331,5 | ex. high | 4,41 | low |
| 15-30cm Meadow | 10,5 | 206,2 | 197,4 | High-very high | 4,41 | low |
| 0-15cm Under Trees | | 234,6 | 92,4 | medium | | |
| 15-30cm Under Trees | | 209,2 | 67,2 | Medium-low | | |
| 85-95cm Peat | 26,7 | 189,2 | 54,6 | low | | |
| 30-70cm Sand | 0,8 | 53,1 | 79,8 | Low-med. | 0,87 | very low |
| 70-90cm Sand/Peat | | 42,6 | 50,4 | low | | |
| 90-100cm Peat | | 181,1 | 33,6 | low | 13,99 | middle |

## 10.4 Summary

### 10.4.1 Intergroup Interferences

The data collected by the project's interactive working groups during the field and laboratory studies are briefly summarised below and appropriate measures derived from them.

*Potassium*

The amount of potassium in the topsoil (0-15cm) of the arable land and in the fallow is higher that in the lower layers (15-30cm, 30-60cm and 60-90cm). In the arable land, the amount of potassium found was higher than that added by fertilisation. This indicates an accumulation of potassium over a period of years or could be the result of other chemical reactions occurring in the soil that alter cation transport. The meadows, however, had more potassium in the lower layers. This may be due to transport of potassium within the soil by seepage water, through flooding and through the groundwater. Also, the influence of extensive management cannot be excluded.

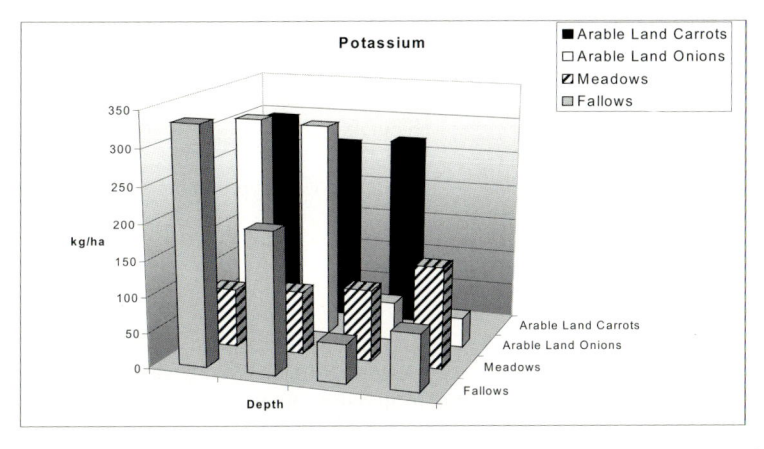

**Fig.10.4.** Potassium Content of Ah-Horizons of Three Land Use Types along the Stahlbach Creek

*Phosphorus*

The samples from the Ah horizon of the arable land show an abundance of phosphorus due to the soil facilities and management. The fallow and the arable land contain sufficient amounts of phosphorus in the topsoil for plant growth. The topsoil samples of the meadows have a medium to poor level of P nutrition.

### Nitrate and Ammonium ($N_{min}$)

No great difference in nitrogen content has been observed between types of land use. However, both the arable land and the meadows receive high doses of nitrogen – through chemical fertilisation and cow manure, respectively – during the growing season to maintain this level off season. The peat layer of the fallows (85-95cm), which is quite rich in organic matter, has the highest nitrate and ammonium ratio. Despite fertilisation, the arable land showed a depletion of $N_{min}$ in the uppermost layer due to plant use.

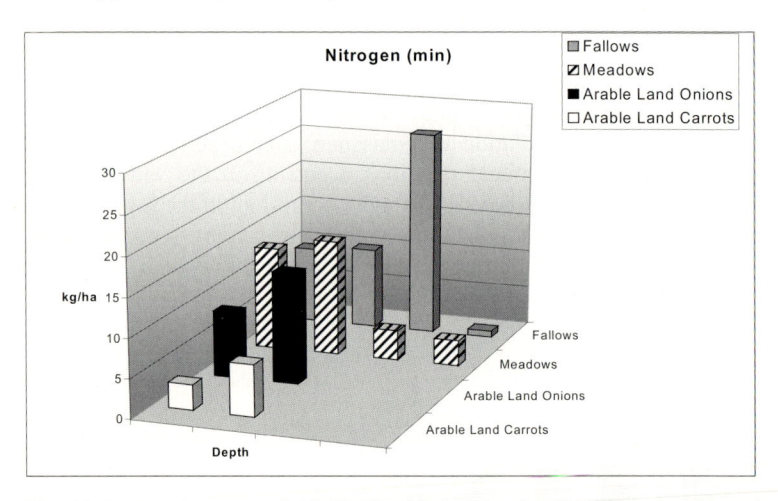

**Fig. 10.5.** $N_{min}$ Amounts of Ah-Horizons of Three Land Use Types along the Stahlbach Creek

### Soil Respiration

A decrease in microbial activity can be observed with increasing land use. This could indicate poorer nutrient quality (due to unbalanced ratios) or possible pesticide use. It also reveals available carbon sources to be decayed by microorganisms. It is natural, however, to have less respiration at greater depths in the soil owing to the decreasing amounts of oxygen and nutrients.

The total amount of heavy metals in all soil samples was below the threshold level set by the Sewage Sludge Regulation for soil content.

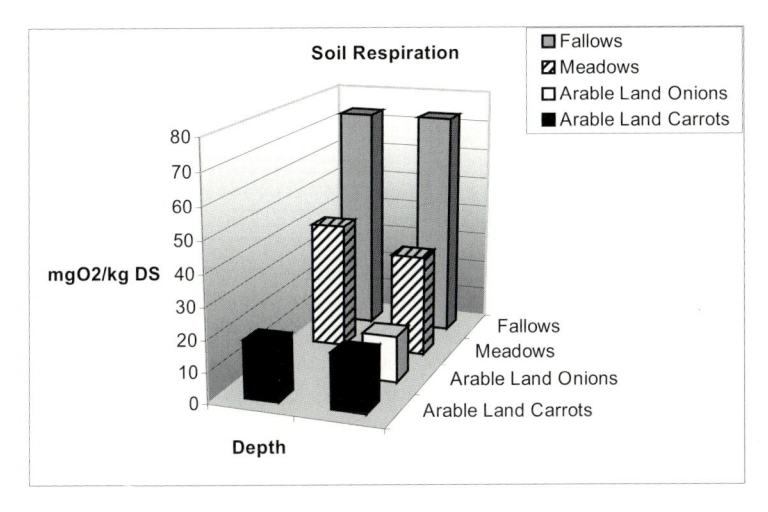

**Fig. 10.6.** Soil Respiration of Ah-Horizons of Three Land Use Types along the Stahlbach Creek

## Investigation of Soil Seepage and Groundwater

The three described and investigated sites are connected by the Stahlbach creek. The fallows are situated upstream and surrounded by arable land. The groundwater flowing from the arable land to the meadows and then into the creek connects the meadows to the arable land.

**Table 10.20.** Soil Water Samples, Arable Land

| Sample | pH | EC (µS/cm) | Chloride (mg/l) | Nitrate (mg/l) | Sulphate (mg/l) | Ammonium (mg/l) |
|---|---|---|---|---|---|---|
| 20 (field) | 6,94 | 1977 | 45,42 | 54,26 | 638,22 | 0,009 |
| 21 (bushes) | 6,87 | 1874 | 47,25 | 0 | 529,25 | 5,48 |
| Limit | 6,5-9,5 | 2000 | 250 | 50 | 240 | 0,5 |

**Table 10.21.** Soil Water and Creek Water Samples of the Meadows

| Sample | pH | EC (µS/cm) | Chloride (mg/l) | Nitrate (mg/l) | Sulphate (mg/l) | Ammonium (mg/l) |
|---|---|---|---|---|---|---|
| 22 meadows | 5,93 | 280 | 6,39 | 0 | 84,38 | 2,4 |
| 26 meadows | 7,3 | 473 | 20,89 | 37,70 | 45,36 | 0,01 |
| Creek water | 7,37 | 351 | 30,177 | 9,406 | 66,056 | 0,02 |
| Limit | 6,5-9,5 | 2000 | 250 | 50 | 240 | 0,5 |

**Table 10.22.** Soil Water and Creek Water Samples of the Fallows

| Sample | pH | EC (µS/cm) | Chloride (mg/l) | Nitrate (mg/l) | Sulphate (mg/l) | Ammonium (mg/l) |
|---|---|---|---|---|---|---|
| 27 fallows | 7,3 | 458 | 5,96 | 1,34 | 23,53 | 0,01 |
| 28 fallows | 5,78 | 416 | 4,551 | 3,249 | 171,76 | 0,21 |
| Creek water | 7,5 | 353 | 30,55 | 10,366 | 65,155 | 0,01 |
| Limit | 6,5-9,5 | 2000 | 250 | 50 | 240 | 0,5 |

The soil water samplers collected soil seepage water at a depth of 100cm below the surface. The soil water quality under arable land. The tables are showing at the time of analysis a high salt content, deduced from the electrical conductivity, which comes close to the upper limit set by the Federal Regulations on Drinking Water. As there are no regulations for soil seepage water to go by, we referred to the above regulations. Especially in river catchment areas, where high groundwater levels occur, this procedure seems to be the appropriate one, resembling groundwater conditions.

The nitrate measured downslope of carrots exceeds the limit set by the regulations. Both sampling areas, the field and bushes downslope of the field, exceed the maximum sulphate concentration laid down the same regulation. There is an ammonium peak that cannot be explained so far. In summary, it can be stated that land management has an impact on the soil seepage water at the arable land site (e.g. in terms of nitrate content). Whether leaching from the remediated waste dump site occurs cannot be deduced from the preliminary results and would require more detailed analysis (sulphate, electrical conductivity). At the sampling site close to the creek below the meadows, nitrate concentration is higher than for both samples (Site B and C) below the fallow. There was no difference found between the water quality of the creek near the fallow and near the meadows.

## *Land Management and Sustainability*

These measurements reflect soil and soil-seepage parameters as well as those of creek water sampled for a short period in late September. They cannot therefore provide a general description of the soil and soil/groundwater conditions of the area. Generally, no huge imbalances of nutrients or other indicators of soil quality have been observed. The same is true of the quality of the crops, which reveals a very good uptake of the available nutrients and trace elements in the soils. No exceeding of the limits (German Foodstuffs Ordinance 1992) for cadmium and lead in plant residues was observed.

The junior scientists' final report reached the following conclusion:

The influence of land management on soil and creek-water quality obviously follows the order of extension (arable land, meadows, fallow). Most of the identified plant species are indicative of moist, loose sandy or sandy clay, sandy loam or sandy silty soil texture and characteristic for soil pH values ranging from neutral to mild to slightly acidic. Some species point to base-saturated and also sandy soils. *Cirsium arvense, Urtica urens, Trifolium repens* and *Rumex* species indicate nitrogen-enriched locations close to the creek.

Landscape analysis, plant determination, soil mapping and analyses of soil and water as well as knowledge about land management of the investigated locations served as an information base on the filtering and transformation capacity of soils and their corresponding susceptibility to degradation and generally provided a more holistic view of sustainable land and water use.

From these preliminary results it can be deduced that the types of land management investigated contribute to sustainable, integrative soil and water protection. Hence, further long-term investigations should be carried out to monitor the status of development.

As indicated by the project examples, an extensive land use type can be recommended to meet the environmental goal of nature protection, emphasising its abiotic resources (soil and water) as demanded by the Agenda 21 (UNCED – United Nations Conference on Environments and Development, Rio de Janeiro 1992) and in the Action Plans of the International River Catchment Commissions. The development of aims and skills should be issue-focused. If arable land is managed sustainable, there is less threat to nutrient loss and water pollution, even on sandy parent material. The ecological methods learned and used, thanks to the skills and competence contributed by the junior scientists from a variety of professional backgrounds, to identify the threshold values for water, soil and plants are transferable to any other environment world-wide.

The results of the work from this project can be used as basis for future investigations focusing on environmental issues relating to sustainable land use and integrated river catchment management as defined in the ISCO (1998) proceedings. This would facilitate transboundary ecoregional land conservation and basin-wide watershed development under the terms of international conventions (e.g. AGENDA 21), action programmes or regional frameworks. Woman play a key role as users of land and water and in related programmes of land use, environmental protection, health care and education. Their participation in development, planning and decision-making processes must be strengthened and translated into public policy.

## References

Alfred Toepfer Akademie für Naturschutz (NNA) (2001) Syntheseberichtdes Forschungsvorhabens Leitbilder des Naturschutzes und deren Umsetzung mit der Landwirtschaft. Ziele, Instrumente und Kosten einer umweltschonenden und nachhaltigen Landnutzung im niedersächsischen Elbetal. - BMBF-Forschungsvorhaben (1.11.1997-31.03.2001), Schneverdingen

ARGE Arbeitsgemeinschaft für die Reinhaltung der Elbe (ARGE-ELBE). http://www.ARGE-elbe.de

Bergsma E ed. (1996) Terminology for Soil Erosion and Conservation. ISSS Wageningen

Blume H-P (1992) Handbuch des Bodenschutzes. 794, ecomed

Bodenübersichtskarte von Niedersachsen, 1:50 000 - digital, L3128 Uelzen, NLfB Hannover 2000

Bundes-Bodenschutzgesetz und Altlastenverordnung (BBodSCHG) (BBodSCHV (1999)

EarthColors Soil Color Book (1997) A Guide for Soil and Earthtone Colors. Color Communications, Inc. Poughkeepsie, NY

EU-Wasserrahmenrichtlinie. http://213.198.25.84/wasser/themen/5_99_wrrl.htm

Fink A (1989) Dünger und Düngung. VCH Weinheim

Finck A (1982) Pflanzenernährung in Stichworten. Hirts Stichwortbücher

Futtermittelverordnung (1992)

Gebrenegus Beyene T (2001) Risk Assessment of Heavy Metals in Topsoils and Fodder Plants of Flood Plains along the Middle Elbe River (Lower Saxony, Germany). MSc Thesis: 98, University of Applied Sciences, Suderburg, unpublished

IKSE (1997) Bericht über den Stand der Umsetzung der "Ökologischen Sofortmaßnahmen zum Schutz und zur Verbesserung der Biotopstrukturen der Elbe". (Internationale Kommission zum Schutz der Elbe), Magdeburg

ISCO (1998) Towards Sustainable Land Use. Furthering Co-operation Between People and Institutions. 1625, Catena Verlag

ISSS, ISRIC, FAO (1999) World Reference Base for Soil Resources. Wageningen, Rome www.fao.org/waicent/FaoInfo

Kahl K, Sänger H, Urban B (1994) Bodenreinigung durch Pflanzen. Landschaftsarchitektur: 45-47

Klärschlammverordnung (AbKlärV) (1992): Bundesgesetzblatt. (Sewage Sludge Regulation, Federal Law Gazette)

Kuntze H, Roeschmann G, Schwerdtfeger G (1994) Bodenkunde. Eugen Ulmer, Stuttgart

NNA Elbe-Projekt (2000) Leitbilder des Naturschutzes und deren Umsetzung mit der Landwirtschaft. Zwischenbericht zum Forschungsvorhaben (01.01.-31.12.1999)

Oldeman LR, Hakkeling RTA, Sombroek WG (1991) World Map on Status of Human-Induced Soil Degradation. Global Assessment of Soil Degradation GLASOD. ISRIC and United Nations Environmental Programme. CIP-Gegevens Koninklijke Bibliotheek, Den Haag

Quartärgeologische Übersichtskarte von Niedersachsen und Bremen 1:500000, NLfB Hannover 1995

Schäfer B (1999) Estimation of Soil Loss for Kulfo Watershed in South Western Ethiopia Using the Universal Soil Loss Equation in Combination with Geographic Information System. Thesis,7, unpublished, Suderburg

Singer MJ, Munns DN (1999) Soils, An Introduction. Prentice Hall, NJ

tvo, Datenbank der umwelt-online, http://www.umwelt-online.de, 2001

UNCED – United Nations Conference on Environments and Development (1992) Konferenz der Vereinten Nationen für Umwelt und Entwicklung im Juni in Rio de Janeiro – Dokumente, Agenda 21. - Ed.: Bundesministerium für Umwelt, Naturschutz und Reaktorsicherheit. 289, Bonn

Urban B, Brenner-Herrenbrück, C, Heins A (2001) Untersuchungen zu nachhaltiger Bodennutzung und zum Grundwasserschutz im niedersächsischen Elbetal. Ein Beitrag zum BMBF-Forschungsvorhaben „Leitbilder des Naturschutzes und deren Umsetzung mit der Landwirtschaft". 279, University of Applied Sciences, Suderburg

Wischmeier WH, Smith DD (1978) Predicting Rainfall Erosion Losses - A Guide to Conservation Planning. Agr. Handbook: 537, USDA, Washington DC
Williams MAJ, Dunkerley DL, De Deckker P, Kershaw AP, Stokes T (1998) Quaternary Environments. Edward Arnold, London

# 11 Evaluation "There is No Unanimous Judgement on *ifu*"

Sigrid Metz-Göckel

*ifu* was not only a German-style postgraduate programme, but also a combination of international academic conference and world women's conference – a comparison used by some lecturers during *ifu*. The analogy to an international academic conference was based on the numerous presentations, presenters, the introductory lectures and workshops. The character of a world women's conference was reflected more in the many informal discussions during which academic topics were debated and expanded on, as well as in the networking activities. Nevertheless, it is true that *ifu* was conceived as a postgraduate programme. The women junior scientists had to conduct research, study and submit results – not an easy task given the tight schedule. *ifu* was a global community of educated women working in the context of a global communication project. This was bound to cause a lot of conflict and problems, but it also produced a high level of identification with *ifu*'s concept and objectives on the part of most participants.

In the preparatory phase, *ifu* was seen as an open-ended experiment. The step from the conception to the implementation phase of *ifu* was, in a way, a quantum leap. It was only achieved through the tremendous commitment of all those involved. For evaluation purposes, not only the basic conditions for the implementation of the programme needed to be accurately recorded, but also had to be considered during assessment. For this purpose, an evaluation team was put together. In the concept phase, the potential effects of a weak organisational structure and insufficient personnel, technical and financial resources on the implementation of the curriculum were given relatively little consideration. However, in the execution phase and during the evaluation, such factors played an eminent role and indicated that there were many more interfaces between organisational and academic work than was previously thought.

## 11.1 Evaluation Concept

*ifu* should be measured according to its own objectives, i.e. in terms of how the implementation succeeded primarily in relation to the institutional framework and the structure of the women junior scientists. Additional criteria for determining its level of success are:

- The expectations of the junior scientists
- The achievements of the study period
- The self-assessment of the project areas

### The Expectations of the Junior scientists
The *ifu* application brochure created high levels of expectation among prospective applicants. Over 90% gave as their reason for applying the desire to improve their academic knowledge and know-how, to better their understanding of gender-related topics and to gain practical experience of interdisciplinarity and intercultural co-operation. Nearly half of them were looking for supervisors for their academic work.

### Contextualisation of Critique
The programmed equality of all cultures, nations and individuals involved in *ifu* did not prevent the establishment of a manifest but hidden "speaking order", which provoked criticism, especially in relation to a North-South dialogue. These areas of dispute are an important outcome of *ifu*.

That such critique was formulated at all, and with disarming ease and occasional abruptness, is not to be wondered at in the light of the ambitious targets set by *ifu*. The critique cannot merely be attributed to the context and concrete background of the critics, but should also be seen as an expression of the junior scientists' independence and, above al, their commitment to *ifu*. In referring to the contextualisation of the critique, what is impled is the scientific procedure of relating critical comment on *ifu* to the circumstantial context and personal background from which the critique comes.

### The Evaluation of Research Groups and Their Research Fields
*ifu* benefited from professional competence for evaluating the highly complex composition of the many areas of study and research. These included international comparative university and student research, educational management, university teaching as well as evaluation research. The evaluation group – comprising four teams – formulated subprojects which further crystallised thematic-organisational aspects, creating an integrated concept.

The evaluation concept of *ifu* was finally divided into the following four research fields, each with detailed targets:

- The content of the study and research programmes in each of the six project areas
- The junior scientists, visiting scholars and lecturers as actors in the study, teaching and research programmes
- The organisation and management of *ifu*, including the virtual *ifu* (vifu)
- The projects spanning all project areas, such as the Service Centre and the Open Space

This section will focus on the evaluation results for the project area Water. The evaluation is based on several visits to the Northeast Lower Saxony University of Applied Sciences at Suderburg, which was the study location for this project area; on junior scientists' observation of the lectures; and on interviews with a selection of junior scientists and visiting scholars.

## 11.2 Bridging the Gap Between Mutually Unfamiliar Disciplines and Socio-Technical Innovation

*"One could never travel through the whole world and get to know all these women, I find that is really great"*[1]

The project area Water had the most applicants from the so-called "developing countries" and was the project area focusing most strongly on natural and engineering sciences. At the same time, it endeavoured to link engineering and natural science know-how to the perspectives of social and cultural scientists, in addition to fulfilling the requirement of imparting the latest knowledge from the fields of water technology and biology. Because of its laboratory component, it took place in Suderburg, in co-operation with the Northeast Lower Saxony University of Applied Sciences. It was here, too, that the junior scientists were accommodated, which created some special conditions during the residency phase. During these three months, *ifu* had sole use of the whole university, including the computer room which quickly became an important centre of communication. The Service Centre was centrally located and the spacious conditions at the newly equipped university were very satisfactory.

Through the dean, Professor Kunst, the project area Water submitted an application for funding to the German Federal Foundation for the Environment (DBU) as part of the Agenda 21. This application emphasised the targets of environmental education as well as the importance of the project area for international environmental protection as a "novel experiment in the area of feminist ecology on a practical level".

### 11.2.1 Curriculum of the Project Area Water

"Water is life" – such was the motto of this project area. This statement of what should be is put here in particularly apposite terms and was realised in concrete examples and recorded references.

The structure of the curriculum was problem-oriented and conceptually cut across boundaries in two senses: first, because of the transfer of ideas on water technology to other disciplines; and second, because of the attempt to integrate global and regional perspectives and also to present the latest technological know-how. It was, per se, a bold and innovative enterprise. Unlike the other curricula at *ifu*, this project area built bridges between various mutually unfamiliar disciplines with the objective of "bundling traditionally different techniques and different know-how of hygiene, health, irrigation, drinking-water processing, rainwater use and wastewater cleaning."

---

[1] From an interview with a Canadian junior scientist

Water is a problem of survival in many regions of the world and the suggested solutions and procedures are becoming more and more technologised. The high level of technology in water science, which is implemented in the hope of improving conditions in many regions or not implemented through lack of capital, leads to a multifaceted conflict relationship with the users whose perspective was constitutive for the multidisciplinary concept of the curriculum.

As many examples show, modern technology can destroy existing knowledge about the handling and use of water, create new dependencies and produce environmental problems which in the long run make the original problems worse. Also, the introduction of new technology for irrigation or water cleaning is dependent on the users, and therefore on people. Water problems are always a concern of environmental education and depend on the message getting across.

The challenges and characteristics of this project area were consequently related to the connection between transferring knowledge of state-of-the-art technology and simple water treatment and practice-oriented teaching methods in environmental education. The project area was concerned not only with problem analysis, but also with concrete practice, e.g. how to make NGOs (Non-Governmental-Organisations) less dependent on sophisticated technology and how to work with available knowledge about culture. Environmental education and teaching methods are concerned here with making knowledge accessible by applying it to the existing situation of the users.

Four key fields define the concept of the curriculum:

- A specific interdisciplinarity resulting from the problems set
- A transfer of theory with practical relevance to concrete water projects
- A transfer of know-how with the potential for political activity (environmental education, NGO mobilisation)
- An interdisciplinary syllabus and innovative research in the natural and engineering sciences

## 11.2.2 Evaluation of the Curriculum from the Perspective of the Junior Scientists

The interviews with junior scientists on the curriculum were analysed in relation to the following:

- Interdisciplinarity including the dimension of gender
- Multiculturality
- Assessment of their own learning processes
- Style of teaching and teaching methods
- Aspect of self-organisation and the social environment in Suderburg

During the first two weeks of lectures, the programme should have been organised with particular emphasis on linking the overall topic to the cultural and social sciences. There are very few comments from the junior scientists about further implementation of interdisciplinarity especially in their projects.

Remarks on multiculturality were almost entirely concerned with the encounter between junior scientists as well as the informal communication and teaching/learning processes between junior scientists.

It should be understood that the junior scientists' critique in the initial phase was due in part to the "culture shock" they experienced on being transplanted to the "flatlands". They felt they had not been fully informed and were not then properly prepared. These "teething problems" were so overwhelming that they dominated discussions during the first three weeks at Suderburg.

What is interesting is how the junior scientists assessed the programme at the end of *ifu*, in the light of their experience with the projects, the excursions and a self-organised and varied cultural programme. At the farewell party, the study location was celebrated as "the paradise of Suderburg".

Owing to the intervention of junior scientists, some changes were made to the curriculum and, as the German dean put it, "the level of the study programme was lifted so high that it was impossible to keep up."

## Type of Interdisciplinarity, Different Levels of Knowledge and Other Differentiation

According to the prepared curriculum, during the first two weeks the junior scientists were to attend lectures on culture, religion and social sciences, all of which were related to the topic water and which sought to take into account the junior scientists' varied cultural backgrounds. What happened in practice was that West European cultural and social scientists gave lectures to an audience that for the most part had different expectations. In addition, the programme was conducted in such a way that different disciplinary perspectives were lined up one after the other.

Opinions on the introductory phase varied widely. One participant felt: "It was basically positive that many different cultures came together and that is was interdisciplinary from the beginning, and also that it stimulated questions about religion and culture. Through this, one got into discussions with one another and also asked about personal things. Whenever one met, there was a common reference point, although initially that was overemphasised. But basically I found the beginning really good. Only there could perhaps have been more multidisciplinarity". (Biochemist, Germany)

The multidisciplinary encounter between junior scientists produced a heterogeneous audience at the lectures and also caused problems with understanding and being understood. The project area Water brought together junior scientists with narrowly focused engineering science expectations and without any interest in interdisciplinarity; junior scientists with social and cultural science interests without any ties to the engineering sciences; junior scientists with very different academic backgrounds, qualifications and professional demands. Some junior scientists complained that others were not familiar with the elementary vocabulary of environmental biology or had an insufficient knowledge of the subject. Others were very advanced in their knowledge and expected teaching sessions to take them even further. Some lacked discipline-specific knowledge. For different reasons, the level of the introductory lectures was considered unsuitable. One of the visiting scholars shared this view: "The problem with the lectures was the level. I think it is difficult for the lecturers when they have no idea what kind of women are sitting there. Ok, there are women sitting there who have no idea. Where do I set the level then? Do I really start very low? Then half of them are sitting there saying 'I know all this already, I'm leaving'. If I pitch it in the middle, perhaps some won't understand. I simply don't know where the middle is. When I don't know the group, it's difficult. There were lectures that were so highly pitched that actually only women who had a little background in this area could understand... there were some lectures which were two times one and a half hours. They organised it so that there was a little introduction in the first part, and in the second part they went into more detail about this project or their area. Then it was ok..." (Biologist, Austria)

In the second half of the study phase, the integration of different scientific perspectives and demands into practice-oriented projects worked significantly better. One Western participant highlighted a further heterogeneity: "I think the biggest difference is the one between the women who come from the West and those from Africa and Asia. Those from the West are more willing to fight for their rights and to say something negative. From the others one only hears, 'Oh, how nice,' while those from the West think, 'No, we have to do something, we have to improve it.' I've learnt at *ifu* that those who come from North America, from South Africa and from big cities are definitely more Western in their world-view... They know better what they are worth and have then expectations in accordance with their worth." (Environmental Biologist, Canada)

### *Multiculturality as a Key Feature*

Of course, initially there were plenty of problems and difficulties on the everyday practical level with a multicultural group of junior scientists, but this soon developed into the most interesting side of the programme.

"I think the multicultural aspect is ok, because it is a chance to learn from others. I mean it is a real chance for the whole world to get together in one place and learn about other cultures and their ways of thinking... It is more than ok, it is good to have this." (Chemist, Sudan)

The direct comments and implied opinions on the interculturality of the junior scientists were invariably enthusiastic. The variety of cultures and regional differences were unanimously perceived as an exciting innovation and an intellectually rewarding element of the project area. These assessments by women with backgrounds in the natural and engineering sciences clearly signalled their interest in and openness to other cultures and also indicated that the accompanying cultural and social science programme was highly appreciated.

## The Dimension of Gender from the Perspective of the Different Disciplines

The talk of a gender perspective and gender relevance in the context of the topic Water was completely new to many junior scientists and initially produced a very ambiguous response. While it was relatively easy for those with a background in cultural science to think in these terms, others felt rather disconcerted. Nevertheless, their curiosity and interest were finally awoken.

In one interview, the conversation turned to a Pakistani lecturer who had spoken about water and gender and women's access to the irrigation system. She is an anthropologist, had studied in New York and was interested in water irrigation because there is a drought area south of Lahore and there the water supply is of vital significance because the farmers' livelihood depends on it. The shortage of water in this area reaches a minimum level of 30%, and the distribution system – all water comes from four rivers, particularly the Indus – is supported and supervised by the government. The water shortage is so severe that even water theft is resorted to. Control of irrigation is in the hands of the male members of the country's elite.

This lecturer was one of the people who, with their competence and personality, were able to establish links between the gender perspective and the issues of water and soil and build bridges between these disciplinary perspectives. The junior scientists constantly expressed their admiration for and amazement at this fact. However, this was the exception rather than the rule. Some lecturers were considered to be very good when their personality came through and they were very enthusiastic about their subject. Competence was always taken for granted; lecturers who gave the impression that they were unable to meet the expectations of the junior scientists in terms of form, content and language were criticised mercilessly.

## Curricular and Cultural Self-Organisation by the Junior Scientists

In the project area Water, self-organisation by the junior scientists played a more extensive role than in all other project areas and was also differently accentuated – in one respect as a supplement and corrective to the curriculum, but above all as a leisure programme. Junior scientists certainly made their own contributions to the study and research programme, but this was particularly true of the cultural and leisure activities.

Examples of self-organisation among the women were language courses; a choir; a dance group for salsa and Arabic dance; a varied film programme featuring and discussion groups, e.g. "Women in Development". The junior scientists organised regular meetings in the *ifu* café, and there were numerous private invitations to, and parties held in, the rooms and student hostels they were staying in. There was a profound interest in getting to know other women and an extraordinary willingness to help each other.

"We had the whole morning only for junior scientists presentations about the environmental situation in Cuba and also the Mangroves in Ecuador. That was very interesting because they presented their own work (because a lecturer cancelled at the last minute). There are also different social projects. Many people who are here want to learn German. I think at the moment almost all the German participants are giving German courses." (Environmental Biologist, Canada)

"We have movies on the weekend, which I appreciate very much. We have junior scientists' presentations where they talk about problems in their countries. That is very interesting for me, because it is good to hear reports from people who live there, to hear about their lives." (Mathematician, Russia)

During the last *ifu* week, the following remark was made by a student: "We are a little bit like a big family here... However, it was somehow quite varied. A network group was started and an NGO group has been developed and that took quite a lot of time." (Biochemist, Germany)

The computer room was an important working place for Internet research, for e-mailing and using programmes, but also a place with 24-hour accessibility and a centre communication.

## Critique and Suggestions for Improvement by the Junior Scientists

According to the participant questionnaire completed before the start of *ifu*, most people expected stimulating content and inspiring lectures. Most, but not all, were used to clearly focused communication of knowledge. There was also a clash between different teaching and learning styles in this project area. As one visiting scholar commented: "One student would require the information in text form, whilst the other wanted it as a diagram."

The critique of the junior scientists was geared to their own scientific disciplines and specialisation. While assessment of the lectures forms only a part of the study programme, it nevertheless shows how difficult it is for an interdisciplinary and innovative curriculum based on the criteria of the *ifu* programme to win acceptance. As in other project areas, the form of communication and scheduling of lectures was initially not optimally suited to the participants needs and should have been done differently. For the next *ifu*, they themselves made many suggestions for improvement – both spontaneous and solicited.

The way the junior scientists formulated their critique differed from individual to individual, but there were also cultural differences and perhaps even different critique cultures. In the introductory phase, most of the junior scientists were deeply interested in and highly committed to introducing changes of particular concern to them. While there was no clear consensus here, the following pattern could be identified:

- The introductory phase should last one week at the most.
- Then disciplinary groups should be created.
- After this phase, there should be interdisciplinary groups again.
- Qualified, older and experienced lecturers should be invited.
- The overall organisation should be much better at the next session.
- There should be less junior scientists.
- There should be less repetition.

- The syllabus should be co-ordinated by members of the teaching faculty. Lecturers should send their texts to one another in advance to allow them to refer to one another.

These suggestions are basically concerned with operational matters, not with the fundamental conception of the curriculum. This would appear to indicate that – despite all the criticism – the basic problems lay in the area of communication and organisation and that the innovative design principles of the curriculum were widely accepted.

### 11.2.3 From the Perspective of the Visiting Scholars: "You Can Feel It in the Air"

There were nine visiting scholars in the project area Water as well as an art visiting scholar. The art visiting scholar was chosen by the art council, the others by the curricular working group. One criterion – besides academic and professional qualifications – was to ensure a broad range of visiting scholars in terms of nationality, which was ultimately achieved. Of the nine visiting scholars, two were German, two Greek (one from a second-generation "guest-worker" family), two African, one Indian, one Bolivian and one Thai.

The visiting scholars and the tutorials played an important role in knowledge transfer and academic dialogue, the tutorials functioning to some extent as "home groups" responsible for orientation and adjustment.

The following is an excerpt from a discussion with two visiting scholars during the third week:

"The first weeks were very hard, the weekend was worked through and only now are we experiencing the first hours in which we can be a little more relaxed. We had a lot to read. The visiting scholar training was good and interesting and the international collection of visiting scholars is very important.

The first turbulent 'teething problems' have been coped with, the junior scientists have now also begun to help and organise themselves, to share and to adapt." (Group discussion with visiting scholars)

In the visiting scholar groups, a great many subjects were discussed and group bonds were established. One visiting scholar reported that at the first meeting each participant spoke only for herself largely avoiding eye-contact with the others. On the second day, they conducted interviews and presentations with partners. "Then a group dynamic got into gear..."

The perspective of the visiting scholars was focused on the programme as a whole. As both mediators and teachers, they were crucial to the programme. The impression they gave was one of competence and commitment.

The acculturation of the junior scientists to the locality and their adaptation to the programme in this project area did not always proceed synchronously, though they did more or less converge in the end. The status of neither student nor visiting scholar nor lecturer was verbally questioned, although there were arguments about competence and scientific standards.

In the final interviews with four of the visiting scholars, differences in emphasis were observable. One visiting scholar remarked critically that not enough attention had been paid to experience in the choice of personnel, and towards the end the "friendly mask", as she put it, was taken off in their culturally heterogeneous visiting scholar group and a cultural dissonance emerged which she did not wish to go into in more detail.

Another described plainly the heavy demands made on her time: as a mother of two small children, she had to repeatedly reorganise her household. She was highly impressed by her visiting scholar group and closely identified with it, without losing an overall perspective. She spoke about spontaneous group-building processes according to regional background and mentioned that she had found a notepad on the street containing a list of addresses, almost all of them African: "Most of them, I think, stick together. Also the Indians stick together... They didn't know each other before, but they were feeling more comfortable with each other. And also at this multiculti party you could see that they were sitting, the Asians together or the Indians or the Africans... Although now in the project period they are more together from different backgrounds. Now there is an eagerness to get to know the other cultures also." Other visiting scholars confirm that the group-building between the junior scientists started as couple-building; then small groups were formed, initially often by nationality, but later the situation loosened up.

One visiting scholar brought up the matter of the junior scientists' dissatisfaction during the first weeks and attributed this to the heterogeneity and organisation of the curriculum. "They organised the lectures in the first part and the projects in the second part. And I think that was not so good because the junior scientists got a little bit tired only listening the whole day and not being active and participating. Sometimes we had people only from English literature or linguistics and we were talking about wastewater equipment planning and they had no idea, they couldn't participate. But now they are also active, most of them. We also did an evaluation of the first few weeks and most were pleased with this period, with the workshop period more than with the lectures. Because we had people from different backgrounds and you can't please everybody.

In the first week we had more about social things and gender, so all the people from the social sciences were pleased, but not the scientific engineers. And the last weeks we had more scientific lectures for them - and the social workers were just sitting there. So that was a little bit a problem. That's why maybe there should have been projects from the beginning, not to have this atmosphere in here...

Well, it's very interesting always to get people together from different backgrounds and from different countries *You can feel it in the air*! Because many conflicts came also from people living together in the same apartments." (Physicist, Greece)

Like the other visiting scholars she also saw positive things in the course of the programme, particularly through the common project work which brought together women from different cultures and disciplines, though there was also tension. For her the various complaints about the different phases of the programme were quite normal. She reflected on the way the programme developed and formed her own opinion about it by questioning the participants herself as well: "I asked some junior scientists if they had participated in another summer school like this before. And most of them hadn't. So it was their first experience like this. Because also in a conference when you go you hear so many boring things and it is only one hour that is worth while for you. And you go for that special hour not for the whole thing. It's like this here also. You are here to hear some things. It may not all be interesting for you. And also only two had some summer school experience – in France, for example, one participant, but that was only for four or five students. Here it was three months, one hundred junior scientists. You can't please everybody. And some of them had also an English problem."

She found it worthwhile introducing a gender perspective into the project area Water – on the one hand, because the problem begins in the family "where the manager is a woman in most families", but also because women in other parts of the world are very active in environmental politics. "Because you can see in which countries the women are active. For example in India, the women are really active. And I would say they are more active in the results than the women from the developed countries. That was for me very surprising. They are so active in NGOs those women, also in Africa and India and Indonesia. The problem with those women is they have the qualification but it is so difficult to get the higher jobs in the government."

Her project group was a very colourful mixture, with women from Indonesia, Nepal, Bangladesh, South Africa, Greece, Bolivia, Brazil and Cameroon. For her, it was an interesting group that worked very well together and had worked on an environmental plan. "We are very happy. We had breakfast together, we did everyday warming up for five minutes, for example meditation, jogging a little bit. We are feeling very good together now. And I think in other groups you can feel it also. We also had evaluations and I read the evaluations. You could see that they were really very pleased with the second part. The only problematic part was, for example, that they needed more information in English. This is the problem with the library, for example."

Question: "Do they still complain about Suderburg?" Answer: "No, no, not at all. And they are marvellous, they are pleased – because if we were in Hannover they would be lost. But now they stay out here so they are pleased. So you have to see all the different aspects of it."

Another visiting scholar explained in detail that, contrary to expectations, as regards intercultural communication, it went very well – apart from minor problems – and actually better than the interdisciplinary communication. In the workshops, junior scientists switched off and did not say when they had failed to understand something; language problems were also a factor. "I think, this communication with specialist words and these technical explanations in English, which was for most the first foreign language or the second or the third, particularly when it came to the technical level, that many participants had problems in clarifying something. 'Water Processing' would also not be easy to explain off the cuff."

And she said something else about the different levels of knowledge, that the young women who were still doing their master's were not so experienced but "more open". I mean, maybe not the knowledge, but more ideas or more spontaneous. Although that of course also depends on the particular person."

One visiting scholar with a master's degree in Agriculture in the Tropics and Sub-tropics dealt with the SIRDO project (cf. Chapter 8), which had a complicated genesis in Suderburg. This project was a kind of key project in the development of the curriculum. However, its transfer was difficult. During the planning phase, the conduct of the project was, for a long time, not clear because the international dean did not attend the second planning meeting, but sent a colleague instead[2]. Also the transfer of data collected in Mexico was not accomplished in time, so the analysis and subsequent calculations in Suderburg were not really possible, as the dean reported in her final interview. One visiting scholar made the following comment about the junior scientists participating in this project: "They then did microbiological research, and also soil and compost research. We collected several samples ourselves and then it worked really well. Then there was also something for them to do. Because earlier they had stood in the laboratory and didn't know what they should do." (Visiting scholar, Biology)

### 11.2.4 Description of the Study Venue – the Environment from the Perspective of the Junior scientists

Suderburg is a small city in a rural area with about 5,000 inhabitants. The nearest cities are Lüneburg in the North and Hannover in the South, both an hour's train ride away. The location itself does not offer any cultural or consumer amenities that might make it attractive to women from large cities around the world. The study sites at Suderburg and Lüneburg were affiliated to the Northeast Lower Saxony University of Applied Sciences. The university is built in the modern northern German style of architecture and has over 100 years of tradition in the area of tropical water science.

When asked for her opinion of the project area, one participant immediately replied: "Ambivalent, good for learning, bad for shopping." She had friends in Frankfurt, Munich and Oldenburg, but could probably not visit them because the trip would be too expensive. She also wanted to shop and to see more life. By contrast, another engineer from Seoul enjoyed the peacefulness and fresh greenness of the environment. For most, the area was just boring at the beginning. During the first weeks, the project area underwent a hard probationary period, until the participants had become acclimatised to their unexpected environment.

#### *Social and Cultural Differences between the Junior Scientists*

The junior scientists were accommodated either in private families or in student hostels with cooking facilities. Most of the junior scientists borrowed a bicycle from their hosts or from a member of the community. There was a large area for

---

[2] The German dean mentioned that this was due to the changing political situation in Mexico.

bicycles in front of the entrance hall to the Service Centre. After the plenum meetings, the women went out to their bicycles and rode away en masse, dressed as they were in saris, European or other national dress. One group of North African junior scientists always went on foot. When asked why these women did not use bicycles, one of them said that women in the Sudan were not allowed to ride bicycles. Even young girls are forbidden to ride a bicycle in public, so they never learn how. This little insight is a perfect example of how easy it is to make gendered activities transparent in a multicultural environment. While in the Sudan cycling is apparently a gendered mode of transport, in other cultures it is gender-neutral. One participant from the Sudan, a leading chemist in the Ministry of Energy, was fascinated by this and inspired to initiate a comparative study of the situation of women in the United Arab Emirates and the Sudan.

Some junior scientists were disappointed about the size and style of their apartment, others found the same accommodation both spacious and good. It depended largely on what one was used to, as one visiting scholar explained in an interview. Some women from African, Arab and Asian countries came from a social stratum where cheap service personnel is available, and in their shared apartments in Suderburg they first had to learn how to take care of themselves. As one visiting scholar explained, her two flatmates had learn that they were responsible themselves for their own garbage disposal and cleaning.

## 11.3 Incongruity of the Perspectives: A Summary

One aspect of *ifu*'s introductory phase in Suderburg was the friendly environment, as mentioned above, and the university; the other was the disappointed expectations of the junior scientists. Initially, the project area Water was influenced by its location in the country, and although this was a dominant factor at first, it eventually came to be accepted as a challenge. The dull environment did not serve to reinforce the participants study habits, and an additional problem was that the library was insufficiently stocked with English literature. Reading became less important than the discussions and communication exchange between the junior scientists, which transcended cultural differences.

The atmosphere in the project area changed dramatically as the junior scientists began taking advantage of the opportunity for self-organisation. From the existing deficiencies they developed their own shadow programmes outside the tutorials and tapped their own potential.

A second key aspect of the evaluation was bridging the gap between the different and mutually unfamiliar disciplines of the natural and engineering sciences on the one side and the arts and cultural sciences on the other. Also of key importance was the question of how far the gender perspective could play a negotiating and clarifying role. From the perspective of the junior scientists, the integration of these two aspects into the practical projects went more smoothly than in the introductory lecture phase, with some qualifications. The engineers reacted rather sceptically, if not dismissively, to the social and cultural science lectures.

What came across well was the "holistic" approach most clearly evident in the presentation of women as end consumers and users of water and in the analysis of their dependence on social rights of accessibility and resources. It is possible that here the NGO activists played an important role in communication.

An evaluative insight into the projects must remain fragmentary, so evaluation of the projects' academic achievements was left to the deans, lecturers and visiting scholars. At the very least, the junior scientists became familiar with a new understanding of interconnectedness and initiated further professional networks.

It is not possible to meaningfully analyse the incongruity of perspectives between the lecturers and the heterogeneous audience in this phase and at this time. The impression is, however, that the teaching faculty succeeded in bridging the gap between the different cultures better than that between the different disciplines, in so far as such a gap exists between the disciplines. This may have been a different matter for related disciplines, but it depended primarily on the topics addressed by the projects, it was not always possible to incorporate a gender dimension into these.

## References

Kunst S, Burmester A (2000) Water is Life. In: Neusel A (Ed.): Die eigene Hochschule, Internationale Frauenuniversität 'Technik und Kultur'. Opladen

Kunst S (1999) Water is life. In: DUZ (Deutsche Universitäts Zeitung)

Neusel A (2000) Die Internationale Frauenuniversität 'Technik und Kultur', das Besondere des Konzepts. In: Neusel A (Ed.) Die eigene Hochschule, Internationale Frauenuniversität 'Technik und Kultur'. Opladen

# 12 Future Perspectives for Sustainable Water and Soil Management

Sabine Kunst

*ifu*'s project area Water, featuring projects on controversial issues from the fields of science and technology, focused on the problem of sustainable water and soil management. The management of resources played a primary role. Currently available knowledge was compiled in condensed form and documented in the various chapters presented here. By combining traditional technological, scientific, ecological and economic considerations with sociological, gender-related, cultural and political aspects, it was possible to develop innovative concepts for wastewater treatment, the provision of potable water, rainwater harvesting, management of catchment areas (rivers) and give proper consideration to the interaction between water and soil. The novel aspects of this approach are

- orientation of project work to pressing current problems
- consideration of gender-specific aspects in planning and management
- international and intercultural co-operation
- intercultural discourse and interdisciplinary work on concrete topics (cf. chapters 4–10)

This yielded new potential for knowledge transfer, providing fresh impulses for planning and ecological and social action and leading to a "globalisation of thinking".

Environmental problems do not stop at national borders, and given the already existing problem of providing enough water for the world's population, a problem with enormous conflict potential, it was decided to discuss water as a political issue. The seriousness of the water issue is felt most acutely in the countries of the so-called Third World, where the social divide continues to grow, also with regard to the supply of clean drinking water. While the countries of the North, with their ample supply of water, are increasingly suffering from storms and floods, the situation in the southern hemisphere is characterised by water shortages. Declining groundwater levels are causing whole regions to subside. Monsoon rainstorms are leading to huge tidal waves that trigger landslides.

The international composition of the junior scientist body in the project area Water promoted an intercultural discussion, encouraging junior scientists to take a critical view of their own culture, values and practices. Combining the expert knowledge and experience in the field of water and gender of a large number of scientists and practitioners was seen as the best way to develop concepts for sustainable management and more balanced and just distribution of water and soil. The hope was to integrate diverse technologies and traditional knowledge about hygiene, health, irrigation, potable water recovery, rainwater harvesting and wastewater treatment in order to bring greater influence to bear on future planning decisions concerning the environment.

## 12.1 Internationality and Intercultural Work

With junior scientists from 44 different countries in project area Water, *ifu* was a highly international and intercultural event. Interest in the topic "sustainable water and soil management" was shown predominantly by junior scientists from the developing countries of the South, who accounted for 85% of all junior scientists in the project area Water (cf. chapter 1). For the women from Southeast Asia, Africa, South America and Europe, particularly Eastern Europe, the issues addressed in the project area Water were of practical relevance and considerable urgency.

In actual fact, the junior scientists' interests were as diverse as the cultures they came from. In Southeast Asia, for instance, integrated treatment of the issues of water and gender is very important, perhaps even crucial, in order to develop sound water- and soil management concepts for coming generations. For Eastern Europe, on the other hand, it is of little interest – at least that appeared to be the perception of the junior scientists from that region. Many women in Southeast Asia subscribe to the view that water and soil management is a political issue. They are therefore much more aware of the instruments of political work than are, say, the women of Eastern Europe, who tend to focus more on academic issues. NGO work thrives in many Latin American countries and is playing an increasingly prominent role in African countries. A crucial prerequisite for such work is the provision of personal computers and Internet access, important tools in setting up world-wide networks (cf. section 12.3).

Internet research enabled contacts to be established with women's networks and organisations. Particularly helpful here was the Women's World-Wide Web AVIVA, which lists, in alphabetical order, all organisations, initiatives, groups and information sources for women world-wide. Simple and swift access to detailed information on specific organisations – e.g. the IWA, its members, congresses and representatives in Germany – was a key factor in developing an international curriculum.

Discussions centred on water as a political issue, incorporating a feminist perspective and focusing on selected regions of the world. Sociological, cultural, and political aspects were treated along with technological, scientific, ecological and economic considerations.

Access to a wide range of information was a basic prerequisite for exploiting the potential offered by the international expertise assembled here to initiate an intercultural dialogue. Such information must cover not only scientific content, but must also give proper consideration to cultural differences, especially those reflected in different attitudes towards water. This provides the basis for developing innovative concepts by an interdisciplinary discourse. What also became evident in the course of the *ifu* project is that intercultural exchange on the meaning of resource management, habits, myths and norms calls for a framework transcending traditional forms of knowledge transfer such as lectures and discussions. The most fruitful exchange of ideas and results took place during the junior scientists' leisure time, during informal encounters on the campus, "incidentally", so to speak. This "incidental" exchange was fostered by the junior scientists' openness and willingness to share their own personal experience with others – by talking about their own backgrounds, by comparing attitudes on a wide range of topics, by providing details of cultural idiosyncrasies, by dancing, singing and doing artwork, by cooking and eating together.

The international composition of *ifu*'s junior scientist body and faculty had a definite impact on the choice of topics addressed. This reflected not only the acuteness of the problems facing the developing countries, but also the fact that this was a congress of women experts only. Priority was given here to such topics as Water and Health and User Participation, not exactly the usual fare at international events of this kind. Just how different *ifu* was from other comparable meetings of experts is clearly reflected in the title of the contribution by Mary Lindsay Elmendorf, recipient of the Nobel Peace Prize for 1947: "Water is Life – A View Through the Eyes of Women". This title is indicative of *ifu*'s focus on the holistic development of resource-management concepts. Instead of adopting an exclusively academic approach to the problems of water-resource management, discussions in the project area took into account the specific concerns of women with regard to water (health education, availability, utilisation, treatment).

Rural Women and Water Problems, Women, Water and Health and Feminisation of Water Management were some of the other topics that attracted considerable interest. By reference to the situation in India, it proved possible to demonstrate how the largely poor rural population of the so-called developing countries can participate in planning and decision-making processes, and which roles women in particular can and should play here. Discussions addressed methods of water acquisition and simple ways of processing drinking water without using chlorine. A key focus of discussions was the problem of extracting drinking water from stagnant bodies of water and ways of enhancing knowledge about health and hygiene. A great deal of effort went into exploring the options for rainwater harvesting – a technique which, for countries like India, is likely to be the key to ensuring adequate water supplies in the future.

Further typical questions on which discussions centred were: the involvement of rural communities in planning and the importance of women groups' efforts to empower women as well as strategies for avoiding and settling conflicts.

The contribution Integrated Wastewater Management in Urban and Rural Areas Including Source Separation and Reuse of Resources reports on the work of several innovative pilot projects in Germany and Sweden incorporating and putting to the test new conceptual approaches to the separation and reuse of wastewater, which were successfully further developed. The general consensus in all the discussions was that in countries with an underdeveloped water supply infrastructure the wastewater disposal systems must incorporate sustainable concepts for separating faeces and grey water. "Planning for recycling" was the motto chosen for this issue. The report considers the problems and chances of such a policy being accepted by decision-makers and the public and discusses strategies for extensive information campaigns on this topic.

The impact of the project area's international composition can be summarised as follows:

- International resource management is an across-the-board task that can only be tackled by an interdisciplinary approach. It is impossible, for instance, to consider the resource water in isolation; its interaction with soil must be taken into account.
- International dialogue is fostered by the intercultural composition of junior scientist body and faculty.
- Analysis of the goals of sustainable resource management results in a critical rethinking – or even rejection – of European concepts.
- The innovative concepts developed here reflect a holistic approach, integrating ecological and economic concerns with gender perspectives.

## 12.2 Interdisciplinary Work and Gender Perspectives

Freshwater resources are scarce, and those available are distributed unevenly across the different regions of the world. Today, some two billion people are still without access to clean drinking water, and experts are agreed that this number will continue to increase over the next century, along with the growth of the world's population. The pollution of water supplies by industrial and domestic wastewater, which is often discharged untreated into the environment, the leaking of contaminated leachate from unofficial waste dumps and the often improper use of pesticides and fertilisers are impairing the quality of surface waters and the groundwater. The enormous challenges faced here call for both international and intercultural communication and interdisciplinary research as well as the conducting of projects to develop technological concepts, facilities and management strategies. Interdisciplinary research and exchange can yield the knowledge needed to manage water resources such that water supplies are used economically, distributed evenly and restored when depleted. "Management" includes recording data on water resources (their quantity, quality and location), planning, exploration and distribution as well as quality control and conservation. If all elements of the hydrological cycle and all levels of water utilisation are covered, the term Integrated Water Resource Management is used.

Establishing truly interdisciplinary co-operation, research and development would appear to be the hardest task. While an international and intercultural exchange of ideas and results can make scientific work more exciting and attractive, the conflict potential in practical interdisciplinary co-operation is relatively high. The scientific cultures vary considerably from one world region to another.

A general observation made at *ifu* was that most natural and engineering scientists are "conditioned" to accept as good science and valuable research whatever is recognised as such by the mainstream of their own particular discipline. The subjectivity of the natural and engineering sciences is not generally acknowledged. In particular junior scientists from countries dominated by a scientific elite, though themselves excluded from such circles, often take a totally uncritical attitude towards the scientific establishment. More highly rated than the overall social utility of a scientist's work is the number of his/her publications on a specific topic. By contrast, many of the junior scientists from countries where knowledge about issues like resource management is not available at universities or other higher-education institutions are self-taught. In Africa, for example, there are scarcely any educational opportunities for engineers, but only for chemists, mathematicians, biologists, etc. Self-taught scientists who apply their knowledge to questions of holistic resource management have much more affinity with those who use traditional knowledge and methods, which are proven and much less susceptible to scientific bias.

The practical orientation of such work also gives a political dimension to technological projects. Scientists from problem-oriented scientific cultures are used to taking an integrated view of science and economics, and they use this perspective to develop politically feasible projects. Others regard taking such a view as a betrayal of scientific principles. Especially in the developing countries, however, it is being increasingly recognised that the importance of a given resource transcends its monetary value. The consumption of a vital resource like water is soon regulated by its price. This was clearly illustrated in Germany following the fall of the Berlin Wall in 1990. Previously, water consumption in East Berlin had been 500 litres per capita per day. After the infrastructure had been improved, and the price of water put up, consumption dropped to approx. 145 litre per capita, the normal level for countries in Northern Europe. Efficiently regulating water consumption by introducing water charges is certainly an important instrument for achieving a sustainable conservation of resources. Critics of this plan, which is already being implemented in several countries today, point to its negative social impact. It would make it impossible to provide the poorer sections of the population with clean water. The result: clean water for the "rich", dirty water for the "poor". Even though such developments cannot be ruled out, this practical approach has achieved some success (NGO work).

No satisfactory concepts are yet available to reconcile the differing perspectives on scientific work – and this also applies to the interdisciplinary consideration of gender issues and expert topics. *ifu*'s success in this area is reflected in the project results presented in chapters 4-10.

The junior scientists developed concepts for adopting a holistic approach to issues, integrating social, ecological and economic concerns.

**Gender Perspectives**

The adoption of a gender perspective was a common feature of all the projects conducted at *ifu*. Efforts were typically made to directly integrate gender aspects in the projects' objectives. A good illustration of this is the conceptual approaches developed to tackle the problem of rainwater harvesting – a technique that will make a significant contribution to meeting drinking-water requirements in the foreseeable future. Before implementation of the various rainwater-harvesting programmes in recent years, women were the ones responsible for applying this technique at the domestic level. Little attention was paid, however, to their demands with regard to, say, system design or their experience with mosquito infestations. By contrast, the conceptual work done at *ifu* specifically focused on women's demands, leading to changes in the design of such systems and ideas for improving storage conditions. The impact of integrating gender perspectives, for instance in villages where rainwater-harvesting techniques are used, will need to be examined by further sociological studies. The studies by Zwarteveen et al. 2001 have shown, for example, that implementing user requirements has not only boosted women's empowerment; women who gained increased competence and skills in irrigation matters also had to renegotiate their role in the overall social structure. For some women, however, this was stressful and not exactly conducive to emancipation.

The discussion of gender perspectives was also found to have another highly positive effect on project work in the areas of technology and the natural sciences. The disastrous global situation with regard to natural resources is forcing people – men and women – to take matters into their own hands and start actively shaping their own destiny. The NGOs of Latin America, for instance, have very high proportion of women members, and many NGOs are headed by women. Self-help is the only means still available to achieve sustainable solutions. Across the globe, people are giving up waiting for radical changes and improvements, e.g. wastewater treatment concepts that are ultimately unsatisfactory. Instead, they are taking an active role in developing decentralised solutions that do not require expert knowledge. The high proportion of women involved in these projects is not surprising, given their clear perception and awareness of the problems faced. And this close involvement in turn encourages the adoption of a pragmatic approach to solving the problems. Women's clearly defined requirements – often based on their experience in caring for their families – lead to approaches that integrate gender perspectives. The recognition that such approaches are feasible (and readily implementable) and capable of yielding impressive solutions has had a huge impact, especially on the thinking of European engineers who are often incapable of tackling problems from novel or unconventional perspectives. *ifu* was, then, a highly successful exercise in transcultural communication and mutual learning. The establishment and maintenance of international contacts via women's networks like WINS offer the promise of an even more successful future.

On the other hand it soon became evident how promising this approach was. Particularly challenging was the intercultural and interdisciplinary nature of the work, which spawned a number of innovative, forward-looking projects. An impressive result was the networking of women experts from all over the world, thus enabling a permanent transfer of knowledge: the founding of the non-government organisation WINS (Women's International Network for Sustainability). WINS incorporates the guiding principles of the project area Water and defines its objectives as follows: "The goal of WINS is to achieve sustainable development based on gender equality emphasising women's empowerment, community participation, international networking and the integration of art and science".

## 12.3 Women's International Network for Sustainability: A Post-*ifu* Initiative Promoting Equitable and Ecologically Sound Alternatives to Mainstream Development

Dolly Wittberger

### 12.3.1 "Development is Well-Being – Concerning the Individual as well as the Community Level – for the Past, Present and Future." (Andrea Heckert, U.S./Mexico)[1]

The Women's International Network for Sustainability (WINS) is an initiative of over 50 women from 17 countries, primarily in the southern hemisphere, that began during the International Women's University (*ifu*) held in 2000 in Germany. Inspired by academic discussions on gender and development as well as by individual reports on development projects in their home countries, junior scientists in the project area Water discussed the effectiveness of "sustainable" initiatives and voiced their desire to maintain and build upon the networks and collaborations that began with *ifu*.

WINS is a continuation of the synergistic connections and collaborations established during *ifu* and will maintain and extend a project-driven network linking junior scientists, members, experts and organisations. WINS provides assistance in designing, implementing and funding projects with a clear focus on women and sustainable development. WINS believes in the synergistic interplay between all aspects of the global environment and envisages a world in which economic development is balanced with cultural diversity, biodiversity, justice, freedom of choice and spirituality. WINS advocates more holistic and sustainable approaches to development by promoting awareness of gender issues, by empowering women, campaigning for women's participation in policy-making and lobbying for women's and girls' rights at the national and international levels.

WINS believes that the key to true sustainability lies in actively promoting responsible and ethical relationships between human beings, natural resources, ecosystems and other species. Based on myriad case studies and the personal anecdotes and expertise shared by women during *ifu*, alternative, community-driven approaches to mainstream development practices have emerged at the local level. One of the key requirements for WINS projects, then, is that they are developed within local communities – even beginning at the personal level through affinity groups – and are based on the perceived needs of the affected communities and/or individuals. WINS projects maintain a clear focus on empowerment, leadership

---

[1] The heading quotes used throughout the text were made by WINS members during discussions on gender and development that took place at the Northeast Lower Saxony University of Applied Sciences, Suderburg, Germany, July-October, 2000.

and gender equity, particularly for rural and indigenous women. At every level of decision-making, planning and implementation, WINS seeks to ensure that women participate as equal partners and have equal access to rights, opportunities and resources.

### 12.3.2 "What Should I Say? Now We Are Developed?" (Christobel Chakwana, Malawi)

Especially in socio-economic and political climates dominated by popularised images of globalisation and free trade, the current rhetoric of international development initiatives revolves around the concept of sustainable development. WINS is deeply concerned about the impact of this new terminology. It questions, for example, whether sustainable development necessarily translates into a socially and ecologically sound lifestyle. Does sustainability promote and ensure equitable participation in decision-making, political processes and education for women and girls?

WINS is interested in whether sustainability, as it is currently defined, can foster more ethical relationships between human beings, ecosystems and other species. WINS members feel the urgent need to nurture alternative approaches to mainstream development projects. As a result, WINS and WINS projects work towards a future of "development", or sustainable development, that truly meets human – specifically women's – needs and values all aspects of the global environment. The three basic principles of *ifu* – internationality, interdisciplinarity and gender – provide a sound basis for current and future WINS projects.

An essential component of WINS's work is to maintain and facilitate networking between women involved in and/or concerned with issues of development and sustainability world-wide. WINS seeks to accomplish this by maintaining discussion groups, databases and archives as well as supporting initiatives to provide women with access to and training in communications technology. The need for and incentive to establish the network resulted from actual projects promoting gender equity and sustainable development that junior scientists designed during *ifu*.

An important aspect of WINS work is supporting and funding these projects, which focus on topics ranging from social forestry in Nepal to health education in the slums of Mumbai to rainwater harvesting in Tanzania. While WINS will initially have its headquarters in Western Europe (Austria and Germany), it plans to maintain field offices for specific areas of WINS projects, especially in the southern hemisphere where many of the currently submitted projects will take place. The current members of WINS work in a variety of fields: they include engineers, scientists, journalists, field workers, activists and artists. This diversity, coupled with WINS's international focus, is the strength of our organisation. Despite our differences and geographical distances that separate us, WINS members share an affinity for and a dedication to the principles of an ecologically sound way of living and gender equity. In fact, at WINS, the concept of equity begins within our organisation by acknowledgement of the often unequal access to resources – including funding, information and educational opportunities – and the willingness to deal with ethnical and cultural differences between WINS members in both the global South and North. Although there are no easy solutions to these issues of inequity, WINS aims to harness the resulting tensions and work towards creative answers.

### 12.3.3 "...I Would Like to Have a Computer, this Would Empower Me." (Arig Bakhiet, Sudan)

During the *ifu* Water project in Suderburg, women were strongly encouraged to develop their skills in the information and communication technologies. The Women's International Network for Sustainability helps to strengthen a vital communication network for women, enabling an exchange of concerns, experience, knowledge and ideas. It seeks to establish links between women engaged in different projects and working in different disciplines, women activists and field workers, women scientists within and outside academia, women's initiatives and NGOs, NGOs and isolated activists, and field-worker projects and donors.

The resulting manifold interactions will empower women and enable them to build strong coalitions, be a motor for change in their individual environments, and promote social and ecological justice at the local, national and international level.

This article (chapter 12.3) was written in collaboration with Kathy Becker, Anne Dabb and Jennifer Reddig.

## References

Zwarteveen M, Meinzen-Dick R (2001) Gender and Property Rights in the Commons, Examples of Water Rights in South Asia. Agriculture and Human Values 18: 11-25

# Appendices

# 13  Manual for Analysis of Soils and Related Materials

Brigitte Urban

With the assistance of Christiane Hillmer

## 13.1  Introduction to Soil Exploration and Soil Sampling

The biological characteristics of soil offer an indication of its fertility. The following experiments mostly cover soil microflora which, in regard to the number of organisms/g of soil and the live weight (kg/ha), is substantially higher than the soil fauna. In addition to characterisation of the organism groups - both quantitative and qualitative - tests of carbon and nitrogen conversion as well as a number of different physical and chemical parameters are significant for evaluation of biological cycles in soil (Urban in Kuntze et al 1994).

In order to eliminate external influences such as those that occur in field tests (varying distribution of organisms on the testing surface, changing water balance, etc.), standardised in vitro experiments have been developed that allow laboratories to compare their findings with others. Careful sample preparation and sampling are imperative.

Samples are taken with the aid of a soil auger that produces soil samples of 10cm/30cm. The testing area should be paced off diagonally. A sample should be taken every meter or, depending on the size of the area, every 1.5 to 2.0 meters. More detailed information can be found in the federal soil protection regulations. The soil is collected in a bucket carried by the person conducting the test. The soil from the entire testing area is mixed together. This ensures homogenisation of the soil samples, compensating for any differentiation. Depending on the moisture content, one to two kilos of soil are required for the following tests. The sample material should be processed immediately in the laboratory or stored in an airtight container in the refrigerator.

### 13.1.1  Preparing Samples in the Laboratory

This material is mixed well in the lab and then passed through a 2mm sieve (exceptions: organic soils, compost, sewage sludge, etc.). The sample can also be strained after it is dried. Part of the sample (approx. 60 to 100g) is placed in a bowl to air dry; it can also be dried in a drying chamber at max. 60°C. The rest of the sample is stored in the refrigerator (or elsewhere under 10°C) until it is to be processed.

## 13.1.2 Conversion

To be able to compare the results of the analysis with values from published sources, some of the values must be converted.
Normal conditions for soil exploration according to VDLUFA:

Conversion from mg/kg to mg/l soil:

$$mg/kg \cdot density\ (kg/l) \tag{13.1}$$

Conversion from mg/100g to kg/ha:

$$mg/100g \cdot layer\ density\ (cm) \cdot raw\ unit\ weight\ (kg/l) \tag{13.2}$$

Normal conditions for soil exploration according to LUFA (German Agricultural Testing and Research Agency), (org. substance < 8%):

$$layer\ density:\ 0.30\ cm,\ raw\ unit\ weight\ (dry):\ 1.4\ kg/l \tag{13.3}$$

Other examples for density in kg/l:

- sand: 1.2-1.8
- cohesive soils: 1.1-1.6
- compost: 0.5-1
- bog soil: 0.05-0.5 (depending on rate of decomposition and drainage)

## 13.2 Moisture Content and Dry Weight

| DIN 19 683, page 4 | for soil |
|---|---|
| DIN 38 414, part 2 | for compost, peat, mud, sediments and materials that can be filtered out |

Moisture content is the quantity of moisture contained in a soil. It is expressed in percent in weight, based on soil dried at 105°C.

*Dry sample = 100%*

When the moisture content is calculated for compost, peat, mud, sediments and materials that can be filtered out, the moist sample serves as the reference quantity (DIN 38 414, part 2, see below).

*Moist sample = 100%*

When the dry substance is calculated using either of the stated formulas, the result is the same.

## 13.2.1 Procedure

**Material:** moist sample

1. Label and weigh glass bowl: $A$
2. Place approx. 20g moist soil in the bowl and weigh (with bowl): $B_{moi}$
3. Dry in the drying chamber at 105°C until constant weight is reached (overnight)
4. Place the sample in the desiccator to cool
5. Weigh dry sample with bowl: $C_{dry}$

## 13.2.2 Calculation

$$\text{Moisture content \%} = ((B_{moi} - C_{dry})/(C_{dry} - A)) \cdot 100 \tag{13.4}$$

$$\text{Dry substance \%} = (100/(100 + \text{moisture cont. \%})) \cdot 100 \tag{13.5}$$

### Calculation of Moisture Content of Compost, Peat, Mud and Sediments (DIN 38 414, Part 2)

$$\text{Moisture content \%} = ((B_{moi} - C_{dry})/(B_{moi} - A)) \cdot 100 \tag{13.6}$$

$$\text{Dry substance \%} = 100 - \text{moisture content \%} \tag{13.7}$$

If only the dry substance is to be calculated, the following formula can be used: (applies to both DIN's)

$$\text{Dry substance \%} = ((C_{dry} - A)/(B_{moi} - A)) \cdot 100 \tag{13.8}$$

## 13.3 Determination of Organic Matter

DIN 19 684, part 3

Determination of organic matter. The components of a sample that are volatile at 550°C are designated as ignition loss. The remaining mineral components are considered residue on ignition.

Humus content is the organic matter in the soil. It consists mainly of excretions from live organisms and remains of dead organisms as well as of newly formed compounds of organic and inorganic components.

The C content can be determined by the ignition loss when the samples are free of clay and lime:

$$\text{Ignition loss \%} \cdot 0.47 = C\% \qquad (13.9)$$

If the sample contains clay or coarse clay, a higher amount of sesquioxide (e.g. iron oxide, etc.) or lime, the C content must be determined using the Lichterfelde method, because constitutional water escapes from the clay and lime glows away at high temperatures, or other gaseous losses may occur, meaning that the weight loss from glowing is not of an organic cause.

### 13.3.1 Procedure

**Material:** 105°C dry soil, strained to < 2mm (Cohesive soils should be finely crushed.). An analytical balance is used for initial weighing (0.0001g).

1. Make a note of pan number
2. Weigh pan $A$
3. Place dry sample in pan then weigh $B_{dry}$
4. Allow sample to glow away for four hours in the muffle furnace at 550°C
5. The next day, place the pan in the desiccator to cool
6. Weigh pan again $C_{comp}$

### 13.3.2 Calculation

$$\text{Ignition loss \%} = ((B_{dry} - C_{comp}) / (B_{dry} - A)) \times 100 \qquad (13.10)$$

### 13.3.3  Evaluation

**Table 13.1.** Evaluation of Humus Content in Soils (Pedological mapping instructions 1994)

| Humus content % | Designation |
|---|---|
| < 1 | very slightly humic |
| 1-2 | slightly humic |
| 2-4 | somewhat humic |
| 4-8 | strongly humic |
| 8-15 | very strongly humic |
| 15-30 | extremely humic |

## 13.4  Determination of pH

DIN 19 684, part 1

The pH value (Latin: potentia hydrogenii) is the negative decadic logarithm of the effective hydrogen ion concentration. It is determined electrometrically. The pH value directly or indirectly influences chemical, physical and biological soil properties and plant growth.

Using a 0.01mol/l $CaCl_2$ solution, a soil solution such as that found in agricultural land is simulated. Seasonal fluctuations are compensated for.

The pH value ($CaCl_2$) is usually around 0.3 to 0.7 lower than the pH value ($H_2O$). This is due to the fact that $H^+$ and $Al^{3+}$ ions are replaced by Ca ions. The pH value in water reacts more quickly to some fertilisation measures and also shows stronger seasonal fluctuations.

### 13.4.1  Procedure

**Reagents:** Dist. $H_2O$ or 0.01mol/l $CaCl_2$: dissolve 1.47g $CaCl_2 \cdot 2H_2O$ and add distilled water to 1000ml.

**Material:**
- Air dry or 60°C dry soil, strained to < 2mm
- Cohesive soils may need to be crushed
- Soil rich in humus, e.g. compost or peat: moist
- Mud samples, see below

1. Weigh 10g soil in a 50ml glass beaker
2. (Samples rich in humus: 25ml moist sample)
3. Add 25ml 0.01mol/l $CaCl_2$ (or dist. $H_2O$)
4. (Samples rich in humus: 75ml)
5. Allow to stand at least 1 hour, stirring occasionally
6. Measure with a calibrated pH meter. The unit indicates the pH value

pH value in *mud samples* according to DIN 38 414, part 5

- If the sample is sufficiently liquid, the pH electrode is directly submerged to measure.
- If the sample contains little water, dist. $H_2O$ is added to form a paste such that homogenous wetting of the electrode is possible

### 13.4.2 Evaluation

Table 13.2. Classification of Soil Reaction (Pedological mapping instructions 1994)

| Classification of soil reaction | | | |
|---|---|---|---|
| 7.0 | neutral | | |
| 6.5-7 | very slightly acidic | 7-7.5 | very slightly alkaline |
| 6-6.5 | slightly acidic | 7.5-8 | slightly alkaline |
| 5-6 | somewhat acidic | 8-9 | somewhat alkaline |
| 4-5 | strongly acidic | 9-10 | strongly alkaline |
| 3-4 | very strongly acidic | 10-11 | very strongly alkaline |
| < 3 | extremely acidic | > 11.0 | extremely alkaline |

## 13.5 Salinity of Soils (Electric Conductivity, EC)

VDLUFA (German Agricultural and Research Testing Agency) Book of Methods, Volume 1, 1991, Soil Exploration, Method A 10.1.1

### 13.5.1 Basics

Different types of plants show different degrees of sensitivity to salt. When the salt concentration is too high, it impairs the root functions and thus the growth of the plant. (Chhabra 1996).

The water soluble salts dissociate in their ions, which can be characterised by measuring conductivity. The salt content is calculated as potassium chloride. Absolute concentration values can not be obtained this way, because the conductivity of the ions varies.

## 13.5.2 Procedure

**Material:** Air dry or 60°C dry soil, strained < 2mm, possibly crush cohesive soil
Bog soils and compost : moist, strained < 5mm

1. Weigh 10g of soil in a conical flask
2. Add 100ml dist. $H_2O$
3. Shake on mechanical shaker for 1 hour
4. Measure conductivity: The electrode must be completely submerged in the eluate
   (Values indicated in mS/cm:    m-mhos/cm, µS/cm: 1000 = mS/cm)

## 13.5.3 Calculation of Salinity

Salinity in % (calculated as KCl) = electr. conductivity (mS/c ) · 0.528
(0.528 applies for autom. temperature correction of conductrimity device of 25°C and the above extraction ratio)

$$\text{Conversion to mg/100g: } \% \cdot 1000 = mg/100g \tag{13.11}$$

$$\text{Conversion to g/l soil: } (mg/100g \cdot density\ in\ kg/l) / 100 \tag{13.12}$$

Examples for density in kg/l:
  arable soil: 1.4
  sand: 1.2-1.8
  cohesive soils: 1.1-1.6

## 13.5.4 Evaluation

The salinity in compost is usually around 1-2% of the dry substance (LAGA (Intrastate Waste Commission) code of practice 10).

For soils free of carbonate the threshold for plant groth is considered < 200mg salt / 100g soil. As the carbonate content rises, the plants can tolerate up to twice the amount of salt without harm. (VDLUFA, Book of Methods 1)

**Table 13.3.** Salt Sensitivity of some Agricultural Plants (summary of various sources, refer to VDLUFA, Book of Methods 1)

| Tolerance in g/l soil | Type of plant examples | Reaction of plant to salt |
|---|---|---|
| 0.5-1 | All types of germinating plants: lettuce, radishes, cucumbers, heather, berries, fruit trees | sensitive |
| 1-2 | Wheat, oats, corn, potatoes, grapes, tomatoes | somewhat tolerant, tolerant |
| 2-3 | Asparagus, spinach, cabbage, barley, sugar beets, rape | very tolerant, tolerant |
| Over 3 | Few halophytes, no cultivated plants | salt-loving |

## 13.6. Cress Test (Germinability of Lepidium sativum)

Certain plants have particularly strong reactions to pollutants in the soil. One of these plants is cress. A certain number of cress seeds is planted in soil, then the growth is monitored. The soil that is used as a comparison is either uncontaminated soil or potting soil.

The soil should be kept sufficiently moist during the entire experiment. The plants are observed for 10 days. A plant lamp that radiates wave lengths conducive to plant growth can be used to artificially extend the days to 16 hours, reducing the time required to one week.

### 13.6.1 Procedure

**Material:** Moist soil, not strained (Remove rocks or similar matter)

1. Fill flower pots with testing soil (3 parallel samples)
2. Put comparison soil in other flower pots (3 parallels)
3. Moisten if necessary
4. Sow the same amount (e.g. 30) of cress seeds in each pot, press slightly
5. Allow to stand under plant lamp for one week (16 hours of light); water if necessary

## 13.6.2 Evaluation

The following parameters can be evaluated:

- Leaf colour compared to comparison soil
- Leaf deformations
- Leaf discolourations (chlorosis)
- Growth
- Number of plants
- Size (length) of seedling, above soil and the root
- (This is done by cutting the plants with a scissors, or extracting carefully, when measuring the root, placing them on millimetre paper and measuring them)
- Fresh weight
- Dry mass

## 13.6.3 Calculation and Rating

The sample is evaluated by comparing it to the comparison sample (average of parallel samples), e.g. number of plants: 87% of number of plants in potting soil. Number of plants, weight, size, etc. are inserted one after the other in the following form as A.

$$(sample \cdot 100) / A \text{ comparison sample} \qquad (13.13)$$

Parameters that can not be evaluated with such a formula (e.g. leaf colour), are evaluated descriptively.

## 13.7 Determination of Total Amount of Microorganisms in Solids (Microbial Number)

The quantity of soil organisms can be determined directly or indirectly.

## 13.7.1 Direct Methods

(E.g. fluorescence microscopic method according to STRUGGER)

Count bacteria under the fluorescence microscope after treating with acridine orange (live bacteria appear green, dead organisms and organic substances red). E.g. determine cell count with a THOMA chamber. A highly diluted germination solution is applied to a microscope slide divided into sections. After colouration and suppression of movement and growth (toluol), organisms are counted.

## 13.7.2 Indirect Methods

(The KOCH plate casting method)

This method was introduced in 1881 by Robert Koch. Adding agar to liquid nutritive media creates solid nutritive media in which macroscopically visible local colonies derived from one micro-organism and remaining stationary can grow.

Nutritive media that fulfil the organisms' needs are used for this procedure. The liquefied sterile nutritive media is mixed with an exactly measured microsuspension and, after stabilisation, forms colonies that can be counted.

## 13.7.3 Composition of Nutritive Media

Micro-organisms generally need the following substances to grow:

1. Carbon source: e.g. sucrose (Czapek agar), starch, protein (gelatin), cellulose, meat extract or peptone (anaerobic bacteria), mannite (azotobacter), etc.
2. A source of nitrogen: e.g.: $(NH_4)_2SO_4$, $KNO_3$, atmospheric nitrogen (azotobacter)
3. Nutrients: P, Fe, S, K, as $KNO_3$; $Fe_2$; $SO_4$; $KH_2$; $PO_4$, etc.
4. Oxygen for organisms that respire aerobically

Agar (also agar-agar, Malay) is a polymeric carbohydrate derived from red algae. It is added 1.5-3% to the nutrient solutions as a stabiliser. Agar melts at 95-98°C and hardens at temperatures under 45°C. Few micro-organisms are capable of utilising agar or its cleavage products (D or L galactose).

Nutrient bouillon containing the necessary substances are commercially available. e.g. 15g Standard 2 nutrient bouillon (manufactured by Merk) + 15g agar are mixed with sufficient distilled water to equal 1 l Standard 2 nutrient bouillon is used to breed less demanding bacteria. Components in g/l:

- peptone from meat 4.3;
- peptone from casein 4.3;
- sodium chloride 6.4

The nutrients required differ so substantially for the individual micro-organisms that it is impossible to breed all micro-organisms with one medium. The nutritive can become a selective medium by adding certain substances. When indicators are added, it can be used for biochemical differentiation.

## 13.7.4 Principles of Sterile Working Conditions

Work with micro-organisms should occur in a low-germ content laboratory or at a work station in a clean room. Avoid drafts in the laboratory; keep all doors and windows closed. Instruments (pipettes, Petri dish covers, etc.) may not be placed on the table. Pipette storage vessels are to be closed immediately after pipette or Eppendorf pipette tips are removed. Do not open vessels any longer than necessary. Avoid talking, coughing and sneezing; miniscule droplets with micro-organisms could be transmitted.

### *Sterilisation*

Microbiology work requires sterile nutrient media and sterile instruments. All beakers, pipettes, etc. must be germ-free (sterile). They are sterilised in a drying chamber at 160-180°C for at least two hours. They must be cold to be used. The following are required:

- 100ml measuring cylinder with an aluminium foil cover,
- conical flask with plug,
- 1ml and10ml pipettes (sterilising box).

The diluting water is sterilised in an autoclave (steam steriliser): 20 minutes at 121°C. The autoclave should not be opened until the temperature has fallen below 80°C. The diluting bottles (baby bottles) are filled with 104 or 95ml tap water. The cover may not be screwed completely closed (approx. 5ml evaporates in the autoclave).

The first diluting step (in the conical flask) additionally requires sterile water. If the Eppendorf pipette is used instead of a normal pipette, the blue 1ml pipette tip must be sterilised in the sterilising box.

When the bottles are removed from the autoclave, they must be screwed closed. The water must be cold before it can be used. The glass instruments can also be sterilised in the autoclave instead of in the drying chamber.

### *Sterilisation of Nutritive Medium*

Approx. 35ml of nutritive solution is required for a Petri dish.

- Prepare nutritive medium in a 500ml bottle:
- 7.5g standard 2 nutrient bouillons + 7.5g agar are mixed with sufficient water to equal 500ml; swivel.
- Sterilise in autoclave at 121°C for 20 minutes (do not screw bottle completely closed).
- Place nutritive medium in 60°C water bath to cool.

## 13.7.5 Creating Soil Suspension

**Material:** moist soil, (strain < 2mm); determine moisture content at the same time

- Weigh in 10g of moist soil in a sterile conical flask (= dilution $10^{-1}$)
- Add 90ml sterile water ( sterile measuring cylinder). Close conical flask and shake for 1 hour

## 13.7.6 Creating Dilution Levels

Before final dilution, any number of dilution levels can be created. Since the total number of germs is usually around $10^6$/g of soil, it makes sense to select a high level of dilution for determining total bacterial germ count (e.g. for arable soils $10^{-5}$ to $10^{-6}$; compost samples should be diluted to $10^{-6}$ or $10^{-7}$).

E.g. creating soil suspension with dilution level $10^{-6}$:

- 10g soil + 90ml water:                         dilution $10^{-1}$
- 1ml dilution $10^{-1}$ to 99ml water:          dilution $10^{-3}$
- 1ml dilution $10^{-3}$ to 99ml water:          dilution $10^{-5}$
- 10ml dilution $10^{-5}$ to 90ml water:         dilution $10^{-6}$

Each pipette may only be used once! The diluting bottles must be shaken before each additional dilution step.

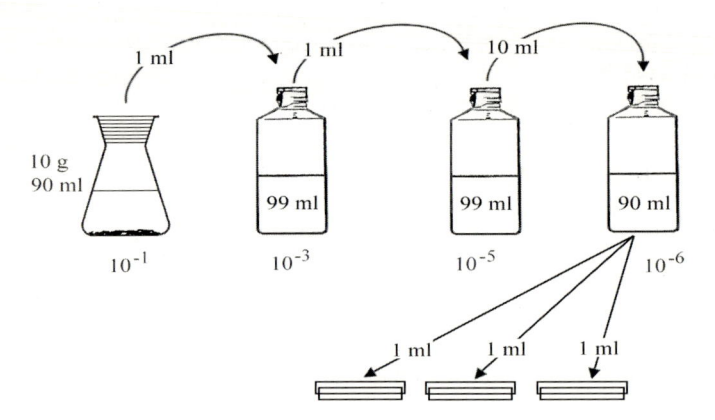

**Fig. 13.1.** Creation of Elution Levels and Seeding the Nutrive Media

## 13.7.7 Seeding the Nutritive Media

1. Label covers of disposable Petri dishes (3 parallel samples)
2. Shake diluting bottle
3. Measure dilution suspension with a sterile pipette and place in Petri dish. When doing this, open the cover only briefly and close quickly
4. Pour lukewarm nutritive medium into Petri dish, again taking care to open cover only briefly
5. Prepare a blank sample to check the nutritive medium
6. Allow nutritive media to cool (This takes approx. 10-30min.)

## 13.7.8 Incubating Nutritive Media

The Petri dishes are placed in an incubator with the cover on the bottom (so no condensation can drip onto the sample). Incubation : 1 week at 27°C

## 13.7.9 Counting the Colonies

To count the colonies, position the Petri dish with the cover on the bottom. Mark each colony with a felt-tip pen to ensure that each colony is counted only once. The germ development process can be disrupted by the formation of fungus. When this occurs, the Petri dish should be discarded.

## 13.7.10 Calculating the Germ Count

- Number of colonies/g DS:

$$(colony\ quantity \cdot 100) / DS\ \% \qquad (13.14)$$

- Consideration of dilution level: The dilution $10^{-x}$ becomes $10^x$ in the result
- The result is stated with one digit before the decimal point.

**Example:**
126 colonies were counted in a $10^{-5}$ dilution,
DS % : 92      (126 · 100) 92 = 136.96
colony quantity /g DS = 136.96 · $10^5$
The result is stated with one digit before the decimal point: 1.3696 · $10^7$
The sample has a colony quantity of 1.4 · $10^7$ / g DS

## 13.8 Soil Respiration, Biological Oxygen Demand (BOD)

Determining oxygen demand of a soil sample with the SAPROMAT

In the course of demineralising carbohydrates through the soil microflora, the carbon source in aerobically functioning micro-organisms is broken down to $CO_2$ and water. The electron or hydrogen acceptor is oxygen. Complete decomposition of a glucose molecule:

$$C_6H_{12}O_6 + 6\,O_2 + 6\,H_2O \rightarrow 6\,CO_2 + 12\,H_2O \qquad (13.15)$$

$$\text{Generally: } (CH_2O) \rightarrow CO_2 + H_2 \qquad (13.16)$$

The glucose is decomposed via gylcolysis to pyruvate, a $C_3$ substance which, in activated form as activated ethanoic acid with separation of a $CO_2$ molecule, is transferred into the citric acid cycle. The last two C-atoms are then eliminated. Beginning with the pyruvate, this procedure occurs twice for each glucose molecule, thus decomposing the entire molecule. The coenzymes active in the citric acid cycle transport the absorbed hydrogen to the oxidation-reduction system of the respiratory chain where it (or electrons) releases energy and is oxidised to form $H_2O$. The energy gain per mol glucose is 36mol adenosine triphosphate (ATP), which is 1100kJ of productive energy.

### 13.8.1 The Sapromat

The Sapromat is a device used to measure the biochemical oxygen demand of water and soil samples. The micro-organism activity creates $CO_2$, which is bound by the soda lime. This creates a vacuum in the container, triggering the pressure indicator. A switch amplifier is used to activate oxygen generation, up to pressure compensation.

The oxygen demand is continuously compiled by the computer. The carbon dioxide that forms is bound by soda lime such that the ratio of nitrogen to oxygen remains stable in the gamma space above the sample for the duration of measurement. The reaction vessels are in a temperature-controlled water bath (temperature 20°C).

A measuring unit consists of one each of the following:

- reaction vessel with $CO_2$ absorber (A) built into plug
- oxygen generator ( B )
- pressure indicator ( C )

The components are linked to one another with hoses and form a closed unit.

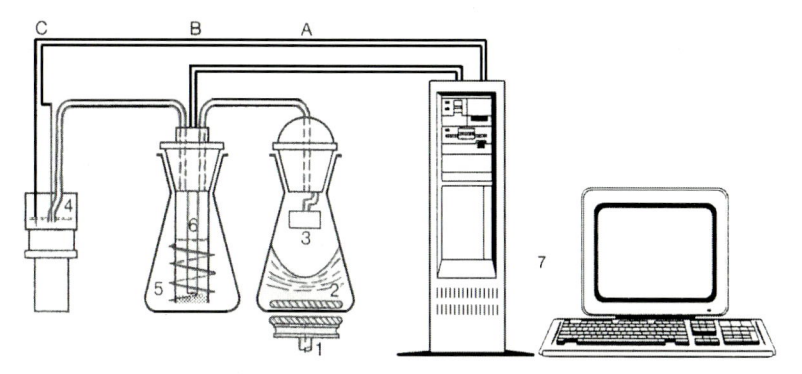

**Fig. 13.2.** Measuring Unit of the Sapromat

| | |
|---|---|
| A | Reaction vessel |
| B | Oxygen generator |
| C | Pressure indicator |
| 1 | Magnetic stirrer f. water samples |
| 2 | Sample |
| 3 | $CO_2$ absorber |
| 4 | Pressure indicator |
| 5 | Electrolyte |
| 6 | Electrodes |
| 7 | Compiler |
| 8 | Computer |

When set to BODEN (= soil), the Sapromat indicates the $O_2$ demand of the sample.

## 13.8.2    24 Hr. Soil Respiration

**Material:** Moist soil; determine moisture content at the same time
1. Weigh approx. 100g of moist soil in a reaction vessel. Make a note of the initial weight. Use a funnel; the lip of the reaction vessel is greasy.
2. Place soda lime in the greased $CO_2$ absorber
3. Place plug on reaction vessel and seal with a clamp
4. Place in Sapromat and switch on (Observe manual!)
5. End measurement after 24 hours

### 13.8.3 Calculation of Results

mg $O_2$/kg TS/24h:

$$( O_2\ \text{final value} \cdot 1000) / ((TS\% \cdot \text{initial weight in g}) / 100) \qquad (13.17)$$

## 13.9 Respiration Activity of Compost

(Determining Rotting Degree) LAGA code of practice 10 (News from the Intrastate Waste Commission LAGA)

The respiration activity offers an indication of the biodegradable components contained in compost. The rotting degree is determined on the basis of oxygen consumption. This occurs in the Sapromat.

**Material:** Approx. 50g of fresh compost, strained to < 10mm
Moisture content and ignition loss are determined at the same time.

### 13.9.1 Procedure

1. Weigh approx. 30g of compost in a Sapromat vessel
2. Set moisture content to 50%, (dry sample at 40-50° if necessary) Water added in g:

$$(\text{initial weight in g} \cdot ( 50 - \text{water content \% sample})) / 50 \qquad (13.18)$$

   Add calculated amount of tap water and mix well
   Allow to stand approx. 1 hour, turning several times
3. Determine oxygen demand during four days in the Sapromat
   The testing temperature is 20°C.

When set to BODEN (= soil), the Sapromat indicates the $O_2$ demand of the sample.

### Evaluation

For evaluation purposes, the final value after four days in mg $O_2$/g $DS_{org}$ or the maximum respiration intensity in mg$O_2$/g $DS_{org}$ per hour can be used.

- Converting initial weight to organic dry substance

$$DS_{org} = (\text{initial weight (g)} \cdot DS\ \% \cdot \text{ignition loss \%}) / 10\ 000 \qquad (13.19)$$

- Respiration activity after four days in mgO$_2$/g DS$_{org}$)

$$mg\ O_2/g\ DS_{org} = O_2\ final\ value\ after\ 96\ hours\ /\ DS_{org} \quad (13.20)$$

- Maximum respiration intensity in mgO$_2$/g DS$_{org}$ · h
  Calculate differences between O$_2$ values after each six hours;
  Use the greatest difference as value A in the formula:

$$mg\ O_2/g\ DS_{org} \cdot h = A/DS_{org}/6 \quad (13.21)$$

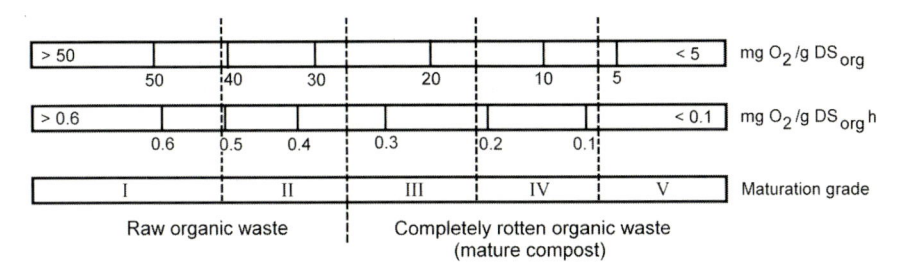

**Fig. 13.3.** Breakdown of Decomposition Process in Rotting Degrees (LAGA code of practice 10 1984)

**Table 13.4.** Soil Respiration Analysis by the Sapromat

| Time h | mg O$_2$ indication | Difference |
|---|---|---|
| 0 | 0 | |
| 6 | 1.5 | 1.5 (1.50-0 = 1.5) |
| 12 | 2.5 | 1 (3-1.5 = 1.5) |
| 18 | 4 | 1.5 |
| 24 | 6.5 | 2.5 |
| 30 | 9.5 | 3 |
| ... | | |
| 96 | 24.5 | |

**Example:**
Initial weight:                39.57g
DS:                             49.88%
IL:                              19.91%
Greatest difference:        03
Final value after 96 hours: 24.5
DS$_{org}$ =                       03.93g

**Calculation Example:**
(39.57 · 49.88 · 19.91) / 10 000
Respiration intensity after 4 days  mg $O_2$/g $DS_{org}$ = 6,23  → 24.5 / 3.93
Maximum respiration intensity  mg $O_2$/g $DS_{org}$ · h = 0,13  → 3 / 3.93 / 6
Rotting degree IV

## 13.10 Carbon Content

Lichterfeld method (determining carbon content via acid digestion) in compliance with DIN 19 684, part 2

The humus content is calculated by multiplying by the factor 1.72 (recently also using the factor 2).

### 13.10.1 Basics

Potassium dichromate in a sulphuric acid solution (amplified oxidation) is used to determine carbon content. Potassium dichromate in an acidic solution destroys organic substances through oxidation. Potassium dichromate is reduced to chromium sulphate. Equivalent amounts of $Cr^{3+}$ ions dyed green emerge from the dichromate. The ions are measured photometrically. The following reaction occurs:

$$8\ K_2Cr\ (VI)_2\ O_7 + 16\ H_2SO_4 + C_6H_{12}O_6 = 4\ Cr(III)_2\ (SO_4)_3 \qquad (13.22)$$
$$+ 4\ K_2SO_4 + 14\ H_2O + 6\ CO_2 + 4\ K_2Cr(VI)_2\ O_7 (10.1)$$

**Material:** air dried or 60°C dried soil, strained to < 2mm, crushed
**Reagents:** $H_2SO_4$ conc. (d = 1.84)
dissolve sulphuric 2n $K_2Cr_2O_7$: 98,07g $K_2Cr_2O_7$ in approx. 500ml dist. $H_2O$, add 100ml conc. $H_2SO_4$, then add water to 1000ml
Calibration Curve: 6.7mg sodium oxalate ($Na_2C_2O_4$) correspond to 0.3mg C (saved in photometer)

### 13.10.2 Procedure

(time needed: 4-5 hours)
1. Weigh sample in a 100ml conical flask; make a note of initial weight
2. Initial weight: arable soil approx. 1g , compost approx. 0.3g
3. Prepare a blank sample (without soil weigh in)
4. Moisten with a small amount of dist. $H_2O$; swivel
5. Add 8ml conc. $H_2SO_4$; swivel

6. Let stand 10min.
7. Add 5ml sulphuric 2n $K_2Cr_2O_7$ (under fume hood); swivel
8. Place in drying chamber 45min. at 150°C; swivel after 15min. and 30min.
9. Remove from drying chamber, then add dist. $H_2O$ to approx. 40ml (under fume hood)
10. Allow to cool 2 hours, then fill to 50ml in measuring cylinder (place in flask again)
11. Swivel sample, then place some of it in a centrifugal flask
12. Centrifuge 10min at 3000rpm
13. With the photometer, compare leftover liquid to blank sample at 578nm

### 13.10.3 Calculation of Carbon Content

The photometer indicates 100x the carbon content of the sample; this value is included in the formula.

$$C\% = photometer\ value\ /\ initial\ weight\ in\ g \qquad (13.23)$$

## 13.11 Determination of Nitrogen (Kjeldahl Procedure)

EAWAG (Swiss Waste Commission), K-3005 (EAWAG - Swiss method)
Nitrogen determination according to Wieninger

Kjeldahl nitrogen is the mass concentration of organically bound nitrogen and ammonia nitrogen in a sample after a certain disintegration. Nitrate oxygen and nitrite nitrogen are not determined; not all organic nitrogen compounds are disintegrated by using this method.

### 13.11.1 Basics

When suitable catalysts in a sulphuric acid solution are used, nitrogenated organic compounds that tend to form ammonia can be reduced to ammonia that can be measured via titration.
When acid digestion in concentrated sulphuric acid with a catalyst (e.g. selenium reaction mixture) is applied, the nitrogen in soil or in plant matter is converted to ammonia. With the sulphuric acid in the excess, the ammonia forms ammonium sulphate.

$$2\ NH_3 + 2\ H^+ + SO_4^{2-} \leftrightarrow 2\ NH_4^+ + SO_4^{2-} \qquad (13.24)$$

In the second step, the ammonia from the sulphuric acid is expelled as $NH_3$ by adding soda lye until becoming alkaline (colour indicator). Distillation conveys the ammonia through the cooler and into the distillation receiver of 2% $H_3BO_3$.

The introduction of ammonia produces ammonium borate, causing H⁺ ions to be bound.

$$H_3BO_3 + NH_4OH \rightarrow NH_4H_2BO_3 + H_2O \qquad (13.25)$$

Titration with 0.025mol/l $H_2SO_4$ and the resulting displacement of the weaker boric acid by the stronger sulphuric acid cause the ammonium borate in the distillation receiver to produce ammonium sulphate.

$$2\, NH_4H_2BO_3 + H_2SO_4 \rightarrow (NH_4)_2\, SO_4 + 2\, H_2BO_3 \qquad (13.26)$$

When the ammonium borate has been completely converted, the excess acid causes the indicator to turn from green to pink. The sulphuric acid utilised is equivalent to the amount of $NH_3$.

**Reagents**:
- Sulphuric acid conc. H2SO4   d = 1.84  (95-97%)
- Selenium reaction mixture (to determine oxygen according to Wieninger)
- For the distilling device: soda lye 30% (300g NAOH + 700ml dist. $H_2O$) boric acid 2%, dyed with mixed indicator 5 for ammonia titration
- For titration: 0.025mol/l sulphuric acid (add dist. $H_2O$ to 50ml 0.5mol/l with
- dist.$H_2O$ to 1 l)

**Material:** Air dry or 60°C dry soil, strained to < 2mm, crushed

## 13.11.2 Procedure

1. Weigh soil in a Kjeldahl disintegration beaker; make a note of initial weight (compost approx. 0.5g, arable soil approx. 2g)
2. Add approx. 1g selenium reaction mixture (one spatula tip)
3. Add 10ml conc. sulphuric acid $H_2SO_4$ (under fume hood)
4. Place disintegration beaker in the preheated disintegration block and cook at 360°C until the disintegrated material is yellowish green (1-3 hours). To exhaust and neutralise the sulphuric acid fumes use water and gas scrubber
5. Allow disintegrated material to cool
6. Place sample holder under the fume hood and add 90ml dist. $H_2O$
7. Turn on distilling device and cooling water; wait for indication of readiness
8. (approx. 10min.). Run program once with dist. $H_2O$ (to rinse hoses)
9. Place conical flask under the right hose. Take care to ensure that the boric hose is directly over the beaker
10. Clamp sample beaker under the left hose

11. Start Kjeldahl soil program, the following sequence automatically occurs:
    50ml of 2% boric acid is pumped into the conical flask
    30% NAOH is pumped into the sample beaker (→ the sample becomes alkaline)
    Hot steam is conveyed into the sample
    In 3.45min. distillation time, the ammonia produced is distilled over to the boric acid; the indicator changes from lavender to green
12. Remove conical flask: Titrate with 0.025m $H_2SO_4$ until colour changes to pink

### 13.11.3 Calculation of Results

- 1ml 0.025m $H_2SO_4$ corresponds to 0.70035mg N
- N% = (consumption $H_2SO_4$ · 0.070035) / initial weight in g

Schroeder 1983: The total N contents (in the DS) in mineral soil is between 0.03 and 0.3%; in peat it is (according to Davis and Lucas 1959) between 0.3 and 4.0%.

## 13.12 C/N and C/P Ratio

Microbial decomposition can be derived from the pH value, the C:N ratio and the C:P ratio of the organic substance.

**Carbon C**
Determined e.g. by acid digestion, with the TOC analyser or based on theignition loss (IL % · 0.47 = C %)

**Nitrogen N**
Total N, e.g. determined according to Kjeldahl or with the N analyser; The extraordinarily low nitrate and nitrite contents in the soil that are not determined during disintegration do not cause any measurable errors. When mineral fertilisation with substances containing nitrate occurs, large amounts of nitrate can be expected. With compost samples, the Kjeldahl nitrogen is the total nitrogen.(Book of Methods for Analysing Compost, Federal Quality Group for Compost, 1994)

**Phosphorus P**
Total P, nitrohydrochloric acid disintegration, photometric determination or ICP
   Example: C = 4.2%, N = 0.3%, P = 0.04%
   4.2 : 0.3 = 14 → C/N ratio = 14
   4.2 : 0.04 =105
   0.3 : 0.04 = 7.5

   0.04 : 0.04 = 1 → C/N/P ratio 105 : 7.5 : 1

## 13.12.1 Evaluation

**Table 13.5.** C/N Ratio from Kunze, Roeschmann, Schwerdtfeger: Pedology, 5th edition

| | |
|---|---|
| under 10 | With a close C/N ratio under 10, the supply of C is quickly exhausted; humus may decline. |
| 10 to 20 | With a C/N ratio between 10 and 20, carbon mineralization through soil micro-organisms occurs relatively quickly. The C/N ratio of good arable soil is between 10 and 13. |
| over 20 | If the C/N ratio in tilled soil is greater than 20, C decomposition is hindered. Micro-organisms consume nitrate, meaning there is none for the cultivated plants. There is a deficiency of N, and the development of the micro-organisms is increasingly hindered. When crop residue containing a great C/N ratio is added to soils, the C/N ratio is even greater after the decomposition process. |

## 13.12.2 C/P Ratio

Phosphates play an important role in the metabolism of micro-organisms. However, their P requirement is lower than the N requirement by a power of ten.

- The C:P ratio of good arable soil is 100-200.
- The optimum is C:N:P = 100:10:1

## 13.13 Determination of Carbonate

VDLUFA (German Agricultural Testing and Research Agency) Book of Methods, Vol.1 (1991), Soil Exploration, Method A 5.3.1

The term carbonate generally refers to calcium carbonate and calcium magnesium carbonate (calcite and dolomite).

### 13.13.1 Method

Dissolution of $(Ca)CO_3$ with HCl; gasometric determination of produced $CO_2$, taking into consideration the influence of temperature and pressure according to Scheibler. The $CO_2$ formed in the reaction vessel fills the Scheibler bulb in the bottle and displaces an equivalent quantity of air. The displaced air forces the confining liquid out of the measuring tube and into the reservoir. The level tube ensures that the result can be read clearly.

**Preliminary Sample**: Put a few drops of 10% HCl on the sample. If it foams, the sample contains lime.

## 13.13.2 Procedure

**Material**: air dry or 60°C dry soil, strained to < 2mm, crushed;
**Initial weight**: 0.5-3g, depending on lime content; make a note of initial weight

When pouring the HCl, touch only the neck of the bottle. A change in temperature can distort results. Temperature and air pressure must be recorded. Place approx. 5ml HCl (approx.18.5%) in the plug insert. The unit has three three-way cocks. Verify that these cocks are in the correct position.

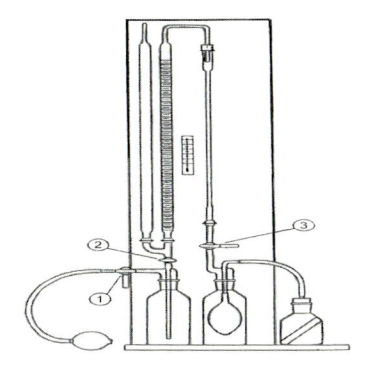

**Fig. 13.4.** Scheibler (Carbonate) Measure Equipment and Instructions

**Table 13.6.** Instructions

| Task | Cock 1 left | Cock 2 center | Cock 3 right |
|---|---|---|---|
| Fill U-tube: slowly press bulb; regulate with 2 | ⊥ | –↗| | ⊢ |
| Balance liquid level; regulate with 2 | ⊢ | –↗| | |
| Wipe dry outside of HCl insert, then place on generating vessel Close system | | | ⊣ |
| Gradually add HCl (by tilting) to the sample. Regulate liquid level by draining with cock 2. The liquid level in the left tube should be somewhat higher | | –↗| | |
| After 5-10 minutes: swivel glass again; balance liquid level; Read $CO_2$ volume in ml | | –↗| | |

### 13.13.3 Calculation

CaCO$_3$ % =

$$(CO_2 \text{ in ml} \cdot \text{air pressure in mm HG} \cdot 0{,}1605) / ((273 + \text{temp. in °C}) \cdot \text{initial weight in g}) \quad (13.27)$$

**Example**
CO$_2$ : 23ml - air pressure : 760mm HG - temp. : 20°C - initial weight : 4.00g
CaCO$_3$ = 2.39%

### 13.13.4 Evaluation of Carbonate Content in Soils

Pedological mapping instructions, 1994

Soils with a carbonate content greater than 10% are referred to as marly soil.

**Table 13.7.** Classification of Carbonate Contents in Soils

| Designation | Carbonate content in % |
|---|---|
| carbonate-free | 0 |
| minimal carbonate | < 0.5 |
| little carbonate | 0.5-2 |
| some carbonate | 2-4 |
| moderate carbonate | 4-7 |
| substantial carbonate | 7-10 |
| rich in carbonate | 10-25 |
| very rich in carbonate | 25-50 |
| extreme amount of carbonate | > 50 |

## 13.14 Determination of Plant-Available Phosphorus and Potassium

Calcium-Acetate-Lactate (CAL) Method; VDLUFA (German Agricultural Testing and Research Agency) Book of Methods (1991), Volume 1, Soil Exploration, Method A 6.2.1.1.

Determination of portion of the nutrients phosphorus and potassium that is available for plants. K and P are washed out and filtered out with the aid of calcium-acetate-lactate solution. K is determined directly with the flame photometer. Phosphorus is determined photometrically as phosphorus molybdenum blue at 578nm.

**Material**: Air dry or 60°C dry soil, strained to < mm, cohesive soil, possibly crushed. (Bog soils: moist)

## 13.14.1 Reagent

**Extraction Solution**
- Stock solution A
- Dissolve 77.0g calcium lactate ($C_6H_{10}CaO_6+5H_2O$) in hot dist. $H_2O$
- + dissolve 39.5g calcium acetate (($CH_3$ $COO)_2$ $Ca.H_2O$ in hot dist. $H_2O$
- + 89.5ml pure acetic acid $CH_3$ COOH
- + Add water to 1 l
- Blank extraction solution: Add water to 200ml solution A to 1 l

**Reference Solution for K**
- 3.5mg and 2.0mg and 1.0mg K/100ml blank extraction solution
- Blank extraction solution as blind value

**Colour Reagents for P**
- Solution A: Store in a dark bottle
- Add 35ml conc. $H_2SO_4$ to 250ml dist. $H_2O$ (first water, then acid)
- + dissolve 3g ammonium heptamolybdate $(NH_4)6$ $MO_7$ $O_{24}.4H_2O$ in 75ml dist. $H_2O$+ 25ml potassium antimonyl tartrate solution
- (0.2743g (KSbO)$C_4H_4O_6$ · 0.5 $H_2O$ in 100ml dist. $H_2O$; 25ml of this)
- L(+) ascorbic acid $C_6H_8O_6$: prepare fresh every day
- Dissolve 0.88g ascorbic acid in 50ml dist. $H_2O$
- Mixed reagent B : Mix solution A and ascorbic acid in a ratio of 7:3.
- ascorbic acid (ml): ((quantity of samples + 1 blind. · 8) / 10) · 3
- mixed reagent B (ml): ((quantity of samples + 1 blind. · 8) / 10) · 7

## 13.14.2 Procedure

1. Weigh 5g soil in a 500ml plastic bottle
2. Add 100ml blank extraction solution
3. Shake two hours
4. Filter; discard first filtrate

**Determining K**
The filtrate is measured on the flame photometer
Standard 0: blank extraction solution
The K content is stated in 100ml (= 5g soil)

**Determining P**
- Pipette one aliquot (portion) into a 50ml measuring vessel
  (e.g. with compost 2ml, arable soil 5ml, no more than 10ml)
- Prepare blind value (5ml Gebrauchs-Extraktions-Lösung)
  Add 8ml mixed solution B
  Add dist. $H_2O$ to 50ml; swivel well
- Allow to stand 1 hour
  Photometric measurement at 578nm relative to colourless blind value.

### 13.14.3 Calculation of Results and Classification

$$K\ mg/100ml\ soil = 20 \cdot mg\ K\ (flame\ photometer\ value) \cdot 1.4 \qquad (13.28)$$

$$P\ mg/100ml\ soil = 100/aliquot\ in\ ml \cdot 20 \cdot mg\ P\ (photom.\ Value) \cdot 1.4 \qquad (13.29)$$

(The value 1.4 applies to arable soils, raw unit weight 1.4kg/l soil, conversion factors: $K \cdot 1.21 = K_2O$,  $P \cdot 2,29 = P_2O_5$)

**Table 13.8.** Classification of Nutrient Provision in Soils (LUFA, Agricultural Testing and Research Agency 1995, Guidelines for fertilisation, values in mg/100ml soil (CAL))

| Nutrient content class | Soil nutrient provision | Arable soil | Home and allotment gardens | Home and allotment gardens | Grassland |
|---|---|---|---|---|---|
| | | P and K | P and K (K soil: S, LS) | K sL, L, C | P and K |
| A | low | 0-5 | 0-4 | 0-8 | 0-2 |
| B | medium | 6-10 | 5-8 | 9-16 | 3-5 |
| C | high (optimum) | 11-15 | 9-12 | 17-25 | 6-10 |
| D | very high | 16-20 | 13-20 | 26-40 | 11-15 |
| E | extremely high | over 21 | over 20 | over 40 | over 15 |

S = Sand, LS = Loamy sand, sL = Sandy loam, L = Loam, C = Clay

## 13.15 Determination of Plant-Available Potassium and Magnesium (diluted with Calcium Chloride)

### 13.15.1 Method

K     Scheffer-Schachtschabel 1982
Mg   VDLUFA (German Agricultural Testing and Research Agency) Book of Methods (1991), Volume 1, Soil Exploration, Method A 6.2.4.1.

The *nutrient content* of the soil is examined to determine application of fertiliser required. This is done by shaking a soil sample that is a representative composite sample of the arable soil with an extraction solution; the quantity of nutrient released in the solution is determined in the extract. The *potassium content* is determined with a flame photometer. The magnesium content is determined with an atom absorption spectrophotometer (AAS).

**Reagents:** Add dist. $H_2O$ to 0.0125mol/l $CaCl_2$ : 1,838g $CaCl_2 \cdot 2\ H_2O$ to 1 l

**Material:** Air dry or 60°C dry soil, strained to < 2mm; cohesive soil must first be crushed

### 13.15.2 Procedure

1. Weigh 10g soil in a 500ml plastic bottle
2. Add 100ml 0.0125mol/l $CaCl_2$
3. Shake 2 hours on horizontal shaker
4. Filter sample; discard first filtrate
5. K is determined with the flame photometer and Mg with the AAS

### 13.15.3 Calculation of Results

$$mg\ K\ (or\ Mg)\ /\ 100\ g = concentration\ in\ mg/l \quad (13.30)$$

Conversion factors: *mg/100g · 1.4 = mg/100ml* (applies to arable soils, raw unit weight 1.4kg/l soil)

### 13.15.4 Evaluation

**Table 13.9.** Classification of Nutrient Provision in Soils (LUFA, Agricultural Testing and Research Agency 1995, Guidelines for fertilisation, values in mg/100ml soil. Sampling depth: arable soil: 0-25 (30)cm, grassland: 0-10cm

| Nutrient content class | Soil nutrient provision | K arable soil | Mg arable soil | K and Mg grassland |
|---|---|---|---|---|
| A | low | 0-2 | 0-3 | 0-2 |
| B | medium | 3-5 |  | 3-5 |
| C | high (optimum) | 6-10 | 4-6 | 6-10 |
| D | very high | 11-18 | 7-10 | 11-15 |
| E | extremely high | over 19 | over 11 | over 16 |

## 13.16 Determination of N – min (NO$_3$ and NO$_2$)

Method: VDLUFA (German Agricultural Testing and Research Agency) Book of Methods 1 1991, 2$^{nd}$ partial delivery 1997, A 6.1.4.1

Extracting nitrates from a soil sample with a weak saline solution. Determining nitrate nitrogen contents by examining the difference in UV absorption of the extracts at 210nm directly and after reduction with nascent hydrogen.

Usually only the nitric nitrogen in the soil is taken into consideration, because the main time frame of the soil exploration is in early spring. At this time of year, only negligible quantities of soluble nitrogen - mostly ammonium nitrogen - are present. If the ammonium nitrogen content is greater than 10kg/ha, it is added to the nitrate nitrogen. When the soil has recently been fertilised with semi-liquid or liquid manure, ammonium contents are higher.

The samples are usually taken in layers of 30cm: 0-30 cm, 30-60 cm, 60-90cm. The values of the layers are added together.

### 13.16.1 Basics

The method applies absorption of the nitrate ion at 210nm. But since several other materials (especially organic materials) also absorb UV light, a second measurement is conducted on a sample in which the nitrates have been reduced with the aid of copper-plated zinc granules in an acidified medium.

## 13.16.2 Material

Moist soil, strained to < 5mm. The soil must be fresh. If it can not be processed immediately, it should be strained and mixed well then stored in a freezer. The frozen soil should then be used for the analysis. The soil can be stored in a refrigerator at 4°C for up to 8 days. The soil can also be spread out and dried at 105°C for 2-6 hours. When the filtrate has been gained, the nitrate content no longer changes.

## 13.16.3 Reagents

- Extraction solution 0.0125mol/l calcium chloride solution:
  Dissolve 1.84g $CaCl_2 \cdot 2H_2O$ with dist. $H_2O$ to 1 l
- Dissolve sulphuric acid approx. 5% : 5ml conc. $H_2SO_4$ in 100ml dist. $H_2O$
- Copper-plated zinc granules (Zinc, granulated 3-8mm)
  Dissolve copper sulphate solution 2.5g $CuSO_4 \cdot 5H_2O$ with dist. $H_2O$ to 100ml
- Creating copper-plated zinc granules:
  Mix 125g zinc in a beaker with 75ml dist. $H_2O$ and 7.5ml sulphuric acid approx. 5% and stir until the surface of the zinc is completely clean
- Decant liquid. Wash 2-3 times in portions with 75ml dist. $H_2O$
- Add 75ml dist. $H_2O$
- While constantly swiveling, add approx. 12.5ml copper sulphate solution by drops, until the zinc granules are completely covered with a black coating of copper
- Decant liquid
- Rinse with dist. $H_2O$
- Allow to air dry, then store is a closed glass vessel

## 13.16.4 Creating a Calibration Curve

A new calibration curve should be created before each measurement.

- 112ml extraction solution + 1ml $NO_3$ parent solution (1g/l)
- 1ml contains 2µg nitrate N (The solution must be prepared daily)
- Calibration series:
  Pipette 0-2.5-5-10-15-20ml of this solution in one 25 ml measuring flask each. Fill with extraction solution.
- The calibration series contains 0-5-10-20-30-40µg $NO_3$-N in 25ml
- Add 1ml 5% sulphuric acid, mix
- Create calibration curve at 210nm

## 13.16.5 Procedure

(Dry substance identification is done at the same time.)

1. Weigh 50g soil in a 1000ml plastic bottle
2. Add 500ml $CaCl_2$ extraction solution (or 200ml)
3. Shake 1hr.
4. Filter (e.g. folded filter 597 1/2); discard first portion of filtrate
5. Pipette 25ml of the filtrate or an aliquot into a test tube; (Add water to 25ml with aliquot)
6. Prepare blank value: 25ml extraction substance
7. Add 1ml 5% sulphuric acid to each test tube
8. Read $NO_3$-N content on photometer (1$^{st}$ measurement)
9. Place 3-4 zinc granules in each test tube (except blind value), then let stand overnight, free from dust
10. Measure samples on photometer (2$^{nd}$ measurement)

## 13.16.6 Calculating Results

$NO_3$-N in aliquot =

$$\text{(photometer conc. 1}^{st}\text{ measurement)} - \text{(2}^{nd}\text{ measurement)} \tag{13.31}$$

$NO_3$-N mg /100g DS =

$$(\mu g\ NO_3\text{-N in aliquot} \cdot \text{ml extraction volume} \cdot 10) / \tag{13.32}$$
$$(\text{aliquot in ml initial weight in g} \cdot DS\ \%)$$

$NO_3$-N kg/ha =

$$NO_3\text{-N mg/100 g} \cdot \text{layer thickness in cm} \cdot \text{raw unit weight in kg/l} \tag{13.33}$$

Layer thickness 30cm, raw unit weight 1.4kg/l (normal arable soil, organic substance under 8%)

## 13.17 Determination of N-min (NH₄)

According to DIN 38 406 E5

### 13.17.1 Basics

Ammonium ions at a pH value of approx. 12.6 react with hypochlorite ions and salicylate ions in the presence of sodium pentacyano nitrosylferrate (2) (sodium nitroprusside) as a catalyst to turn the substance blue. The hypochlorite ions are formed in an alkaline medium by hydrolysis of ions of the dichloroisocyanic acid.

### 13.17.2 Reagents

**Solution 1:**
Disodium pentacyanonitrosylferrate sodiumsalicylate solution
200mg sodium nitroprusside $Na_2 (Fe(CN)_5NO) \cdot 2 H_2O$
+Dissolve 17g sodium salicylate $C_7H_5NaO_3$ in dist. $H_2O$ and fill to 100ml. The solution will keep only for a limited time.

**Solution 2:**
Citrate solution
100g trisodium citrate $C_6H_5Na_3O_7 \cdot 2 H_2O$
+ Dissolve 10g sodium hydroxide NaOH in dist. $H_2O$ and fill to 500ml

**Solution 3:**
Oxidation reagent
Dissolve 0.030g dichloro isocyanic acidic sodium $NaCl_2(NCO_3)$ in 5ml dist. $H_2O$
+ add 20ml solution 2
The solution must be made fresh every day.

**Material and Extraction**: Refer to Determining N-min

### 13.17.3 Procedure

1. Pipette 25ml 0.0125mol/l into a test tube as blind sample
2. Pipette 25ml of the filtrate into a test tube
3. Add 1ml solution 1 and swivel
4. Add 1ml solution 3 and swivel
5. After allowing to stand 90min. measure with photometer at 690nm; compare to the yellow blank value

The values are stated in mg/l $NH_4$. 0.0125mol/l $CaCl_2$ should be used for dilutions. The calibration curve must be created with 0.0125mol/l $CaCl_2$.

### 13.17.4 Calculation

1mg NH$_4$ is equivalent to 0.78mg NH$_4$-N

NH$_4$-N mg/100g DS =

$$\textit{(mg/l NH}_4 \cdot 0.8 \cdot \textit{extraction volume in ml} \cdot 10) \; / \; \textit{initial weight in g} \cdot \textit{DS \%} \tag{13.34}$$

NH$_4$-N kg/ha =

$$\textit{NH}_4\textit{-N mg/100 g} \cdot \textit{layer thickness in cm} \cdot \textit{raw unit weight in kg/l} \tag{13.35}$$

Layer thickness 30cm, raw unit weight 1.4kg/l (normal arable soil, org. substance under 8%)

## 13.18 Determining Exchangeable Cations at Soil pH

(Cation Exchange Capacity, CEC) According to Meiwes et. al. 1984
Exchange using percolation method with 1mol/l NH$_4$Cl

The exchangeable cations in the soil are replaced by NH$_4$ ions. The exchange occurs with the percolation method and results in an excess of ammonium ions. This method is suitable only for soil free of carbonates. The cations are exchanged at a pH value close to that of the soil. The NH$_4$ solution is not buffered; it has a pH value of 4.1 to 4.6. All cations of quantitative relevance are measured in the solution: Na, K, Mg, Ca, Al, Mn and H; NH$_4$ is not measured.

When measuring cations containing soluble salts, oxides or hydroxides in the soil, these substances may be partially dissolved and also measured. The sum of the absorbed cations is greater than the cation exchange capacity, because - in addition to the replacably bound ions - the soil solution (dissolved salts) is also determined.

**Material:** Air dry or 60°C dry soil, strained to < 2mm; cohesive soil may need to be crushed

**Reagents:** 1mol/l ammonium chloride solution: Dissolve 53.49g NH$_4$Cl in dist. H$_2$O and add water to 1 l
(Filter wadding: extra fine glass wool)

### 13.18.1 Preparing Percolation Tube

Percolation tube: 50ml Eppendorf Combitips

1. Fill tube with 0.7–0.75g filter wadding
2. Rinse approx. 3 times with 1mol/l $NH_4Cl$; push wadding firmly back each time with a glass rod
3. Add 10g quartz
4. Add 4-10g soil sample (record exact weight), Sand 10g, less cohesive soil; silica sand is added to 10g
5. Add 10g quartz
6. Close the bottom of the tube with a plug (yellow Eppendorf standard tip with fused together tip)
7. Add 5ml 1mol/l $NH_4Cl$
8. Allow to stand overnight
9. Prepare blind sample : 0.7g filter wadding, 30g quartz

When a sample is difficult to moisten, the sample is carefully mixed with the quartz using a glass rod.

### 13.18.2 Percolation

1. Place percolation tube on 100ml measuring flask
2. Position filtering stand with winchester (with inserted infusion drip) over tube
3. Close drip all the way
4. Pour 90ml 1mol/l $NH_4Cl$ in winchester
5. Set drip speed: one drop flows every 12sec. (90ml in approx. 5 hours). If at first no drops are emitted, carefully press the drip together once. The dripping may cease, so it should be checked periodically
6. Upon completion: Fill 100ml flask to calibration mark with 1mol/l $NH_4Cl$, then swivel

### 13.18.3 Identifying Elements

**Identifying $H^+$:** Pour a portion in a beaker and measure pH value. The pH value of the 1mol/l $NH_4Cl$ must also be determined

**Calculable $H^+$:** The pH values of the percolate ($pH_x$) and of the pure $NH_4Cl$ ($pH_o$) are measured potentiometrically (with the pH-meter). The $H^+$ concentrations in M/l according to $ph = -\log H^+$ are derived from this and then subtracted from one another ($H_x^+ - H_o^+ = H$ – value in M/l ). Multiplying by (100 / initial weight · 100) results in the $H_{eff}$ value in $cmol_c/kg$

**Formula for Excel (program):**
H concentration in M/l $(A_{sample})$ = 1/POWER (10; pH value of sample)
H concentration in M/l $(B_{NH4Cl})$ = 1/POWER (10; pH value NH$_4$Cl)

$$C = A_{sample} - B_{NH4Cl}$$

$H_{eff}$ in cmol$_c$/kg =

$$((C \cdot 100) / \text{initial weight in g}) \cdot 100 \tag{13.36}$$

Identifying Na, K, Mg, Ca, Al, Mn, Fe: Acidify sample with 2 drops of conc. HNO$_3$. Identify elements with ICP, AAS or flame photometer

## 13.18.4  Calculation of Cation Exchange Capacity

cmol$_c$ / kg

$$\frac{(\text{conc. in mg/l}^{1)} \cdot \text{volume in ml} \cdot 100)}{(\text{molar mass} \cdot \text{initial weight in g} \cdot 1000)} \tag{13.37}$$

[1] conc. in mg/l = conc. sample - conc., blank value  Volume = 100, molar mass = molecular weight / valence

**Table 13.10.** Calculation of Molecular Weight, Valence and Molar Mass for Determination of Cation Exchange Capacity

| Element | Molecular weight | Valence [2] | Molar mass |
|---|---|---|---|
| Na | 22.99 | 1 | 22.99 |
| K | 39.098 | 1 | 39.10 |
| Mg | 24.305 | 2 | 12.16 |
| Ca | 40.078 | 2 | 20.04 |
| Fe | 55.845 | 3 | 18.62 |
| Mn | 54.938 | 2 | 27.47 |
| Al | 26.982 | 3 | 8.99 |

[2] It is assumed that Mn is bivalently bound and Al and Fe trivalently bound to the exchanger

### Terms and Conversions

$$cmolc \quad centimol / \text{ion charge } (10_{-2} \text{ mol}) \tag{13.38}$$

$$cmolc / kg \cdot 10 = \mu molc/g \tag{13.39}$$

$$cmol_c / kg \cdot 10 = mmol_c/kg \tag{13.40}$$

(c = charge); ($\mu mol_c/g$ = mmol$_c$ / kg)

Old unit of measure: *mval* (= milli equivalent): mmol$_c$ / kg = mval / kg
Old form: *mmol/z/100g*, now: mmol$_c$ / 100g, cmol$_c$ / kg = mmol/z/100g

Conversion from mg / kg to cmol$_c$ / kg:

$$( mg/kg / 1000) \cdot ((1 / molar\ mass) \cdot 100) = cmolc / kg \qquad (13.41)$$

e.g. 1000 mg /kg K = 2.56 cmol$_c$ / kg K

## 13.18.5 Examples of Cation Exchange Capacity of Soils

**Table 13.11.** Absorption Ratios in Top Soil of Soils of Arable Land and beneath Forest (Scheffer/Schachtschabel 1998)

|  | CEC$_{eff}$ (cmol$_c$/kg) | Saturation (of CEC eff) ||||
|---|---|---|---|---|---|---|
|  |  | Al | Ca | Mg | K | Na |
| Soils under arable land; Central Europe |  |  |  |  |  |  |
| Grey-br. podzolic soil (loess, Straub.) | 14 | 0 | 80 | 15 | 5 | <1 |
| Chernozem (loess, Hildesheim) | 18 | 0 | 90 | 9 | 0.5 | 0.4 |
| Marsh (Wersermarsch) | 25 | 0 | 50 | 42 | 3 | 5 |
| Pelosol (Liaston, Franken) | 17 | 0 | 83 | 8 | 9 | 0 |
| Podzol (sand, Celle) | 3 | 0 | 86 | 6 | 7 | 1 |
| Soils under forest; Central Europe |  |  |  |  |  |  |
| Podzol (granite, Bav. Forest) | 6.8 | 65 | 22 | 6 | 4.6 | 2.6 |
| Stagnosol (loess, Bonn) | 5.4 | 69 | 13 | <2 | 6 | 11 |
| Cambisol (loess, pumice, Vogelsberg) | 12 | 85 | 5.8 | 4.2 | 5 | 0 |
| Soils formed in other climates |  |  |  |  |  |  |
| Vertisol (Sudan) | 47 | 0 | 71 | 25 | 0.4 | 3.8 |
| Oxisol ( Brazil) | 2.6 | 89 | 2.7 | 3.5 | 3.1 | 1.2 |
| Ultisol (Puerto Rico) | 7.2 | 72 | 15 | 8.3 | 2.8 | 1.4 |
| Aridisol (Arizona, USA) | - | 0 | 45 | 5.5 | 2.5 | 47 |

|  | CEC$_{pot}$ (cmol$_c$/kg) | pH CaCl$_2$ | C$_{org}$ (%) |
|---|---|---|---|
| Soils under arable land(Central Europe) |  |  |  |
| Grey-brown podzolic soil (loess, Straubing) | 17 | 6.3 | 1.4 |
| Chernozem (loess, Hildesheim) | 17 | 7.2 | 1.6 |
| Marsh (Wersermarsch) | 37 | 5.1 | 2.7 |
| Pelosol (Liaston, Franken) | 22 | 6.7 | 2.4 |
| Podzol (sand, Celle) | 12 | 5.2 | 2.5 |
| Soils under forest(Central Europe) |  |  |  |
| Podzol (granite, Bav. Forest) | - | 2.6 | 12 |
| Stagnosol (loess, Bonn) | 18 | 3.8 | 5.7 |
| Cambisol (loess, pumice, Vogelsberg) | 60 | 2.9 | 20 |
| Soils formed in other climates |  |  |  |
| Vertisol (Sudan) | 45 | 6.8 | 0.9 |
| Oxisol ( Brazil) | 13 | 3.5 | 2.8 |
| Ultisol (Puerto Rico) | 26 | 3.5 | 3.3 |
| Aridisol (Arizona, USA) | 36 | 9.9 | 0.4 |

## 13.19. Nitrohydrochloric Acid Disintegration

DIN 38 414 - S7

Bound ions are transformed to a soluble state through a very acidic environment and heat reaction. Nitrohydrochloric acid disintegration is applied to identify heavy metals and other elements. Disintegration includes an indefinite percent of the total content, particularly when dealing with flash heated oxides or some silicates. However, this type of disintegration has proven useful for evaluating environmental issues (evaluation of contents in compliance with the sewage sludge ordinance). The disintegration reagent is a mixture of concentrated nitric acid and concentrated hydrochloric acid in a ratio of 3:1.

**Material:** 105°C dried soil, strained to < 2mm, ground with an agate mortar. When HG or Cd is to be determined, the sample should be air dry (or dried at 60°C).

### 13.19.1 Procedure

Disintegration must occur under a fume hood; highly toxic gases are released during boiling.

- Place 3g of soil (analytical balance) in a disintegration beaker and moisten with a small quantity of bi-dist. $H_2O$ (to prevent sample from bursting apart when acid is added).
- Then add 21ml HCl (37%) and 7 ml $HNO_3$ (65%). When the sample has a high lime content, the HCl must be added by drops (it foams).
- The beaker is placed in the disintegration block.
- The reflux condenser is put into place, and 10ml 0.5mol/l $HNO_3$ is added to the absorption vessel.
- The samples must stand at room temperature for at least 12 hours. Then they are slowly brought to a boil (approx. 120°C) and kept at this temperature for 2 hours.
- The condensed acid is returned to the sample through the reflux condenser (→water cooling) and the absorption vessel.
- When it cooled, the contents of the absorption vessel are placed in the reflux condenser, then the absorption vessel and the reflux condenser are rinsed with 0.05mol/l $HNO_3$.
- The disintegrated material is poured through a folded filter - previously rinsed with 0.05mol/l $HNO_3$ - into a 100ml measuring flask. Then bi-dist. $H_2O$ is added and the flask is swiveled.

### 13.19.2 Microwave Disintegration

An alternative to nitrohydrochloric acid disintegration is microwave disintegration. However, classification according to the sewage sludge ordinance requires nitrohydrochloric acid disintegration.

Disintegration occurs in a compression-proof, tightly closed Teflon container. A higher boiling point can be achieved than with a nitrohydrochloric acid disintegration due to the higher energy supply (microwaves) and higher pressure. Substances that are difficult to disintegrate (e.g. Cr) are disintegrated better with this method.

Microwave disintegration is better suited for plant disintegration or soils with high organic material content, because these samples foam excessively with nitrohydrochloric acid disintegration.

**Example of microwave disintegration:**
0.5g soil + 1ml $HNO_3$ (65%) + 3ml HCl (37%) + 1ml $HClO_4$ (60%)
if greater than 30% organic components + 2ml $H_2O_2$ (30%).

Disintegrate sample in the microwave, filter, then add water to 50ml.

## 13.20 Sewage Sludge Regulations

(AbfKlärV = Garbage and Sewage Regulations) Federal Law Gazette 1992

Depositing sewage sludge on soil used for gardening or agricultural purposes is not permitted, if soil exploration shows that any one of the following values for heavy metals is exceeded: Heavy metals are more accessible at low pH values. Thus lower limits apply when the pH value for cadmium and zinc is lower than 6.

**Table 13.12.** Threshold values of heavy metals in soils and sewage sludge

| Element | Symbol | Content in soil (mg/kg TS) Limit | Limit pH value < 6 | Contents in sewage sludge (mg/kg TS) Limit | Limit pH value < 6 |
|---|---|---|---|---|---|
| Lead | Pb | 100 |  | 900 |  |
| Cadmium | Cd | 1.5 | 1 | 10 | 5 |
| Chromium | Cr | 100 |  | 900 |  |
| Copper | Cu | 60 |  | 800 |  |
| Nickel | Ni | 50 |  | 200 |  |
| Mercury | Hg | 1 |  | 8 |  |
| Zinc | Zn | 200 | 150 | 2500 | 2000 |

**Table 13.13.** Conversion Factors

| Given | Required | Factor | Given | Required | Factor |
|---|---|---|---|---|---|
| N | $NO_3$ | 4.425 | K | $K_2O$ | 1.21 |
| $NO_3$ | N | 0.225 | $K_2O$ | K | 0.83 |
| N | $NH_4$ | 1.29 | Ca | CaO | 1.40 |
| $NH_4$ | N | 0.78 | CaO | Ca | 0.717 |
| P | $P_2O_5$ | 2.29 | Mg | MgO | 1.66 |
| $P_2O_5$ | P | 0.44 | MgO | Mg | 0.60 |

(e.g. 24mg $NH_4$ corresponds to ($NH_4 \cdot 0{,}78$) 18.72mg N

**Table 13.14.** Ideal pH Values in Arable Soil

| Soil type | Ideal pH value ($CaCl_2$) at humus contents in % | | | | |
|---|---|---|---|---|---|
| 0 | 0–4[1] slightly humic | 4-8 strongly humic | 8-15 very strongly humic | 15-30 slightly organic | Over 30 peat |
| Sandy soil S | 5.5 | 5.5 | 5.0 | 4.5 | 4.0 |
| Loamy sand | 6.0 | 6.0 | 5.5 | 5.0 | 4.0 |
| Sandy loam | 6.5 | 6.0 | 5.5 | 5.0 | 4.0 |
| Loam and clay | 7.0 | 6.5 | 6.0 | 5.5 | 4.5 |

[1] Most common values in arable soil

## 13.21 Elution with Water

DIN 38 414 - S4

Eluates must be created to identify the soluble components in a material (e.g. soil). This procedure is to be used to identify the components that are dissolved in water. The extracted elements should offer information regarding possible impairment or danger to open waters resulting from storage or deposit of the material. When the results have been evaluated, the material can be classified in a waste dump category. The sample should be examined in the state in which it is deposited.

## 13.21.1 Procedure

1. Weigh in a 2000ml plastic bottle: 100g dry substance corresponding to the quantity of the original sample (10,000 / DS%)
2. Add 1000ml dist. $H_2O$; If the sample has a high water content, the water in the sample should be taken into account, such that the ratio 1:10 is maintained, e.g. 210g moist material corresponds to 100g DS. There is already 110ml water in the sample, so only 890ml (1000-110) dist. $H_2O$ need to be added
3. Shake on overhead mechanical shaker for 24 hours, approx. 1rev/min
4. Filter, then discard first filtrate

Identify elements with AAS, ion chromatography photometer (ICP), photometer, etc. If the samples are to be measured with the AAS or ICP, the filtrate must be acidified with $HNO_3$ (65%); the pH value must be > 2.

## 13.21.2 Calculation

(This formula can also be used for other eluates or disintegrations)

Element (mg/kg) DS = ( c · v ) / w
c = concentration of measured element in mg/l
v = volume of extraction solution in ml
w = dry weight in g
    (conversion of moist samples: w = moist weight in g · DS %) / 100

Example: measured value Cr 1.2mg/l, 1000ml added dist.$H_2O$, weight 100g
(1.2 · 1000) / 100 = 12 → The sample contains 12mg Cr/kg

## 13.22 Soil Moisture Retention Capacity, pF Value

Method: DIN 19 683, page 5, Determining soil moisture tension (p = potential, F = free energy of water)

The *pF value* is a logarithmic designation for the applied tension, e.g. 1000cm water column is equal to pF 3 (log 1000 = 3). *Soil moisture tension* is the tension that keeps water in the soil. The water quantity remaining in the soil at a certain soil moisture tension is measured.

Application: Determination of usable field capacity, information regarding pore size distribution. If a steel sampling ring sample is completely saturated with water and then subjected to overpressure or underpressure, increasing by steps until the sample no longer emits water, the pore portions, in vol.%, correspond to the pressure values can be determined by the water loss. The underpressure that can be applied to the individual steel sampling rings with the ceramic plate corresponding to the soil moisture tension of roots in nature. The underpressure method is used for pF values up to 2.2. The overpressure or centrifuge method is used for pF values between 2.2 and 4.2.

### 13.22.1 Sampling

All of the steel sampling rings are driven into the soil in the upper part (0-60cm) of the test pit. To ensure accurate results, 10 steel sampling rings should be examined for each horizon. To drive the sampling rings into the soil, place a wooden board across the top and hit the board with a hammer. Do not hit the steel sampling ring directly!

The steel sampling rings should be far enough apart that each sampling ring can take an undisturbed soil sample. First drive in all sampling rings, then take them out with sufficient space surrounding them; clean the sampling rings. There should be a bulge at both ends. The ends of the steel sampling rings are sealed with plastic caps then individually wrapped in newspaper. Place all sampling rings in a plastic bag so that they stay moist.

### 13.22.2 Determining pF Value

Preparing steel sampling rings

- Unwrap steel sampling rings and carefully remove plastic caps
- Use a spatula to scrape off any excess from the surface (This surface must be smooth, because it forms the connection to the ceramic plate)
- Position filter paper
- Place spatula on filter paper and turn steel sampling ring around (do not pull out spatula)
- Remove upper plastic cap and scrape off excess with a spatula (the top surface may have gaps; the missing volume must be determined)
- Clean the outside of the steel sampling ring with a brush
- Place the sampling ring on the ceramic plate (the filter paper remains under the sampling ring)

### 13.22.3 Operating the pF Meter

The steel sampling rings are on a ceramic plate. When the plate is damp, it is water permeable but not air permeable. The steel sampling rings must be completely saturated with water at the beginning of pF value determination. This is achieved by pouring water (approx. 1cm deep) on the ceramic plate and then waiting for all sampling rings to be saturated before beginning measurement. Underpressure is applied to suction the water out of the steel sampling ring.

The underpressure is generated by an evacuating pump connected to a so-called bubble tower. Different soil moisture tensions (60cm, 100cm, 300cm) can be calipered one after the other on the bubble tower. The sampling ring is weighed when equilibrium has been established in the steel sampling ring between the soil moisture tension applied and the soil forces that bind water. This usually takes one to five days, depending on the type of soil and the soil moisture tension. The only purpose of the cover is to prevent the ceramic plate from drying out and the moisture in the soil in the steel sampling rings from evaporating.

### 13.22.4 Centrifuging Procedure

The steel sampling rings are placed in special trays. The individual trays are exactly counterbalanced on a tray by adding sand to the drip pan. The samples are centrifuged for 45min at 2000rev/min then weighed. Then they are centrifuged at 4000rev/min for 45 minutes and weighed. Then they are dried at 105°C and weighed.

Conversion from rev/min (centrifuge) to cm water column:

$$cm\ water\ column = (h \cdot rev^2 \cdot r) / 180,000 \qquad (13.42)$$

(steel rings : h = 5.1, r = 12.5)
e.g. 2000rev/min equals 1417cm water column

### 13.22.5 Drawing pF curve

Mean values of the individual values are calculated

**1$^{st}$ dot at pF 0:**
Value from column 16. Total pore volume, in example 38.18. When saturated, the total pore volume is filled with water; this results in the lowest point

**2$^{nd}$ dot at 60cm water column:**
Value from column 6 ($H_2O$%vol). If the water was drained from the sample by soil moisture tension of 60cm water column, the sample still contains 12.5vol.% water (in example)

**3$^{rd}$ and following dots:**
Column 6

## 13.22.6 pF Value Calculation

**Table 13.15.** pF Value Calculation

| Value | Explanation; known values (x) |
|---|---|
| A | (x) Steel sampling ring number (top column, 60cm, 100cm, ...dry) |
| B | (x) Sampling depth of steel sampling ring (test pit) |
| C | (x) Horizon (test pit) |
| 9 | (x) Weight of steel sampling ring (from list) |
| 10 | (x) Volume, every steel sampling ring has a volume of 100cm$^3$ (Ø 50 mm, h 5.1cm) |
| 11 | (x) Enter lacking volume (otherwise 0) |
| 13 | (x) Organic substance: ignition loss % of horizon or of sample examined |
| 1 | (x) Gross weight of steel sampling rings from list of weights 1st section 60 cm water column, etc. |
| 2 | Net weight: (steel sampling ring + soil) - steel sampling ring = weight of soil. Enter values from weighing list : 60cm, 100cm, ...dry |
| 3 | Volume:   steel sampling ring volume - lacking volume = volume |
| 4 | Specific gravity:   net weight / volume = specific gravity (g/cm$^3$) |
| 8 | Dry specific gravity: net weight/volume = dry specific gravity (g/cm$^3$) |
| 12 | Transfer values from column 8 to column 12 |
| 14 | Specific weight:   2.64 - (value from column 13 / 100 · 1.24) Pure humus has a specific weight of 1.4g/cm$^3$, pure mineral 2.64g/cm$^3$; The specific weight of humic soil can be calculated with the above formula |
| 15 | Substance volume: volume of solid material |
| 16 | Total pore volume; When dry it consists of air, when saturated of water Substance volume + total pore volume = 100% |
| 5 | H$_2$O % dry mass : percentage of water in relation to dry mass |
| 6 | H$_2$O % volume : percentage of water in relation to volume |
| 7 | Air % volume : percentage of air in relation to volume (H2O % vol. + air % vol. = total pore volume) |

## 13.23  Soil Texture (Grain Size Distribution)

DIN 19 683, page 1

### 13.23.1  Basics

Grain size distribution indicates the mass fractions of graining groups present in a type of soil. Grain sizes greater than 0.063mm are separated out by straining; smaller grain sizes are separated out by sedimentation.

### 13.23.2  Determining Grain Size Distribution by Straining

Straining means separating a soil into graining groups with the aid of a test sieve. Grain sizes determined with sieves are named after the hole size of square hole sieves or the mesh size of sieve mesh which the grain fell through last. This size is referred to as grain size or grain diameter.

If the soil has no (or only few) grain sizes under 0.063mm, dry straining is done; if it contains grains smaller than 0.063mm, wet straining must be done. If the fraction under 0.063mm is greater than 10%, elutriation analysis must be conducted.

### Procedure, Dry Straining

**Material:** 105°C dried soil

**Quantity:** "Normal" arable soil: approx. 200g

Accdg. to DIN lgst. grain → approx. initial weight:

- 2mm → 150g
- 10mm → 700g
- 20mm → 2kg
- 30mm → 4kg
- 60mm → 18kg

Larger quantities must be strained in batches of approx. 2kg each. (If the sample contains too much organic substance, the organic substance must be destroyed by glowing it away.)

**Positioning of sieves:** drip pan on bottom, above that 0.063mm → 0.1mm → 0.2mm → 0.63mm → 1.0mm → 2.0mm → 6.3mm

If the sample contains larger grains, it must be passed by hand through the following sieves: 100mm → 63mm → 20mm → 10mm

- Weigh in sample (e.g. approx. 200g) in a bowl; record weight
- Place bowl contents on tower of sieves without losing any soil (Use a brush)
- Screw cover on tightly; strain 10 minutes (level 7)
- Weigh strainer contents again:
  -Place a bowl on the scale and set scale to 0.0
  -Use a brush to transfer the sieve contents to the bowl
  -Record weight
  -Set scale with bowl to 0.0 again.
  -Empty next sieve, etc.
- Calculate grain size distribution as shown on form
  The loss should not be higher than 1 of the initial weight; if higher; repeat straining with a new sample.

### Procedure, Wet Straining

If the sample contains no more than 10% fine grains under 0.063mm, wet straining must be done. Wet straining should also be done when the soil "sticks together." When the fraction of fine grains is greater than 10%, elutriation analysis should be conducted.

**Material:** 105°C dried soil

- Initial weight approx. 70g; record weight
- Dump contents of bowl into 0.063mm sieve without losing any soil
- Carefully rinse the soil through the sieve with the aid of a hand-held shower and a brush. Grains under 0.063mm in size are not caught

**Table 13.16.** Calculation of Grain Size Distribution

| Grain size distribution | | | | | | | | |
|---|---|---|---|---|---|---|---|---|
| Sample No. | | | | Initial weight in g (A) | | | | |
| | Total sample | | | Portion < 2mm | | | | |
| Mesch size (mm) d | Residue in g | Residue in % | Portion in % [1] | Residue in g | Residue in % | Portion in % | German Classification | Portion in % [1] |
| | B | E = (B/C) 100 | | B | E = (B/C)·100 | | | |
| 20,0 | | | | - | | | | |
| 10,0 | | | | - | | | | |
| 6,3 | | | | - | | | | |
| 2,0 | | | | - | | | | 100 |
| 1,0 | | | | | | | coarse sand | |
| 0,63 | | | | | | | | |
| 0,20 | | | | | | | medium sand | |
| 0,10 | | | | | | | fine sand | |
| 0,063 | | | | | | | | |
| (Bowl) <0,063 | | | | | | | silt | |
| Sum C = ΣB | | 100 | | | 100 | | | |
| Lost D = A - C | | | | | | | | |

)[1] Portion < d in % of the total sample

If the sample contains a substantial amount of large grains, it should first be strained through a larger sieve (e.g. 1mm) to protect the fine sieve. When there is a substantial amount of large grains, the soil should be rinsed through the sieve in several batches.

- Use a small amount of water to rinse contents into a bowl, then dry at 105°C
- Place sample in the sieve machine; refer to Dry Straining. There should be no residue left in the bowl (if so, discard)
- Column Bowl: Initial weight wet straining - Sum of sieve contents up to 0.063mm, Column Loss:0

(Calculation, please see Table 13.16)

### 13.23.3 Determining Grain Size Distribution with Sedimentation

DIN 18 123 (Elutriation analysis)

## Basics

This method is used when over 10% of the grains are smaller than 0.063mm. Grains of different sizes sink in water at different speeds. The relationship between grain size, density and sink speed is explained by Stokes' law. Since this law applies to spherical objects, only equivalent grain diameters are determined when the law is applied to grains in natural soils.

To separate the grains, the soil sample is stirred in water to form a suspension, then it is left to stand in a glass cylinder. Since the grains sink at varying speeds, depending on their size, the distribution of grain size varies with time; thus the density in the suspension varies between the levels in the glass cylinder.

The areometer method according to Bouyoucos-Casagrande is generally used to measure this change and to determine the mass fractions of the grain sizes. When applying this method, the density of the suspension is measured with an areometer at set intervals. The grain size distribution is calculated based on the densities and the submersion depths of the areometer.

## Dispersing Agent

The finest particles in the suspension tend to coagulate (flocking). If this occurs during sedimentation analysis, the amount of finest grains measured is usually less than the actual amount contained in the suspension. A dispersing agent must be added to prevent coagulation.

Dispersing agent sodium diphosphate (= sodium pyrophosphate)

**Parent solution:** Dissolve 20g $Na_4P_2O_7 \cdot 10H_2O$ in dist. $H_2O$ and add water to 1000ml. Use 25ml of the parent solution for elutriation analysis

## Procedure

**Material:** 60°C dried soil or moist soil
Initial weight 30-60 g (the higher the clay content, the lower the initial weight)

- Dry soil: Record initial weight, moisten with dist. $H_2O$, stir if necessary
- Moist soil: To determine the dry substance, moisture content can be determined at the same time to then be calculated back to the dry substance of initial weighing. Or after elutriation is completed, the cylinder content is dried at 105°C.

1. Place the 0.063mm sieve in a bowl. Then rinse the sample into the sieve with the aid of dist. $H_2O$. Using a brush and approx. 800ml dist. $H_2O$, pass the sample through the sieve. The fine material is collected in the bowl
2. Rinse the soil from the sieve into a bowl and dry in the drying chamber at 105°C. Then strain again when dry
3. Pour 25ml sodium diphosphate parent solution into a 1000ml glass cylinder
4. Pour all of the fine material from the bowl into the glass cylinder, taking care not to spill (use a funnel)
5. Fill glass cylinder with dist. $H_2O$ to 1000ml. The sample can remain this way until the next day
6. Close the glass cylinder with a plug and shake well, shaking the cylinder upside down repeatedly
7. Put down glass cylinder and immediately start a stopwatch. Remove the plug and carefully insert the areometer spindle. It must float freely. Read value: (Reading: Only the last digits are read, e.g. 18.5 (not 1018.5));
Reading times: 30sec. → 1min. → 2min. → 5min. → 15min. → 30min. → 1hr. → 2hr. → 6hr. → 24hr.; After the 5min.
Reading: Remove spindle and measure temperature. Before the next measurement, carefully place the spindle in the cylinder without disrupting the sample. Return spindle and thermometer to storage beaker after each measurement.

(Calculation with computer program)

## 13.24 Grain Fractions and Texture Types

From: Pedological Mapping Instructions 1994

The Pedological Mapping Instructions differentiate between the grain fractions of fine soil $f < 2mm$ and of coarse soil (or coarse fraction) $f > 2mm$. When there is a substantial fraction $> 2mm$, the amount should be stated (Refer to Mapping Instructions). When the soil type is identified (according to Mapping Instructions), the fine soil is determined to be equal to 100%. This differentiation is not made for foundation engineering purposes. The entire grain area is set to 100% (e.g. classifying soils for construction purposes according to DIN 18 196). However, in both field the methods of identification are the same (The only difference is that sieves with smaller mesh sizes are used for foundation engineering purposes).

**Table 13.17.** Grain Fractions that for Identifying Type of Soil

| Grain diameter in mm | Fraction | Abbreviation |
|---|---|---|
| < 0.002 | Clay | C |
| 0.002-0.063 | Coarse clay | U |
| 0.063-2.0 | Sand | S |
| > 2 mm | Gravel | G |

## 13.24.1 Instructions for identifying the soils texture type

When the percentage of grain fractions is known, the type of soil can be identified from the diagram. The percentages can be identified with the graining line (with elutriation analysis) or with the results of straining. If the soil fraction on the abscissa is linked to the coarse clay fraction on the ordinate, the type of soil is reached, e.g. 32% clay and 42% coarse clay → Lc2 slightly clayey loam.

> Read "backwards": 2 = weak, 3 = medium, 4 = strong
> Lower case letter: clayey (c), loamy (l), coarse clayey (u), sandy (s)
> Upper case letter: clay (C), loam (L), coarse clay (U), sand (S)
> cC, uU, sS: pure clay, pure coarse clay, pure sand
> or compare the abbreviations to the table "Limits of Types of Soil"

The soil texture type pure sand (sS) is further differentiated by fine, medium-sized and coarse fractions. With coarse clayey, loamy and clayey sand, the various sand sizes should be stated only when a sand subfraction is clearly the largest amount, e.g. slightly loamy coarse sand.

**Table 13.18.** Ranges of Textural Class "Pure Sand" (Percentages of sand subfractions)

| Designation | Abbrev. | Fine sand 0,063-0,2 | Medium sand 0.2-0.63 | Coarse sand 0.63-2.0 |
|---|---|---|---|---|
| Fine sand | fS | 75-100 | 0-25 | 0-25 |
| Medium sandy fine sand | fSms | 50-75 | 15-50 | 0-35 |
| Coarse sandy fine sand | fScs | 50-75 | 0-15 | 10-50 |
| Medium sand | mS | 0-25 | 65-100 | 0-35 |
| Fine sandy medium sand | mSfs | 25-50 | 40-75 | 0-35 |
| Coarse sandy medium sand | mScs | 0-25 | 40-65 | 10-60 |
| Coarse sand | cS | 0-25 | 0-15 | 60-100 |
| Fine sandy coarse sand | cSfs | 25-50 | 0-40 | 10-75 |
| Medium sandy coarse sand | cSms | 0-25 | 15-40 | 35-85 |

**Table 13.19.** Ranges of Types of Soil Texture (Percentages of Fractions) (Pedological Mapping Instructions 1994)

| Soil type groups | Main soil group | Abbrev. | Grain fractions in % | | |
|---|---|---|---|---|---|
| | | | Clay | Coarse clay | Sand |
| Pure sand | Sand | Ss | 0-5 | 0-10 | 85-100 |
| Slightly coarse clayey sand | " | Su2 | 0-5 | 10-25 | 70-90 |
| Slightly loamy sand | " | Sl2 | 5-8 | 10-25 | 67-85 |
| Medium loamy sand | " | Sl3 | 8-12 | 10-40 | 48-82 |
| Slightly clayey sand | " | Sc2 | 5-17 | 0-10 | 73-95 |
| Medium coarse clayey | " | Su3 | 0-8 | 25-40 | 52-75 |
| Very coarse clayey sand | " | Su4 | 0-8 | 40-50 | 42-60 |
| Coarse clayey, loamy sand | Loam | Slu | 8-17 | 40-50 | 33-52 |
| Very loamy sand | " | Sl4 | 12-17 | 10-40 | 43-78 |
| Medium clayey sand | " | Sc3 | 17-25 | 0-15 | 60-83 |
| Slightly sandy loam | " | Ls2 | 17-25 | 40-50 | 25-43 |
| Medium sandy loam | " | Ls3 | 17-25 | 30-40 | 35-53 |
| Very sandy loam | " | Ls4 | 17-25 | 15-30 | 45-68 |
| Slightly clayey loam | " | Lc2 | 25-35 | 30-50 | 15-45 |
| Sandy, clayey loam | " | Lcs | 25-45 | 15-30 | 25-60 |
| Very sandy clay | " | Cs4 | 25-35 | 0-15 | 50-75 |
| Medium sandy clay | " | Cs3 | 35-45 | 0-15 | 40-65 |
| Pure coarse clay | Coarse clay | Uu | 0-8 | 80-100 | 0-20 |
| Sandy coarse clay | " | Us | 0-8 | 50-80 | 12-50 |
| Slightly clayey coarse clay | " | Uc2 | 8-12 | 65-92 | 0-27 |
| Medium clayey coarse clay | " | Uc3 | 12-17 | 65-88 | 0-23 |
| Sandy, loamy coarse clay | " | Uls | 8-17 | 50-65 | 18-42 |
| Very clayey coarse clay | " | Uc4 | 17-25 | 65-83 | 0-18 |
| Coarse clayey loam | " | Lu | 17-30 | 50-65 | 5-33 |
| Medium clayey loam | Clay | Lc3 | 35-45 | 30-50 | 5-35 |
| Medium coarse clayey clay | " | Cu3 | 30-45 | 50-65 | 0-20 |
| Very coarse clayey clay | " | Cu4 | 25-35 | 65-75 | 0-10 |
| Slightly sandy clay | " | Cs2 | 45-65 | 0-15 | 20-55 |
| Loamy clay | " | Cl | 45-65 | 15-30 | 5-40 |
| Slighty coarse clayey clay | " | Cu2 | 45-65 | 30-55 | 0-25 |
| Pure clay | " | Cc | 65-100 | 0-35 | 0-35 |

## References

Abfallklärverodnung (AbfallklärV) 1992, Part 1
Arbeitsgruppe Bodenkunde der Geologischen Landesämter (1994) Bodenkundliche Kartieranleitung. (Pedological Mapping instructions)
Beck T (1968) Mikrobiologie des Bodens. BLV Verlag München
Blume HB (1992) Handbuch des Bodenschutzes. Ecomed
Brucker a. Kalusche (1990) Boden und Umweltbiologisches Praktikum. Quelle und Meyer, Heidelberg
Brauns A (1968) Praktische Bodenbiologie. Gustav Fischer Verlag Stuttgart
BundesBodenschutz- und Altlastenverordnung (1999)
Chhabra R (1996) Soil Salinity and Water Quality. A.A. Balkema/Rotterdam/Brookfield,
EAWAG (Swiss Waste Commission) K-3005
Engelhardt W (1983) Ökologie im Bau- und Planungswesen. Wissenschaftliche Verlagsgesellschaft, Stuttgart
FAO, ISSS-AISS-IBG, ISRIC (1994) World Reference Base for Soil Resources. Compiled and edited by Spargaaren, Wageningen, Rome 1994
Fiedler JH (1973) Methoden der Bodenanalyse 2, Mikrobiologische Methoden. Verlag T. Steinkopf Dresden
Fink A (1991) Pflanzenernährung in Stichworten.Verlag FL, Hirt
Foth HD, Withee LV, Jacobs HS, Thien SJ (1982) Laboratory Manual for Introductory Soil Science. Sixth Edition, WmC Brown company Publishers, Dubuque, Iowa
Heß D (1991) Pflanzenphysiologie. UTB, Ulmer Verlag
Kunze H Roeschmann G, Schwerdtfeger G (1994) Bodenkunde. Ulmer
LAGA (1984) Mitteilungen der Länderarbeitsgemeinschaft Abfall, Merkblatt 10
Meiwes K-H, König N, Khanna PK, Prenzel I, Ulrich B (1984) Chemische Untersuchungsverfahren zur Charakterisierung und Bewertung der Versauerung in Waldböden - Forschungs-Bericht. Waldöko 7, Göttingen
Rosenkranz, Einsele, Harreß (1988) Bodenschutz. Erich Schmitt Verlag, Berlin
Schlegel HG (1985) Allgemeine Mikrobiologie. Thieme Verlag
Steubing L, Kunze CH (1980) Pflanzenökologische Experimente zur Umweltverschmutzung. Quelle und Meyer, Heidelberg
Scheffer/Schachtschabel (1998) Lehrbuch der Bodenkunde. F Enke Verlag
Schinner, Öhlinger, Kandeler (1991) Bodenbiologische Arbeitsmethoden. Springer Verlag
Singer MJ, Munns DN (1999) Soils An Introduction. Fourth Edition, Prentice Hall, Upper Saddle River, NJ 07458
Schlichting E, Blume PH, Stahr K (1995) Bodenkundliches Praktikum - Pareys Studientexte. Blackwell Wissenschafts-Verlag Berlin, Wien
Sprenger B (1996) Umweltmikrobiologische Praxis. Springer Verlag
Van Reeuwijk LP (1992) Procedures for Soil Analysis. Third Edition, ISRIC, Wageningen
VDLUFA (1991) Methodenbuch 1 - Die Untersuchung von Böden. German Agricultural Test and Research Agency. VDLUFA Verlag, Darmstadt
www/FAO/publ World Reference Base for Soil Resources

# 14 Influencing BOD and N Removal Assessment of Important Parameters

Sabine Kunst

## 14.1 Batch Tests as a Method for Classifying Nitrification and Denitrification Activities in Activated Sludge

Important groups of micro-organisms in activated sludge, e.g. nitrifiers, cannot be identified microscopically. The only way to characterise a given activated sludge in terms of its performance capacity, e.g. as regards nitrogen conversion, is by detailed analysis of the respective metabolism reaction in batch tests. Depending on the approach used for these tests, the actual or potential microbial-substance conversion speed can be measured. Using batch tests, it is also possible to detect inhibiting influences on the nitrogen-conversion reactions.

### 14.1.1 Batch Tests for Nitrification (Aerobic)

The nitrification performance of different activated sludge can be compared. This is done using parallel reactors, each filled with activated sludge from the aerated basin of the respective wastewater treatment plant. The sludge are aerated overnight to ensure that the $NH_4$-N in the sludge is fully oxidised. Using magnetic stirring devices, the activated sludge is mixed in the reactors and additionally aerated. The test can be performed at room temperature or, if the impact of temperature is to be determined, by using a thermostat. The pH value is monitored and corrected if necessary (pH-value range: 7.0-8.0). Then the suspended solids of the activated sludge in the reactors are measured.

At the beginning of the tests, an $NH_4$-N concentration of 50mg/l is set in each reactor. To do so, the already existing $NH_4$-N concentration must first be measured; only enough $NH_4$-N is added to make the amount up to 50mg/l. This done using $NH_4Cl$. The ammonium concentration should not be limited during the test, as this would prevent nitrification from taking place at maximum speed. The tests are monitored by conducting measurements for at least 2 hours at a time; samples are taken after 0, 15, 30, 60, 90 and 120 minutes The samples are immediately filtered using a folded filter and checked for $NH_4$-N, $NO_2$-N and $NO_3$-N. The measured concentrations are entered in a table and subsequently rendered in graph form.

From the measured nitrogen conversions during the test period, the conversion rates are calculated - in mg N/(L·h) and expressed in relation to the suspended solids in mg N/(g SS·h) - as:

- ammonium conversion rate (decrease in ammonium)
- nitrite and nitrate formation rate.

If, in the batch tests, only ammonium is defined, the other parameters (oxygen, pH value) being set to optimum, it is possible to measure the maximum possible nitrification capacity of the mixed biocoenosis. If, on the other hand, the aim is to check the impact of a given wastewater on nitrification performance, the desired wastewater load is defined at the beginning of the test, ammonium being added if necessary, and then the nitrification performance achieved for this particular wastewater is measured. Should the wastewater have an inhibiting effect on the nitrifiers, the nitrification performance for the test with wastewater will be lower than that with ammonium only.

Ammonium is also consumed by biosynthesis for biomass growth (the higher the $BOD_5$ load in the test, the higher the consumption until the maximum growth rate is reached). This ammonium uptake is contained in the ammonium conversion rate, but not in the nitrate (or nitrite) production rate.

## Inhibition of Nitrification by Toxic Substances

Apart from the pH value, temperature and oxygen content, toxic substances contained in the wastewater can also have an inhibiting effect on nitrification. The nitrifiers' high sensitivity to toxic substances is due to the nitrifiers' low growth rate. The ammonium oxidisers appear to react more sensitively than *nitrobacter* because of their more complex enzymatic system. There are a number of substances with a toxic impact, but it is very difficult to identify them more precisely in the overall wastewater stream. Batch tests do, however, enable an inhibiting effect to be attributed to specific industrial discharges or wastewater bit-streams (if separate analysis is possible).

Fig. 14.1 illustrates how batch tests are used to assess the influence of an industrial discharge (in this case from the textile industry) on nitrification performance.

The beginning of the curves shows the state without industrial impact (at 0% of textile wastewater), i.e. the nitrogen conversion rates of municipal activated sludge loaded exclusively with municipal wastewater. In further tests, the ratio of industrial wastewater was increased. The curves clearly demonstrate the inhibiting effect of the industrial wastewater on nitrification performance.

The figure also shows that the nitrification rates of municipal activated sludge are normally between 2 and 2.5mg N/(g SS·h) (here, the ammonium conversion curve starts at approx. 2.25mg N/(g SS·h).

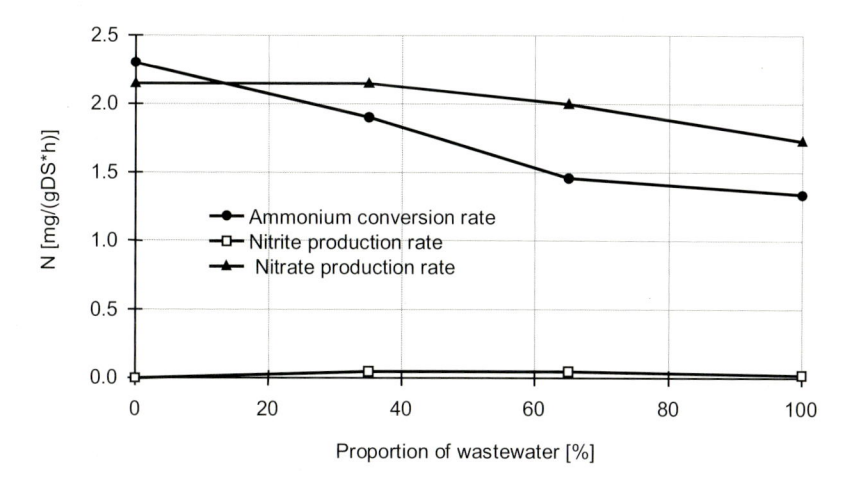

**Fig. 14.1.** The Nitrification Rate for Different Mixtures of Wastewater from the Textile Industry and Municipal Wastewater (Hulsbeek a. Kunst 1994)

It is possible to adapt the nitrifiers to problematic substances in wastewater – provided, of course, the substances are only present in relatively stable concentrations. Peak loads, e.g. as a result of cleaning processes or disturbances in the production process, prevent the formation of an adapted biocoenosis and are often responsible for the impairment of nitrification.

### 14.1.2 Batch Tests for Denitrification (Anoxic)

Anoxic batch tests can be used, for example, to determine the impact of different carbon sources on the denitrification performance of activated sludge. Again, the parallel reactors used are filled with activated sludge that has previously been aerated overnight (to remove COD, in this case). On the day of the test, the sludge in the reactors are stirred without aeration. These tests are also performed at room temperature. The pH value is monitored, and corrected if necessary (pH value: 7.5-8). Consideration should be given to the fact that denitrification – unlike nitrification – consumes protons, so the pH value is supposed to increase. The suspended solids in the activated sludge are measured.

In the reactors, $NO_3$-N is supplied at a concentration of 50mg/l. Here, it is again necessary to measure the already existing $NO_3$-N concentration, only making up the amount lacking by adding $NaNO_3$. Since the denitrification rates depend on the degradability of the carbon sources, it is difficult to compare these tests with an uninhibited denitrification reaction. A control experiment is set up, in which nitrate but no other substrate is added. The endogenous denitrification capacity is measured, as no external carbon sources are available. Here, micro-organisms consume intracellular storage substances as carbon source for nitrate conversion.

In the second reactor, the sludge is blended with wastewater, and in the third with acetate, as carbon source in the following concentrations or loads:

- Reactor 1: Blank experiment – no carbon source
- Reactor 2: Wastewater –load 0.3g COD/g DS
- Reactor 3: Acetate – load 0.3g COD/g DS

Samples should be taken after 0, 15, 30, 60, 90 and 120 minutes. The samples are conveyed immediately to a folded filter and analysed for $NO_2$-N and $NO_3$-N. The measured concentrations are entered into the relevant table and then rendered in the form of a graph. The nitrogen conversion rates measured during the test period are used to calculate the denitrification rates for the three reactors, in mg N/(g SS · h).

If substrates or wastewater are to be checked for their denitrification-inhibiting effect, the sludge is blended, in parallel test set-ups, with the respective wastewater or substrate in increasing concentrations. The following observations can then be made:

1. The denitrification performance increases with the wastewater or substrate load
   $\Rightarrow$ the wastewater or substrate can be used by the denitrifers as carbon source without causing inhibition within the selected load range. The greater the increase in speed, the better the carbon source can be utilised.

2. The denitrification performance increases up to a certain wastewater or substrate load and then decreases again at the next-higher load setting
   $\Rightarrow$ the wastewater or substrate can be used as carbon source for denitrification up to a certain concentration limit; if this concentration is exceeded, the toxic impact prevails.

3. The denitrification performance remains at the low level of endogenous denitrification
   $\Rightarrow$ Either the wastewater contain no carbon source, or the carbon source cannot be used. The tested wastewater has neither a positive nor a negative effect on nitrate conversion.

4. The endogenous denitrification performance decreases after wastewater has been added
   $\Rightarrow$ the wastewater has a toxic impact on the micro-organisms, inhibiting even the endogenous conversion performance.

## 14.2 Respirometry: Determination of the Oxygen Uptake Rate (OUR)

**Fundamentals and Explanatory Remarks**
The micro-organisms contained in the activated sludge are able, among other things, to degrade carbon compounds (BOD, COD). During this process, oxygen is consumed by the micro-organisms through the respiration of biologically degradable substances (BOD). This oxygen consumption is proportionate to the amount of the degraded carbon compounds. By measuring the oxygen consumption over the respective period, it is possible to determine whether and how fast a given activated sludge can consume a defined amount of BOD or COD. The speed of the oxygen consumption thus serves as a yardstick for the respiration activity of the activated sludge and for the degradability of the COD.

The COD can be broken down as follows:

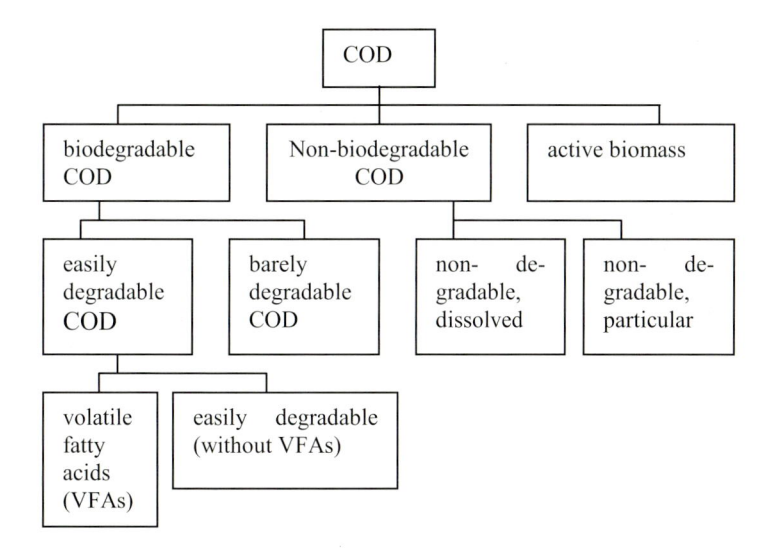

**Fig. 14.2.** Different COD Fractions in Wastewater

This differentiation of the COD is based on the degradability by micro-organisms. Organic substances with a simple molecular structure (such as short-chain organic acids, glucose, starch) are easily degradable; substances with complex molecular structures, however, are barely degradable (e.g. lignin, cellulose, waxes). Non-degradable or very slowly degradable are, for instance, anthropogenic substances such as nylon, polystyrene and halogenated carbohydroxides.

The COD elimination in wastewater treatment plants (WWTPs) is based on a mixture of degradation and adsorption: easily degradable – and to some extent also barely degradable – COD ratios are respired or used for cell building or as a storage substance, whereas non-degradable or very slowly degradable COD ratios can only be eliminated by adsorption to the sludge flocs. The latter may account for up to 50% of the entire elimination capacity of an activated sludge. Here, however, it should be taken into account that an increase in the age of the sludge leads to an increase in the adaptability of the activated sludge, i.e. a high sludge age means that an activated sludge can also respire barely or very slowly degradable COD ratios.

### 14.2.1 Determination of the Respiration Rate of Activated Sludge by Measuring the $O_2$ Utilisation Rate

Before beginning the test, the activated sludge must be aerated for at least 12 hours without substrate feeding. This should guarantee that the COD adsorbed to the sludge flocs has been largely respired.

To determine the respiration rate of a given activated sludge, a glass vessel is filled with the $O_2$-saturated activated sludge and then tightly closed to stop oxygen from the surrounding air getting in. In the vessel is a $O_2$-measuring probe. A recorder records the decrease in the $O_2$ concentration throughout the test period. The sludge must be stirred to provide a sufficient blower-stream velocity for the probe. For longer measuring periods, the reaction container brought to the right temperature in a water bath.

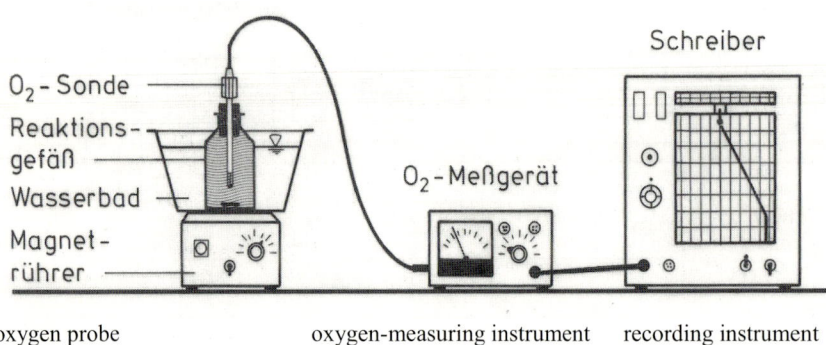

oxygen probe
reaction vessel
water bath
magnetic stirrer

**Fig. 14.3.** Test Set-up

After lengthy aeration without substrate addition, the endogenous respiration can be determined using the measuring set-up. If, however, biologically degradable substrate is supplied to the micro-organisms, it will be immediately metabolised with $O_2$ consumption, i.e. the respiration activity of the activated sludge increases. To determine this substrate respiration, an injection syringe is used to add a defined substrate amount through the sludge during the measuring period.

Since the $O_2$ consumption does not increase linearly with the increased substrate supply but moves towards a saturation value (Michaelis-Menten Kinetics), it is possible, by repeatedly taking measurements as the substrate supply increases, to determine the maximum respiration.

If the injected substance has an inhibiting effect on the respiration activity of the micro-organisms, this is reflected in the decreased oxygen consumption. Toxic substances thus reduce the respiration rate or even inhibit respiration completely.

From the recorded data, it is possible to determine the $O_2$ consumption in mg $O_2/(l \cdot h)$. In order to compare measurements of activated sludges with different concentrations of dry solids, the respiration activity is correlated with the dry solids in the respective measuring container and expressed as $O_2$ consumption in mg $O_2/(g\ SS \cdot h)$.

### 14.2.2 Evaluation of the Recorded Data

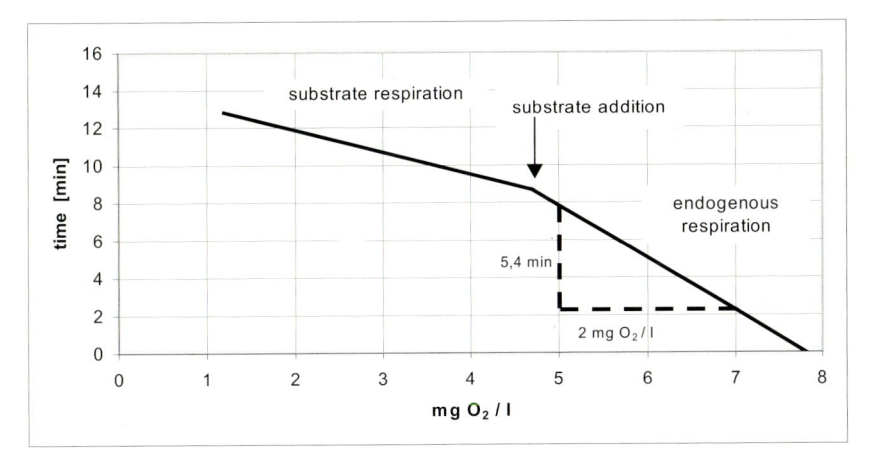

**Fig. 14.4.** Example of Data Recording

**Calculation of the endogenous respiration in the given example**
In their endogenous respiration, the micro-organisms consume 2mg/l $O_2$ in 5.4 minutes. The respiration rate is thus 0.37mg $O_2/(l \cdot min)$ or 22.2mg $O_2/(l \cdot h)$. With a dry-solids content of 4g/l in the activated sludge, the respiration activity is 5.6mg $O_2/(g\ MLSS \cdot h)$.

### 14.2.3 Dependence of Oxygen Consumption on Toxic or Inhibiting Substances in Water

To enable the inhibiting effect of a substance to be clearly detected, the activated sludge must be as close as possible to maximum respiration. First, a maximum-respiration curve of the activated sludge is recorded by adding increasing concentrations of a readily biodegradable substrate. Usually, an aqueous peptone solution is used, but it is also possible to use other organic substances that can be aerobically utilised by micro-organisms (e.g. acetate or glucose) without causing inhibition.

By repeating the test with the same concentrations (as COD) of the substance that is to be tested, it is possible to predict the primary aerobic biodegradability of the substance compared with the reference substrate. Furthermore, the shape of this second curve allows us to draw conclusions about the possible toxicity or inhibitory effect of the substance.

A second test series is used to assess the effects of the substance on the respiratory activity of the activated sludge, i.e. to determine the maximum concentration of this substance that can be allowed to enter a WWTP without affecting the activity of the activated sludge.

Peptone (or another substrate) is thus added at the very beginning of the test, in the concentration determined in Test 1, to obtain the maximum respiration activity. The unaffected $O_2$ consumption (approx. 1-2mg/l $O_2$ decrease) is then briefly measured and the inhibitor substance injected. The test is repeated with increasing concentrations of the potential inhibitor. By comparing the respiration activity without inhibitor with the potentially inhibited tests, a degree of inhibition [%] can be deduced for every concentration.

### 14.2.4 Further Applications for Oxygen-Consumption Measurements

Apart from the application of oxygen-consumption measurements described above (recording the maximum respiration of activated sludge, dependence of oxygen consumption on toxic or inhibiting substances), the presented test set-up can be used – in a slightly modified and possibly automated form – for further examinations of wastewater and activated sludge. Such examinations are conducted using batch reactors. The major difference from the test set-up of the practical experiments is that the oxygen content in the batch reactors is controlled during the test: after an aeration phase, until a defined oxygen concentration is reached, there follows the oxygen consumption proper (consumption rates are calculated automatically) until a defined minimum value is reached. Then, the reactor is aerated again, and so on. This procedure enables the oxygen-consumption rates to be determined throughout the test period. This facilitates, among other things, the quantification of the concentration of easily degradable COD in the wastewater. An idealised curve of the oxygen-consumption rates over such a test period is shown in Fig. 14.5.

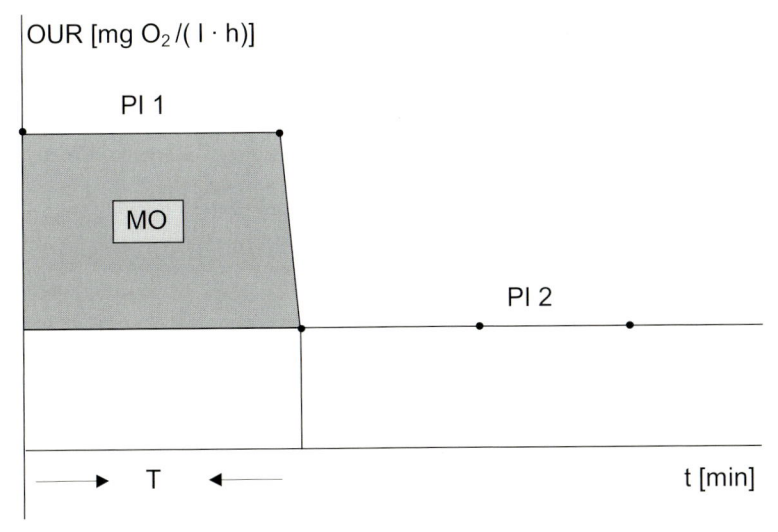

**Fig. 14.5.** Ideal Curve for Determining the Easily Degradable COD by Measuring the Oxygen-Consumption Rate, according to Dold et al 1991

Here, the upper level P1 is equivalent to the oxygen-consumption (-utilisation) rates (=OUR) of the activated sludge as long as easily degradable substrate is available. If this is consumed, the oxygen-consumption rates drop to the lower level P2. The area highlighted in grey shows the MO (= oxygen amount necessary to respire the easily degradable COD, the unit being mg $O_2$/l). This MO is thus proportionate to the easily degradable COD in the wastewater. In this idealised case, it can be determined by multiplying P1-P2 by the time that passes before the drop fall down.

Using the described test set-up, it is possible - at parallel determination of the dissolvable COD in the batch reactor and of the concentration of PHB (polyhydroxy butyric acid, a storage polymer widely used by micro-organisms) - to determine the heterotrophic yield coefficient, which expresses the ratio of the absorbed substrate (i.e. COD) that is used for building up new biomass.

To determine the heterotrophic yield coefficient using oxygen-consumption measurements, the ratio of the remaining COD – i.e. that which is not converted into biomass – is recorded. This can be done indirectly by recording the oxygen-consumption rates because this COD ratio is respired by the bacteria, i.e oxidised with oxygen consumption.

If the entire amount of substrate used is known, the COD balance can be completed, and the heterotrophic yield coefficient calculated, using the measured COD ratio and the analyses described above. Fig. 14.6 shows the results of a test conducted at a municipal WWTP in Hannover.

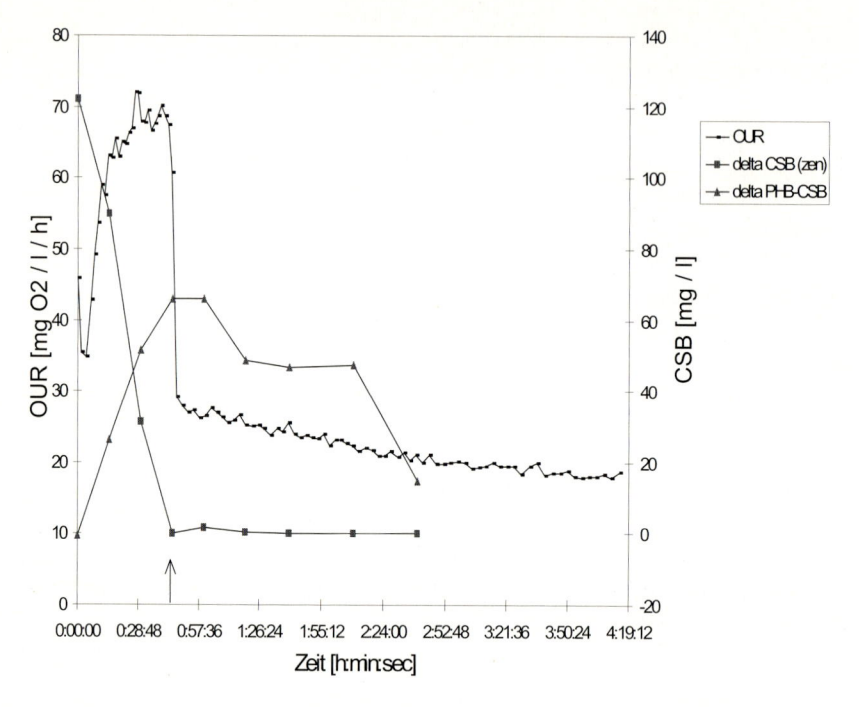

**Fig. 14.6.** Test to Determine the Heterotrophic Yield Coefficient. The Oxygen-Utilisation-Rate (OUR), the Concentration of Dissolvable COD (COD (zen)) [CSB=COD] in the Reactor, and the PHB Concentration in the Activated Sludge (in COD Units) as a Function of Time [Zeit=time]

This test set-up also enables the amount of active biomass in the wastewater and the maximum growth rate of heterotrophic bacteria in the activated sludge to be determined.

# Vitae of Contributors

**Aithal, Vathsala**, M.A. Educational Sciences; researcher; member of the Cornelia Goethe Centre, Johann Wolfgang Goethe University, Frankfurt am Main, Germany; specialised in gender and environment, education and development processes in the Third World as well as intercultural communication; involved in a variety of conferences and congresses focusing on gender, development and the environment.

**Burmester, Andrea**, Dipl.-Ing., Scientific Assistant at the International Women's University, Hannover, Germany; specialised in water management and wastewater treatment with particular emphasis on simple water treatment and irrigation efficiency; in charge of co-ordinating and implementing the postgraduate programme of the project area Water

**Devasia, Leelamma**, Professor, Tirpude College of Social Work, Nagpur, India; specialised in rural development; has published more than 30 studies and nine books, mainly on women's development, empowering women and feminisation of water management; engaged in research on water harvesting by tribal women; Director (Hon.) of Community Action for Development (CAD), an NGO working with women at grass-roots level; has conducted more than 20 internationally sponsored development projects through CAD.

**Kayser, Katrin**, Dipl.-Ing., Scientific Assistant at the Department of Water Quality and Waste Management, University of Hannover, Germany; specialised in wastewater management, water biology and biological processes of wastewater treatment; working in particular on capacity analysis of small wastewater-treatment plants and optimisation of water purification in constructed wetlands and lagoons.

**Kruse, Tanja**, Dipl.-Päd., Scientific Assistant at the International Women's University, Hannover, Germany; specialised in the didactics and methods of educational science with particular emphasis on adult education; in charge of co-ordinating and implementing the postgraduate programme of the project area Water

**Kunst, Sabine**, Prof. Dr.-Ing. habil. Dr. phil., head of the subject group Biology and Wastewater Treatment at the Department of Water Quality and Waste Management, University of Hannover, Germany; specialised in wastewater management, water biology and biological processes of wastewater treatment; in charge of several international projects; has conducted research projects on water and gender; special background in interdisciplinarity and particularism.

**Mennerich, Artur**, Prof. Dr.-Ing., Dean of the Department of Water and Environmental Management, University of Applied Sciences, Suderburg, Germany; specialised in wastewater treatment, sewage-sludge treatment and disposal, technical and energetic optimisation of wastewater-treatment plants.

**Metz-Göckel, Sigrid**, Prof. Dr. phil., Professor at the Faculty of Educational Science and Sociology, head of the Centre for Research on Higher Education and Faculty Development; research on higher education and gender; speaker of the unique Ph.D. programme Women in Gender Studies, funded by the German Research Foundation from 1993-1999; also speaker of the Ph.D. programme Knowledge Management and Self-Organisation, funded by the Böckler Foundation since 2001; in charge of a variety of projects on the technical competence of women and women working in the sciences; member of several federal-state and national commissions on politics in higher education.

**Pathak, Namrata**, Ph.D. in Environment and Biomass; Project Scientist at the Centre for Rural Development and Technology, Indian Institute of Technology, Delhi, India; research interests include rainwater/roofwater harvesting, water-quality monitoring (bacteriological and chemical), treatment methods, gender-related issues, rural environmental/ecological issues, issues of sustainability and sustainable development, biomass-based technologies (e.g. vermicomposting, mushroom cultivation, biogas generation) and use of plant-based extracts as pesticides, antimicrobial agents.

**Töppe, Andrea**, Prof. Dr.-Ing., Professor of Hydraulic Engineering and Water-Resources Management, University of Applied Sciences, Suderburg, Germany; specialised in water-resources management, coastal engineering and hydrology; responsible for the partnership programme with the Izhevsk State Technical University, Russia.

**Urban, Brigitte**, Prof. Dr. rer. nat., Professor of Soil Science and Biology, University of Applied Sciences, Suderburg, Germany; specialised in water and soil management and environmental history; in charge of international co-operation projects with the USA, Australia and Asian countries.

**Wichern, Marc**, Dr.-Ing., Scientific Researcher at the Department of Water Quality and Waste Management, University of Hannover, Germany; specialised in the modelling and simulation of different treatment systems for municipal and industrial wastewater; in charge of development and research for the computer software DENIKA+, which is widely used for the design and optimisation of WWTPs in Germany.

**Wittberger, Dolly**, Mag. rer. nat. Dr.; obtained Ph.D. in Molecular Genetics from Vienna University, Austria; currently working as a scientific advisor for art projects in the controversial field of gene technology; external lecturer at Vienna and Graz Universities, Austria; teaches gender aspects in the natural sciences, particularly molecular biology; member of academic advisory boards for the integration of feminist/gender aspects in natural-science syllabuses; during the International Women's University (*ifu*), co-founded the Women's International Network for Sustainability (WINS) and serves as its current president; in line with her strong commitment to feminist political ecology, co-ordinates the various joint efforts of this NGO to achieve truly sustainable approaches to development.

# Index

abbreviations 122
absorbed substrate 379
academic
   backgrounds 299
   work 295
acceptance 217
access to resources 24
actinomycetes 191
action level 23
activated sludge 100, 371
   model 120
   system 101
acuteness 311
admission 3
adsorption 148
aeration 129
aeration phase 378
agar-agar 330
Agenda 21 139, 168
agriculture 44
alkalinity 104
alternating denitrification 110
alternative
   sanitation 218
   technologies 184
   vision 43
aluminium sulphate 67, 72
ammonium 372
   conversion rate 372
   oxidise 104
analysis
   metal 75
   microbiological 71
animal species 155
application 3, 296
arable land 246, 268, 277
areometer 365
autoclave 331
autotrophic biomass 121
autotrophic organisms 155
average rainfall 84, 85
   Albania 79
   Ibadan 87
   India 81
   USA 82
average temperature

USA 82
awareness 50, 258

bank 246
   erosion 244
batch reactors 378
batch tests 371
beaver 239
benefit 172
benthic 244
biodiversity 174
biofertilizer 217
biological
   activity 277
   lagoon treatment 153
   nutrient removal 125
   phosphorus removal 108
   purification 110
   self-cleaning process 153
biological WWTPs
   costs 131
   dimensioning data 131
biomass growth 372
$BOD_5$ 118, 142, 148
   concentration 177
   loads 165, 372
bottom
   composition 232
   constructions 249
   level 251
   structures 231
Brazil 176
   fishing village 178
   requirements 176

C/N ratio 341
C/P ratio 341
CAD 32
calcium hypochlorite 67, 72
capability 172
carbon 338
carbon degradation 114
carbonate 342
carousel toilet 207
cascade denitrification 110
catchment area 28, 230, 237, 239
cation exchange capacity 352, 354

cations 352
   absorbed 352
   Na, Mg, K, Ca 73, 74, 75
centralisation 91
centrifuging 361
change agents 49
channel 229
chlorination 61
clamps systems 209
climatic conditions 209
coagulation process 60
COD 118, 142, 148, 375
   degradation 121
   fractioning 118
coliform
   faecal 58, 193
   organisms 58
   total 193
colluvium 264
colonial history 29
colony quantity 333
commitment 303
communication 307
communication technologies 6, 317
community 36
   acceptance 198
   awareness 39
   responsibility 41
community-based 218
competence 301, 303
compost 200, 216, 336
   microbiological quality 214
   nutrient analysis 215
   nutrients 194
   population diversity 190
   temperature 186
composting
   bacteria 190
   basins 164
   bed 157
   C/N ratio 188
   environmental factors 187
   moisture content 187
   oxygen 187
   pathogens 189
   phases 186
   temperature 188
composting plant
   experience 163
composting process 186, 215
composting tank 145

composting toilet 189, 196, 207
   acceptance 217
   dimensioning 196
   solar heating 203
compounds
   nitrogen 149
   organic 148
   phosphorus 149
computations 225
computer programs 121
concepts
   semi-decentralised 141
conduit 232
   constructions 251
conference 295
conflict management 16, 27
connecting rivers 230
consensus 311
constructed wetlands 27, 146
contamination 32, 265
cost-effective 92
critical reflection 26
critical tractive force 228
critique 296, 302
Cuba
   biodiversity 174
   coffee production 174
cultural
   factors 45
   obstacles 23
culture 14, 299
current
   meters 224
   velocity 244
curriculum 304

decentralisation 139, 165
decentralised technologies 184
decentralised wastewater
   treatment 27
decision-making process 43, 311
decomposing matter 207
decomposition 200
   process 208
   toilet 204
degradation 148
dehydration 199
dehydration toilet 201, 203
demineralising 334
democracy 50
DENIKAplus 121

DENISIM  121
denitrification  105, 113, 115, 129, 149, 373
  endogenous  374
  intermittent  106
  nitrogen conversion  374
  rate  105
designing  173
desinfection  169
developing countries  3, 126
development
  of landscape  274
  sustainable  137
development plan  236
developmental programmes  32
dewatering  94
differences  15
different
  cultures  308
  disciplines  308
dilution levels  332
discharge  151, 233
diseases  27
  water-based  59
disinfection  61
  boiling  62
  ozone  61
  solar  62
  UV  61
disposal
  sewage-sludge  109
divalent ions  62
diversity  271, 316
drainage pipes  246
drinking water  31, 57, 194
  pathogenic  58
  quality  58
  standards  74
  WHO guidelines  58
dry sanitation  197, 198
dry solids content  158
dry specific gravity  362
dry toilet  201
Dublin Principles  138
dynamic models  118
  matrix notation  118

ecological
  movement  46
  sanitary unit  204
  systems  30

economy  44
ECOSAN toilet  203
ecosphere  52
education  45
effluent concentration  145
Elbe  237
  ecology project  271
electr. conductivity  327
elements
  identifying  353
elimination
  pathogens  200, 206, 214
elution  358
emancipation  45
empowerment  27, 315
endogenous respiration  377
environmental
  degradation  47
  education  297, 298
equality  34
eradication  246
erosion  227
essentialism  16
estimation  176
eurocentrism  17
Europe  18
  legal requirements  101
  technological development  131
evaluation  7
  concept  296
experiment  295
EXPO  12, 68
extension stretches  233

faecal coliform  64, 71, 73, 193
faecal sludge  163
faecal sludge production  157
  surface area  157
  surface load  157
faecal streptococcus (FS)  64
fallow  282
  region  285
family latrine  202
farm  35
fear  39
feminisation
  water management  31
feminist
  agenda  12
  ecology  297
  perspectives  23

planning 170
fences 255
fertiliser 216, 347
field capacity 359
field research 28
filter
  multimedia 70
filtration
  slow sand-filtration 60
fish
  ponds 243
  population 243
fishway 240
flat land 226
flood area 229
floodplain 238, 242, 253, 272
  biota 253
floods 237
flotation 60
flow formula 225
flow rate 178
fluvisols 273
folklorising 15
food supply 44
forelands 226
fungi 192
future perspectives 309

gender
  approach 171
  approach application 171
  equality 53
  equity 316
  inequality 23
  perspective 169, 301, 312, 313
  specific view 171
gender-specific aspects 309
geological material 262
germ count 333
Germany 12
global 2, 295
  dialogue 6
  networking 6
  significance 267
  vision 50
global impact 89
grain size distribution 362
grants 4
Green Card 12
grit chamber 95, 128
groundwater 42, 289

salinity 69
growth 119
  specific rate 104
  term 119
GTA 213

health 31
heating 72, 75
heavy metal 67, 278, 356
  threshold values 357
heterotrophic
  bacteria 104
  biomass 116, 121
  organisms 154
  plate count 64, 71
  yield coefficient 379
higher-education 10
historical maps 242
horizontal soil filter
  process characteristics 150
horizontal-flow
  planted soil filter 147
  tanks 98
horseradish tree 62
household 33
human faeces 183
human resources 131
humus 164, 324
hydraulic load 178
hydraulics 222
hydrolysis 351

ignition loss 324
illiteracy 44
immigration 12
impairing effect 234
impairment 248, 249, 253
  non 234
  small 235
  strong 235
  very strong 236
implementing 141
improvement 302
  strategies 236
in situ
  collection 64
India 13, 29
indicator bacterium 63
indicator germs 151
indigenous knowledge 47
industrialisation 44

industrialised countries 3
infrastructure 36
inhibition 119
inhibitory effect 378
integrated concepts 25
interaction 38
intercultural
   communication 9, 305
   competence 16
   discourse 309
   exchange 27
   learning 9, 15
   research 2
   teaching 2
   training 16
   work 310
interdisciplinary 88, 89, 299
   aspects 27
   communication 305
   discourse 310
   research 312
   work 309
intermittent denitrification 111
international 1, 88
   education 89
   networking 53
   partnership 89
internationalisation 10
internationality 256, 310
inventory 241, 256
investment costs
   evaluation 126
ion exchange 148
iron 74
irrigation 151, 240

junior scientists 2
   countries 5
   disciplines 4

key soil factors 275
knowledge 30, 41
   production 17
   transfer 24, 26, 298

labour 32
lagoon
   biological treatment 153
   systems 167
lagoon plants 27
land management 272, 290

land requirements 126
land use 263, 274
landowner 35
land-use-pattern 253
layer density 322
leaching 290
leakage 166
legal status 34
level 300
limitation
   of the study 76
living conditions 29, 41
load
   pollutant 145
loading rate 177
long-term
   diversity 229
low-cost 166
low-technology 152, 168

magnesium 347
Maharashtra State 32
mahila mandals 34
main rivers 231
mainstream 24
maintenance 246
   framework plan 243
   requirements 172
   work 240
management 22
   demand-side 84
map of soil 245
mapping
   geological 276
   pedological 276
marginalisation 49
master's program 7
maturation phase 186
meadows 238, 245, 279
measure catalogue 28
mechanical retention 148
mediators 303
meeting 34
mesophilic
   bacteria 191
   phase 186
metal analysis 275
metal ions 62
methods
   small-scale 76
   treatment 76

water purification  76
microbial number  329
microfilter  60
microwave disintegration  357
middle Elbe  238
migration  46, 251
mineral particles  261
mobility  274
modelling approaches  27
models
   basics  113
   biological  112
   mathematical  112
moist soils  281
moisture content  322
monoculture  18
moringa oleifera  62, 75
mountainous  226
multi-chamber systems  209
multicultural  88, 300
multilingualism  15
multimedia  6
multiple use  172

national boundaries  89
nature-like state  255
NGO  32, 305, 310, 314
   CAD  32
   GTA  213
   WINS  314
nitrate  348
nitric nitrogen  348
nitrification  103, 112, 121, 149, 371
   reaction  104
   toxic substances  372
nitrification rate  373
nitrifying bacteria
   growth of  104
nitrogen  103, 341
   compounds  149
   determination  339
   gas  105
   removal  103
nitrohydrochloric acid disintegration  356
$N_{min}$  288
North-South dialogue  296
nutrient  101
   analysis  215
   recycling  262
   removal  101

removal development  102
nutrients and trace elements  278
nutritive media  330

$O_2$ consumption  377
operation costs  131
organic
   compounds  148
   matter  324
   substance  362
organisms
   autotrophic  155
   coliform group  63
   heterotrophic  154
   pathogenic  63
   pathogens  152
   total coliform  63
ownership  38
oxidation
   ferrous ions  70
   microbial  149
oxygen
   consuming substances  91
   demand  114, 115, 334
   dissolved  105
   no dissolved  112
   supply  150
   uptake rate  375
oxygen consumption  336, 375, 378, 379

panchayat  35
participation  170, 271
   local  169
   political  170
   public  91
   user  139
participatory
   approach  139, 172
   atmosphere  34
   research  37
pastures  245
pathogen reduction  200
pathogens  152
peak loads  373
percolation  352
permeability coefficient  153
personality  301
perspective  170
pF meter  360
pF values  360
pH values  325, 358

phosphate uptake 113
phosphate-accumulating organisms 107
phosphorus 341, 344
   compounds 149
   incorporation 116
   removal 106
physical discomfort 36
planning ideas 172, 176
plant
   biomass 148
   indicator 281
   samples 275
planted soil filter 168, 178
   control 153
   design size 161
   effluent 151
   gravel 152
   horizontal-flow 147
   influent 151
   maintenance 153
   multistage 161
   operational stability 152
   process characteristics 150
   sand 152
   vertical-flow 147, 159
political issue 26
pollution 32
pollution-prevention 183
polyculture 52
population 29
pore
   size 214
   system 158
   volume 362
pores 261
postcolonial 10
post-denitrification 110
postgraduate 1, 295
potassium 287, 344
potassium dichromate 338
poverty 44
power 44
P-precipitation 113, 116
practical orientation 313
practice-oriented
   projects 300
   teaching 298
precipitation 107, 257
pre-denitrification 106, 110
P-removal
   biological 113

   chemical 113
preservation 231
presses
   rake material 94
pre-treatment 144
   mechanical 146
problem-centred projects 26
problem-oriented curriculum 297
property 24
   biological 273
   chemical 273
   physical 273
protection 257
   system 236
   zones 69
protozoa 192

quality
   chemical 73
   requirements 79

rain-and-snow type 237
rainwater 59, 64
   agricultural use 78
   agroforestry 84
   available 88
   collection 68
   domestic use 77
   drinking water 80, 87
   DRWH 77
   harvesting 27, 42, 70, 76, 80, 81, 87
   harvesting plant 69
   sustainable system 83
   system components 77, 79
   usage 64, 66
   use 77
   vegetable growing 84
rakes 92, 93
reality 51
receiving waterway 151
recirculation 178
recommendations 218
rectangular tanks 99
red list of birds 238
reduction
   microbial 149
reed 147
reed bed 146
reflection 49
reinforcements 232
religion 299

remeandering  255
removal
   carbon dioxide  70
   iron  70
   manganese  70
requirements  27, 314
   energy  140
   utilisation  236
   water  198
research  1, 34, 312
resistant pathogen  63
resource management  23, 310, 312
respiration activity  337, 377
respiration rate  376
restoration  231
   measures  253
retention  68
   mechanical  148
   ponds  68
retention time  144, 177
reuse  198
reverse osmosis  60
river
   characteristics  222
   discharge  223
   hydraulics  253
   morphology  222
   protection  221
   protection system  229
   restoration  253
   stretch  235
   systems  221
river development planning  221
   field method  223
riverbed  226, 246
   characteristics  253
role-play  15
roof washers  88
rotifers  193
rotting degree  336
roughness  226
round tanks  99
rural
   areas  27, 138
   development  29
   regions  141, 164
   social fabric  33
   women  30
rushes  147

saline soil  266

salinity  327
sample preparation  321
sand catchers  94
   aerated  94
   unaerated  94
sand filtration  75
sanitary engineering  91
sanitation  183
   concepts  134, 185
   ecological project  218
saprobia
   index  247
   value  240
sapromat  335
scientific perspectives  300
scope
   of the study  76
scraping
   bridges  96
   lengthwise  96
secondary settling tank  117
sediment  228
   transport  227
sedimentation  60, 365
self-analysis  49
self-organisation  301, 307
self-reflection  16
semi-decentralised  132
   concepts  141
   units  134
seminatural  68
septic tank  144
   dimensioning  145
sequencing batch reactor  130
Service Centre  5
settlement  140
settlement tank
   dimensioning  145
settling tank  97
   dimensioning  98
sewage
   reduction  183
   system  184
sewage sludge  109
sewer network  101, 125
sewer systems  91
shields curve  228
side rivers  231
simple technology  48
simplicity  172
simultaneous denitrification  110

single-chamber systems  207
SIRDO  28, 204, 206
    dry  211
    solar-heated  212
    wet  211
sisterhood  14
skill  30
    development  48
slow sand-filtration  72
sludge  144, 376
    activated  100
    recirculation  100
    surplus  100
sludge composting  156
    dimensioning  156
    fundamentals  156
sludge composting plant  158
    experience  158
    operational stability  158
sludge level  155
soakaways  68
social
    change  48
    forestry  42
    justice  45
    worker  33
social evaluation  216
social practice  14
socio-political context  11
sodic soil  266
soil  261
    absorption ratios  355
    auger  321
    degradation  263, 267
    dry weight  322
    exploration  322
    filters  27
    formation  262, 263
    grain fractions  366
    horizons  280
    horizontal-flow filter  147
    management  25
    mapping  268
    material  268
    methodology  269
    moisture tension  359
    nutrient provision  346
    organisms  329
    planted filters  140, 146
    pollutants  328
    profile  277, 279, 282, 284

    properties  28
    reaction  326
    respiration  288, 335
    sampling  321
    seepage  289
    suspension  332
    sustainable utilisation  28
    texture  362
    texture types  366
    vertical-flow filter  147
    water  264
soil water  273, 289
soil-clogging  152
soil-forming factors  262
solar heating toilet  203
solidarity  14
solids-retention time  104
solution  39
sophisticated technology  298
specific gravity  362
spread of disease  198, 199
spring marshland  242
stereotypes  13
sterile working  331
sterilisation  331
storage tank  82, 84, 85, 86
    oversizing  77
    size  77
    volume  80
strainers  92, 93
straining  362
    dry  363
    wet  363
strategy  16
street plays  39
strontium  74
study
    habits  307
    venue  306
study offices  5
subprocesses  108
substances
    fibrous  92
    settable  92
substrate  244
summer school  305
surface water  103
surplus sludge
    production  114
survey  244
sustainable

development 137
resource management 175
sanitation 185
solutions 218

teaching/learning processes 299
technical
   design 95
   English 305
   fundamentals 141
technological development 131
technologies 137
   small-scale 59
the cost-benefit analysis 41
thermophilic phase 186
Third World 11
time interval 130
toilet 194
topsoil 283
total coliform 64, 71, 73, 193
total discharge 224
total heavy metals 284
tradition 33
traditional
   techniques 297
   technologies 25
training 40
transfer 306
   coefficient 275
   interdisciplinary 25
   knowledge 25
   practice-oriented 25
treatment
   concept 162
   conventional 124
   heating 67
   mechanisms 141, 148
   moringa oelifera 67
   preliminary 97, 124
   primary 124, 128
   products 184
   secondary biological 125, 128
   sewage-sludge 109
   slow sand-filtration 67
   techniques 67
   tertiary 125, 128
   theory 67
treatment methods 117
   efficacy 59
trickling water 164
turns 222

unfamiliar disciplines 297, 307
Universal Soil Loss Equation 265
universalism 17
university
   German 11
urban landscape 133
urine diversion 202, 203, 206
urine separator 198
utilisation 236
UV absorption 348
UV radiation 66, 71, 75

vector management 189
vegetation 246
velocity 224
vertical soil filter
   process characteristics 150
vertical-flow
   planted soil filter 147, 159
vertical-flow tanks 98
viability 214
village 35
   eco-friendly 53
visiting scholars 10, 303

wastewater 138, 371
   agents 142
   biological treatment 100, 151
   centralised treatment 123
   common technology 167
   decentralised treatment 137, 162, 173
   discharge 179
   domestic 132, 143, 185
   evaluation 123
   industrial 372
   lagoons 140, 153
   loads 143
   machine technology 92
   mechanical treatment 92
   municipal 372
   planted soil filters 140
   purification 170
   purified 151
   quality 185
   quantities 142
   treatment 27, 91
   treatment dimensioning 141
   treatment efficiency 141
   treatment plant dimensioning 177
   treatment plants comparison 127

undissolved  92
wastewater lagoons  159, 160, 178
  control  155
  dimensioning  153
  effluent  160
  effluent concentration  155
  fundamentals  153
  influent  160
  maintenance  155
  purification capacity  154
  stability  155
  treatment mechanisms  154
  unaerated  154
wastewater plants
  design parameter  164
wastewater purification
  Vietnam  173
wastewater treatment
  Cuba  174
water
  agencies  257
  availability  23
  charges  22
  depletion  46
  erosion  265
  integrated resource management  22
  level  252
  management  31, 51, 137
  plants  255
  potability  23, 32
  quality  23, 36, 243
  quality survey  253
  quantity  36
  requirements  36
  resources  21
  shortage  21
  soil  267, 272
  sources  34
  supply  21, 40, 86
water consumption  86, 165
  domestic  143
water managers  171
water policy  31
water purification
  low-cost  52
water quality standards  57
water toilets  194
waterless toilets  195
watershed management  267
well  35
  tube-well  40
WINS  9
women
  status  33
  water management  33
women's participation  315
women's position  171
Women's Research Commission  1
women-oriented programmes  48
workforce  44
working hours  38

zinc  74

# Further Publications by the International Women's University in the Year 2001:

Becker-Schmidt R (ed) (2001) Gender and Work in Transition, Globalisation in Western, Middle and Eastern Europe. Verlag Leske + Budrich, Opladen, ISBN 3-8100-3252-2

Floyd C et al. (eds) (2001) Feminist Challenges in the Information Age. Verlag Leske + Budrich, Opladen, ISBN 3-8100-3255-7

Härtel I, Schade S (eds) (2001) Body and Representation. Verlag Leske + Budrich, Opladen, ISBN 3-8100-3254-9

Metz-Göckel S (ed) (2001) Lehren und Lernen an der Internationalen Frauenuniversität, Ergebnisse einer Wissenschaftlichen Begleituntersuchung. Verlag Leske + Budrich, Opladen, ISBN 3-8100-3253-0

**Further Volumes in the Year 2002.**

Printing (Computer to Film): Saladruck Berlin
Binding: Stürtz AG, Würzburg